GEOPOLITICS

and the

GREEN REVOLUTION

GEOPOLITICS
and the
GREEN
REVOLUTION

Wheat, Genes, and the Cold War

JOHN H. PERKINS

New York   Oxford • Oxford University Press   1997

Oxford University Press

Oxford   New York
Athens   Auckland   Bangkok   Bogota   Bombay   Buesnos Aires
Calcutta   Cape Town   Dar es Salaam   Delhi   Florence   Hong Kong
Istanbul   Karachi   Kuala Lumpur   Madras   Madrid   Melbourne
Mexico City   Nairobi   Paris   Singapore   Taipei   Tokyo   Toronto   Warsaw

and associated companies in
Berlin   Ibadan

Copyright © 1997 by Oxford University Press, Inc.

Published by Oxford University Press, Inc.
198 Madison Avenue, New York, New York 10016

Oxford is a registered trademark of Oxford University Press

Library of Congress Cataloging-in-Publication Data
Perkins, John H.
Geopolitics and the green revolution : wheat, genes,
and the cold war / John H. Perkins.
p.   cm.
Includes bibliographical references and index.
ISBN 0-19-511013-7
1. Wheat Breeding.   2. Wheat—Breeding—Government policy.
3. Wheat.   4. Wheat trade.   5. Green Revolution.   6. National
security.   7. Cold war.   I. Title.
SB191. W5 P42   1997
338.1'6—DC20   96-8885

1 3 5 7 9 8 6 4 2
Printed in the United States of America
on acid-free paper

Preface

The Yield Transformation in Cereal Production

The need for food creates a relationship of fundamental importance between people and the environment. If we do not understand this relationship, we remain unaware of the critical dynamics that exist among human populations, culture, and nature. At the foundation of the relationship are the major cereal grains, especially wheat, rice, and maize, and the yields obtainable from them.

Yields of cereal crops went up dramatically during the past 100 years, and especially since 1950. This book is an effort to understand the yield transformation in the basic cereal crops and thus gain insights into the relationship between people and nature. Its starting point was to explore the scientific changes underlying the green revolution, a public relations term referring to the changes after 1960 in the wheat and rice yields obtainable by farmers in less industrialized countries. Use of the word "revolution" suggested that a fundamentally new relationship existed between people and their major food plants. "Green" implied a benign technology and emphasized the positive nature of the relationship.

The term *green revolution* is widely recognized among agricultural experts and development workers. An immense literature analyzes its scientific and technical components, the economic policies needed to promote it and accommodate its impacts, and its consequences. Despite many studies on the subject, relatively little has been written about why and how the science underlying the green revolution came to be. This book is an inquiry into the origins and unfolding of the scientific work upon which the green revolution was based.

Outline of the Argument

This book sketches the development by plant breeders of high-yielding varieties of wheat, which was a major part of the green revolution. The story, however, could not be confined to the traditional borders of the green revolution. Changes in the agriculture of less industrialized areas were linked too strongly to events elsewhere to be understood in isolation. Highly industrialized countries also developed and adopted high-yielding varieties of wheat in ways that had important links, scientifically and politically, to events in the third world.

Wheat production is a large and important global industry, much too vast to examine here in its entirety. For reasons that are explained in chapter 1, this book focuses on selected events in wheat production in the United States, Mexico, India, and the United Kingdom. Thus the research for this book was built around an effort to understand the plant-breeding science behind high-yielding varieties of wheat in four particular countries, during the time period from about 1900 to 1980. As I worked through archival documents, reports, publications, and personal interviews, however, I realized that an originally unanticipated theme emerged and was essential to any explanation of how and why wheat breeders formed their conclusions. This theme was the immense importance of agriculture in general and the cereal crops in particular to the shape of human culture and the security of nations.

Understanding that wheat breeding had something to do with cultures and nations came from the recognition that political support for wheat breeding was linked to national security planning and to the need for countries to manage their foreign exchange.

I concluded that considerations of national security and foreign exchange were really important examples of an even broader concept: that wheat and people are two species that have evolved a complex codependency since their first major encounter in the Neolithic agricultural revolution. In the approximately 10,000 years in which people have intertwined their affairs with the wheat plant, we have created a situation in which neither species has a future independent of the other.

Codependency of people and wheat made my task more complex. In order to explain the importance of national security planning and foreign exchange management in the affairs of wheat breeding, I had first to lay the foundation that codependency had shaped both human culture and the wheat plant for thousands of years. Accordingly, the narrative begins in chapter 1 with an explanation of political ecology, a framework that opens the way to a consideration of codependency. Chapter 2 then outlines the physical nature of the wheat plant and how humans and this cereal have coevolved since the Neolithic agricultural revolution. Codependency sets the stage for an examination of the origins and the socio-political position of plant-breeding science, the subjects of chapters 3 and 4, respectively.

Wheat breeding was fully formed and recognized as an important activity by 1940 in the United States, Britain, and India. Events after 1940, however, sharply accelerated the pace of work and amplified the science's strategic importance. Chapter 5 begins this part of the story by explaining how and why the Rockefeller Foundation launched a major agricultural science project in Mexico, which launched wheat breeding into international prominence. This chapter also recounts how the Mexican government embraced the Rockefeller Foundation program as its way of shaping national security and managing Mexico's foreign exchange. The strategic importance of wheat breeding was rationalized in the United States by a theory I call the population–national security theory, outlined in chapter 6.

Chapters 7, 8, and 9 move to reconstruct how three nations after 1945 each made a strategic decision to embrace wheat breeding as a way of managing its national security and foreign exchange problems. The United States (chapter 7) made commitments to promote wheat breeding as part of the cold war efforts to contain the former Soviet Union. In addition, the critical importance of agricultural exports in the U.S. economy made wheat breeding important for foreign exchange management. India (chapter 8) moved to embrace wheat breeding along the complex pathway it took to recover from the effects of British imperialism and the shattering of the economy of British India at independence. Security and autonomy of the Indian nation and foreign exchange considerations were the prime drivers in the national commitment to wheat breeding. Finally, the United Kingdom (chapter 9) vastly expanded its commitment to wheat breeding as it struggled to reconstruct its postimperial economy. Once again, considerations of national security and foreign exchange management drove the crucial decisions.

Chapter 10 reconstructs the science of high-yielding wheat in the United States, Mexico, India, and the United Kingdom. Mexico and India constitute the heart of what is usually considered the green revolution. At the simplest level, the material in this chapter provides the answer to the question about how farmers in these countries obtained higher yields from their land. My argument, however, is that a fuller explanation of how and why these higher yields came to be requires a larger framework. The scientists sketched in chapter 10 would not have had the support, nor would their products have been embraced as a matter of policy, without the perception of national leaders that wheat breeding provided important avenues to security and management of foreign exchange. This chapter also dramatizes the idea that the green revolution was a global phenomenon, not just an event in the third world.

Significance of the Argument

This book's first objective is to provide an explanation of how humans make use of resources that are exceedingly important to human survival and prosperity, namely, agriculture. Accordingly, it is first and foremost a contribution to environmental history, the effort to understand how human culture and the environment are related to each other. Reconstruction of an episode in the history of plant-breeding science was the major vehicle to write this essay in environmental history.

The story told here, however, has policy implications. In particular, it is relevant to the extensive debates over the social equity, or lack thereof, associated with the agricultural enterprise, and the question of whether agricultural operations are ruining the resources needed for farming. In contemporary terms, these two questions are often phrased in terms of sustainability, a term that often obscures as much as it enlightens.

More specifically, the argument here appeared to be important for answering a series of questions: Why was high-yielding agriculture developed and promoted, if in fact it is inequitable and destructive? Were, for example, the originators and promoters unaware of possibly deleterious features of high-yielding wheat production? Is it possible that the originators and promoters of the green revolution had a different vision of the human condition, in which the allegations of inequity and destructiveness could not be understood? Did the forces that prompted the green revolution leave a legacy that any social or environmental reform efforts will have to address, if the reforms are to be successful? It was beyond the scope of this book to explore all of these issues fully. However, the epilogue sketches some of the more important points. The argument is that reform of agriculture is unlikely to be successful without a broad understanding of how contemporary practices emerged. An appreciation of how agriculture got to be the way it is by no means guarantees the wisdom or success of the reform movement. Reform without an appreciation of history, however, is even more likely to aim at the wrong target and not succeed.

The relationship between national security policy and high-yielding agriculture is the legacy that will hang most persistently over reform efforts to make agriculture "sustainable." In addition, foreign exchange management has tight connections to national security and national autonomy. Personally, I'm not happy that the connections are so strong. I'd much rather see efforts to make farming less destructive of the environment freed from the terrific emotions and fears that emerge from the depths of national security considerations. Unfortunately, the links are there, and powerful forces will keep agricultural reform tightly tied to efforts to keep nation-states strong.

Any quest for sustainable agriculture will therefore be affected by considerations of national security. I hope one modest contribution of this book will be to show that appreciating the nature and complexity of this tie is helpful for those who would reform agriculture to make it more sustainable. I fear that ignoring the tie will shatter the reform efforts.

Inevitably, this book leaves much of interest unsaid. Stories remain to be told about rice, maize, and other crops, and about soil scientists, irrigation specialists, fertilizer producers, mechanical engineers, and other scientists. Most importantly, the book is silent about the person who has to put all of the disparate pieces of knowledge into practice: the farmer. Hundreds of millions of men, women, and children labor daily to produce the food that keeps the billions alive, including those who write books. Some are well rewarded for their work, but many are not. Farmers, however, whatever their status, work at the interface between humans and nature, which is fundamental to the survival and prospects of our own species and the many other species with whom we share the earth. Those of us who do not work at this interface are well advised at least to try to understand what is at stake.

Olympia, Washington   J.H.P.
June 1996

Acknowledgments

Many people inspired, assisted, or in some other way enabled me to complete this work. I have tried to name all of the relevant people, and I offer apologies to anyone inadvertently omitted.

Many people over the years have guided me into the intricacies of the agricultural enterprise. Without their insights and guidance, I would not have been able to complete this work. Several people consented to be interviewed in depth about their own roles in the events recounted here or about the part played by someone they knew. Of particular importance were R. K. Agrawal, Roger Austin, John Baldwin, G. Douglas, H. Bell, John Bingham, Norman E. Borlaug, Peter Day, Scott Findlay, K. S. Gill, Nigel Harvey, H. K. Jain, Virgil A. Johnson, Francis G. H. Lupton, S. P. McClean, A. M. Michael, C. R. Mohapatra, Benjamin Peary Pal, R. S. Paroda, N. S. Randhawa, M. V. Rao, Alan Roelfs, Lyle Sebranek, B. P. Singh, D. N. Srivastava, Ruth Engledow Stekete, M. S. Swaminathan, J. P. Tandon, and Orville A. Vogel. I am also indebted to Helen Weaver for allowing me access to Warren Weaver's private papers.

A number of students at the Evergreen State College provided invaluable assistance in gathering and summarizing tremendous numbers of documents: Bobbie Barnett, Peggy Britt, David Giglio, James Jenkins, Michael Kent, Linda R. P. Knight, Michael MacSems, Ken Steffenson, and Mariusz Twardowski. To each of them, I am very grateful.

Books are mostly words in a sequence, but they are invariably aided by illustrations. I am indebted to Tim F. Knight for his excellent drawings and maps, prepared especially for this text.

Over the years, it has been my privilege to work closely with a number of colleagues, each of whom has taught me a great deal about environmental and agricultural issues, broadly conceived. Inevitably they have influenced this book for the better. Especially important were Mark Abner, Richard Alexander, Wallis Barker, Pamela Bennett-Cummings, Mike Beug, Peggy Britt, Jovana Brown, Paul Butler, Susan Campbell, Barbara Cellarius, Doris Cellarius, Richard Cellarius, Ellie Chambers, John Cushing, Allen Davis, Betsy Diffendal, Ken Dolbeare, Roland Duerksen, Larry Eickstaedt, Barbara Ellison, Curtis Ellison, Hugo Flores, Steve Ganey, José Gómez, Bill Green, Burt Guttman, Jeanne Hahn, Patrick Hill, Virginia Hill, Thomas Johnson, Lou Ellyn Jones, Teresa Koppang, Karel Kurka, Pat Labine, Eugene Leach, Mike Lunine, Joanne Markert, David Marr, Eugene Metcalf, David Milne, Rick Momeyer, Ralph Murphy, Lin Nelson, William Newell, Nancy Nicholson, Andy Northedge, Nicola Ostertag, Barbara Patterson, Terry Perlin, Ron Pratt, Tom Rainey, Karen Riener, Meredith Savage, Lars Schoultz, Darius Sivin, Stan Sloss, Bob Sluss, Barbara Smith, Oscar Soule, Fred Stone, José Suárez, Pete Taylor, Jennifer Thomas, Phil Trask, Jude Van Buren, Barbara Whitten, Hugh Wilder, Denny Wilkins, Tom Womeldorff, York Wong, Ron Woodbury, and Byron Youtz. I am particularly indebted for the stimulation and critical feedback I received from Ralph Murphy and Tom Rainey, for they encouraged and guided me in thinking about political ecology.

A number of environmental studies specialists, environmental historians, and scholars on agriculture, science, and human affairs have stimulated and guided my thoughts on how to approach these topics. Of particular importance were Robert Anderson, John Baldwin, Jerry Berberet, Paul Brass, Terence Byres, Judith Carney, Karen Colligan-Taylor, William Cronon, Al Crosby, Donald Dahlstan, Thomas Dunlap, Richard Garcia, Paul Gersper, Richard Haynes, Douglas Helms, Carl Huffaker, Donald Hughes, Edmund Levy, Everett Mendelsohn, Carolyn Merchant, William Murdoch, Richard Norgaard, John Opie, Robert Paehlke, Paolo Palladino, Dick Perrine, David Pimentel, A. Rahman, Peter Rosset, Margaret Rossiter, Vernon Ruttan, Al Schwartz, Ray Smith, Richard White, Donald Worster, and Angus Wright. I am especially indebted to Richard Haynes, editor of *Agriculture and Human Values*, who took an early interest in this project.

James Cook, Helena Meyer-Knapp, and Tom Womeldorff were kind enough to read excerpts of the text in draft form, and I benefited greatly from their comments.

Gathering material for any extended study is not possible without the expertise and advice of many librarians and archivists. I am particularly indebted to Mrs. Rama Agarwal, Elaine Anders, Hannah Bloom, Claire Collier, Richard Crawford, Marjorie Dalby, Barbara Glendenning, Joan Green, Lois Hendrickson, Michele Hiltzik, Terry Hubbard, Don Jackanicz, Paul Kaiser, A. L. Kapoor, Ernestine Kimbro, Norma Kobzina, Penelope Krosch, Jacki Majewski, Sally Marks, Pat Matheny-White, Frank Motley, Ann Newhall, Emily Oakhill, Harold Oakhill, Carol O'Brien, Neenah Payne, Sarah Pedersen, Barbara Radkey, Sara Rideout, Tom Rosenbaum, Melissa Smith, Darwin Stapleton, Randy Stilson, Sandy Swantz, Carolyn Treft, Roseann Variano, Evangelina Viesca, Teresa Velasco, Valerie Walter, Beth Weil, and Randy Wilson. I am particularly indebted to the Agricultural Research Council Archives, Cambridge University Libraries and Archives, Centro de Investigaciones de Mejoramiento de Maiz y Trigo, Delhi School of Economics Library, Ford Foundation Archives,

Indian Agricultural Research Institute Library, Indian National Archives, Oxford University Archives, Plant Breeding Institute Library, Rockefeller Foundation Archives, the Evergreen State College Library, United States National Archives, University of California, Berkeley, Libraries, and University of Minnesota Archives.

The initial stages of research for this book were conducted while I was an academic visitor at the Centre for Environmental Technology, Imperial College of Science and Technology, London. Gordon Conway graciously made this visit possible and took a genuine interest in the book's content. Gordon later served as the representative of the Ford Foundation in New Delhi, where he also provided encouragement during one of my trips to India. Several other people at Imperial College also made my stay there very enjoyable: Ian Bell, Richard McCrory, Hilary Morgan, John Peachy, Jules Pretty, and Bashra Salem.

I spent a total of ten weeks on three different occasions in India gathering materials for this study. I am indebted to Craig Davis for first interesting me in Indian issues and to Craig and Ed McCrea for making it possible to visit India for the first time. While in India I was assisted in many ways, both professional and personal, by colleagues Desh Bandhu and D. K. Banerjee and their respective families. Staff at the India International Centre provided a convenient, comfortable place to live while in New Delhi.

Over the years, a number of coworkers at the Evergreen State College provided much assistance and support to this project. Especially important were Paula Butchko, Bonita Evans, David Judd, Jane Lorenzo, Judy Saxton, Jan Stentz, Audrey Streeter, Pam Udovitch, Dee van Brunt, Carolyn Walker, and Karen Wynkoop.

Financial support for this study came from the National Science Foundation (SES-8608372; DIR-8911346; DIR-9012722), from the Smithsonian Institution, Special Foreign Currency Program, and from the Evergreen State College. From the NSF, I am particularly indebted to Rachel Hollander and Ron Overmann; Francine Berkowitz from the Smithsonian was very supportive on a number of occasions. While in India, I was aided on several occasions by the American Institute of Indian Studies, particularly by P. R. Mehendiratta and L. S. Suri.

The editorial staff at Oxford University Press were immensely helpful in preparing the final copy of the manuscript. I am particularly indebted to Kirk Jensen and Cynthia Garver, as well as to the copyeditor, Susan Ecklund.

The most sustained support and encouragement for this study, and some of the best intellectual conversations about it, came from my immediate family, Barbara Bridgman Perkins and Ivan Bridgman Perkins.

Despite the help, encouragement, and support I received from these wonderful people, all the errors of omission and commission remain mine alone.

Contents

1 Political Ecology and the Yield Transformation   3

2 Wheat, People, and Plant Breeding   19

3 Wheat Breeding: Coalescence of a Modern Science, 1900–1959   42

4 Plant Breeding in Its Institutional and Political Economic Setting, 1900–1940   75

5 The Rockefeller Foundation in Mexico: The New International Politics of Plant Breeding, 1941–1945   102

6 Hunger, Overpopulation, and National Security: A New Strategic Theory for Plant Breeding, 1945–1956   118

7 Wheat Breeding and the Exercise of American Power, 1940–1970   140

8 Wheat Breeding and the Consolidation of Indian Autonomy, 1940–1970   157

9 Wheat Breeding and the Reconstruction of Postimperial Britain, 1935–1954   187

10 Science and the Green Revolution, 1945–1975   210

Epilogue   256

Notes   269

Index   325

GEOPOLITICS
and the
GREEN REVOLUTION

1

Political Ecology and the Yield Transformation

The Central Issues

Something quite remarkable happened during the past century, and especially since 1950. Yields rose dramatically in the basic cereal crops such as wheat, rice, and maize, and in other crops as well. Casual inquiry to an agricultural expert about the source of the increase is likely to bring a response such as, "Well, farmers now use better plant varieties and more fertilizer than they used to, so the yields went up."

At the simplest level, this response is perfectly adequate and true. Better varieties and more fertilizer have made it possible to get larger harvests from the same plot of ground. Unfortunately, the simple answer immediately provokes yet further questions: How did farmers obtain the new and better plant varieties? Why did they use more fertilizer? When did farmers start changing their practices? Where? Why? Who helped them?

The last question quickly leads the inquiry into the realm of agricultural science, because scientists enabled farmers to change their practices. Especially important were plant breeders and soil fertility experts. Thus a new realm of questions is opened: How did scientists discover the methods for higher yields? When did they do their research? Where? Why? Who paid for the research? Why? What is the significance of this scientific change?

These questions seem simple, but agriculture is a tricky topic to address. It generates an inordinate number of paradoxes, puzzles, and ironies, which makes answering the queries difficult. Consider, for example, just a few:

3

Agriculture was once the place where the vast majority of human beings worked and lived, but now it increasingly provides a place for only a small minority of people.

Agriculture's harvests are the only source from which most people obtain enough food to stay alive, but few nonfarmers understand or care about its workings.

Agriculture is often considered to be a landscape that is alive, verdant, lush, and redolent of wholesome naturalness, but in reality it represents the complete destruction, indeed obliteration, of natural ecosystems and wildlife habitat.

Agriculture is thought, in American political mythology, to have produced the honorable farmers who are the backbone of republican democracy, but in daily life these same farmers are often ridiculed (unfairly) as naive bumpkins from the backwaters of civilization.

Agriculture is often considered primarily a business, but it is also a human-created ecosystem generating a food web of which we are an integral part and without which most of us could not survive.

Agriculture is seldom considered to have much to do with the security of nations, but in reality it may be as important as the military and industry in guaranteeing national independence.

Agriculture is sometimes alleged to be on the verge of or already in collapse, but the human population growth of nearly 100 million per year suggests food is still sufficiently abundant to maintain growth.

Agriculture is often perceived as a romantic, tranquil refuge from the relentless blight of industrial civilization, but it is buffeted by its own relentless technological change and is also the foundation upon which the machinery of urban industry was built and is maintained.

These seemingly endless internal contradictions suggest a complexity of the subject that makes it difficult to answer the questions about the yield transformation. At the very least, attitudes toward agriculture are mixed and inconsistent, which hinders comprehension. How, then, do we begin to construct meaningful questions and answers for an inquiry into the whys and wherefores of the changes in harvest yields?

One useful way to begin is to analyze agriculture as a complex set of technologies that access natural resources to produce food. More specifically, plant agriculture consists of knowledge, such as how to (1) select appropriate plant varieties, (2) plant seeds in properly prepared soils, (3) provide water and soil nutrients in the right amount at the right time, (4) protect the crop plant against pests, (5) harvest and store the crop, and (6) process the harvest for use. These agricultural technologies enable people to make use of the natural resources upon which agriculture is based: sunlight, soil, plants, water, and climate.

Put more generally, this image of agriculture rests upon the notion that technology consists of knowledge by which people use environmental resources in order to satisfy material wants and needs.[1] In the case of agriculture, the materials produced are the biomass of the harvested crop. Technology, in other words, mediates between people and nature in ways that permit human beings to garner enough biomass to survive, reproduce, and form cultures. Without technologies such as agriculture, people would have to find their subsistence in other ways, such as fishing or hunting and gathering. Schematically, the relationship is shown in Figure 1.1.

Figure 1.1   Technology mediates between human culture and nature.

Once agriculture is seen as a technology that mediates between humans and natural resources by producing harvestable biomass, it can be explored from a political ecological perspective: productivity of agricultural land is both an ecological and an economic process. "Productivity," in other words, has two meanings. The first refers to biological productivity, that is, the physical biomass produced in a particular area in a particular time frame, measured in grams and calories. Second is economic productivity, that is, the value in money or utility of the biomass produced in a particular area in a particular time frame. Economic output, in turn, is linked to the power to control the distribution and enjoyment of the harvest. Therefore, development of agricultural resources (e.g., land and water) is inherently both an ecological and a political economic process. Political ecology seeks an explicit integration of the political economic and ecological dimensions of agricultural management in order to describe, explain, predict, and guide change.

Roots of Political Ecology

Political ecology rests upon many previous ideas. Many writers have developed parts of it as they sought answers to how people should interact with the natural world. Most explored the relationships among (1) the numbers of people and their consumption habits, (2) forms of knowledge and social organization, and (3) natural functions and processes. Since the mid-1960s, an especially large literature has developed, motivated largely by a sense of impending crisis from environmental deterioration. Almost all of these recent studies have related environmental impacts to one or more of the factors: technology, population levels, and consumption levels. Unfortunately, the frameworks developed in this literature were usually inadequate to answer a crucial question: How can and should people collectively manipulate the biosphere in order to satisfy the material needs for food for all people?

A few examples will illustrate the variety of themes in this literature. Some writers, such as biologists Paul and Anne Ehrlich, focused on the sheer number of people and the resulting intolerable burdens placed on nature and food supply systems.[2] Others, such as biologist and political activist Barry Commoner, downplayed the role of population and laid more responsibility for environmental crisis on the kinds of technology adopted.[3] A third variant focused on the high material consumption patterns of the industrialized nations as the source of excessive resource exploitation and environmental exhaustion.[4] Against the symphony of doom from those who saw impending environmental collapse was a counterchorus, usually economists, who believed that modern technology enabled a sustainable consumption of high levels of material goods, including food, for a growing proportion of a growing global population.[5]

Other literature explored related subthemes. For example, philosopher William Leiss explored the concept of "domination of nature" and its relationship to technology, emphasizing that those who sought technology for the control of nature often found it necessary or desirable to control their fellow human beings as well.[6] Historian Carolyn Merchant delved into the origins of modern science and the resultant loss of belief in the vitality and female gender of nature, a change that made exploitation of the earth more feasible.[7]

Environmental historians have made major contributions to the understanding of the interactions between people and the natural world. Richard White, for example, studied the different ways in which Native American and Euro-American settlers both changed the ecology of Island County, Washington, in order to satisfy their respective needs for material resources.[8] Fundamental to White's argument is the notion that all people modify the ecosystems they live in as they become integral to those ecosystems. Similarly, in New England Carolyn Merchant studied the integration over time of changes in land-use practices, ideas about nature, and cultural patterns by which people supplied their needs and reproduced.[9] Merchant's emphasis on reproduction was a vital addition to understanding the importance of interactions between humans and nature.

More recently, biologist David Ehrenfeld and philosopher Luc Ferry explored, in different ways, the importance of values in human interactions with the environment.[10] In a different vein, political analyst Norman Myers and others have raised the issue of how environmental problems are major sources of conflict between nation-states.[11] Jonathon Porritt provided a comprehensive articulation of an environmentally based political platform, based on his experiences in the United Kingdom.[12]

A subtheme explored extensively in the early 1970s was a mass-balance approach to the relationship between people and food. Several bouts of famine or near famine between the mid-1960s and mid-1970s stimulated a vigorous debate about whether technology was available to produce enough food to supply all people with an adequate diet. One school of thought, exemplified by Georg Borgstrom's *Harvesting the Earth* (1973),[13] was heavily influenced by the Malthusian image of unending human misery due to the postulated inevitability of reproduction to exceed the powers of food production. Greater optimism for human ingenuity was voiced by such writers as Colin Clark in his *Starvation or Plenty?* (1970).[14] These latter two studies, despite their different conclusions, came close to the approach endorsed here because they emphasized two critical ideas: the role of photosynthesis in the human food supply and the role of agricultural technology as a factor in the levels of the harvest.

A study that uses a framework analysis very similar to the one adopted here was *So Shall You Reap* by Otto and Dorothy Solbrig. They understood that farming was a massive transformation of the environment and argued that life for over 5 billion people was simply not possible without agriculture. They correctly saw that anticipated population growth in the next few decades necessitates increased production. If those increases come through further environmental destruction from agriculture, however, the ultimate hopes for human security and prosperity will be dashed.[15]

Despite the enormous literature on the environment, technology, and agricultural production, the questions asked and the frameworks developed to provide answers have generally not yet integrated all the salient features of political ecology. Specific problems include:

- Too tight a focus on population as the cause of environmental problems has tended to ignore human ingenuity in problem solving and to slide past a critical moral question: What do we do about all the people who currently exist and are very likely to be born soon?
- An emphasis on technology choice as the major source of environmental problems avoids blaming population but runs the risk of downplaying the role of population or ignoring levels of consumption. Focus on technology also may ignore factors, arising from competition between companies and between nations, which push technological change whether or not the individuals who innovate want to change.
- Identification of overconsumption as a source of environmental problems is useful. Unfortunately, this approach tends to ignore the extensive development of infrastructure and ideology in modern society, which do not adjust easily, if at all, to voluntarily reduced rates of consumption.
- A further problem with most of the existing literature that treats the interaction among people, nature, and food is a lack of broad historical sensibilities.

Lack of historical insight is particularly troublesome in critiques of current agricultural practices as environmentally destructive and socially inequitable. Although both criticisms may be well founded, they avoid a crucial question: How and why did countries and farmers adopt the practices now said to be destructive? Were people coerced into doing something unwise? Were they venal or intellectually deficient? Or did they act in ways that were necessary and honorable at the time, even if the changes ultimately proved to be detrimental?

The latter questions are vital for what the political ecological framework seeks to explore. In order to understand the significance of modern agriculture, it is not enough to know that technical change occurred and that the economy of individuals and nations was thereby shifted. In addition, it is not enough to know that the changes led to more food production and thus the ability to support more people on earth. Likewise, it is not enough to know that modern production practices may be associated with significant social inequalities and that they may destroy the ability of agriculture to produce in the future. All of these issues may be necessary to understand modern agriculture but they are not sufficient, either to understand the past or to guide the process of reform in the future. It is also essential to understand why the changes occurred, and political ecology can help with this question.

Political Ecology as an Analytical Framework

Understanding the political ecological framework begins with a few fundamental principles. The key concepts are (1) that humans are components of ecosystems, (2) that of the necessity born of hunger, humans modify and harvest the productivity of the biosphere with agricultural technology in order to obtain food and other materials, (3) that humans create political economic structures to control the production and distribution of materials from the biosphere, and (4) that both the modifications of the biosphere and the political economic structures have a history that affects subsequent efforts to change either the technology or the social structure of agriculture.

Political ecology's roots lie in both ecology and political economy. From ecology comes the concept of biological productivity or the production of biomass on the earth. More specifically, ecology seeks to understand the distribution and abundance of organisms across the face of the earth. It seeks explanation for the common observation that organisms of a specific kind are abundant in some places, scarce in others. In addition, population sizes can fluctuate, up and down, over time. Invariably, the distribution and abundance of organisms, including people, depend upon the biological productivity of photosynthesis and how a particular species is involved with photosynthetic organisms.

Ecologists seek to understand the significance of relationships among different species that live together in the same place. The term *ecosystem* designates the collection of species in an area and their associated physical surroundings. Central to the study of ecosystems are the mutual interactions and linkages among species and between organisms and the surroundings.

Food webs are a major but not the only important interactions among species. Food webs link organisms of different sorts: primary producers (green plants) fix solar energy; herbivores feed directly on green plants; carnivores feed on herbivores or other carnivores; omnivores (such as people) feed on both plants and animals; and decomposers feed on all dead organisms. In these terms, agriculture is the way people generate a food web and thus tap the primary production from solar energy fixed by green plants. The food web supporting people is the key objective of agricultural ecosystems.

In physical terms, ecologists seek to understand food webs through the flow of solar energy into the earth, its fixation in photosynthesis and subsequent flow into animals and decomposers, and its ultimate dissipation as heat into space. Associated with the flow of energy are biogeochemical cycles that circulate chemicals within the biosphere, from living creatures to the physical environment and back again to living organisms. In these terms, agriculture is the way people tap the energy flows from the sun and the associated biogeochemical cycles. Food is merely trapped solar energy and associated minerals, needed for human survival.

Each species in the ecosystem has a population level that usually fluctuates up and down through time. Ecologists seek to understand what determines the population size and its rate of change over time. For many species, ecologists are also interested in carrying capacity, or the maximum number of individuals that can be supported for an indefinite period in a particular area. Estimations of carrying capacity are an important component of ecological inquiry, particularly for species of high interest to people. In these terms, agriculture is the way people have increased the carrying capacity of the earth for humans. Agriculture permits people to capture a larger amount of solar energy than they could through hunting and gathering, which in turn permits a larger human population.

Ecology, and its concepts of ecosystems, populations, carrying capacities, communities, food webs, energy flows, and biogeochemical cycles, has increasingly become a part of everyday language. Much of the modern environmental movement rests upon the idea that industrial civilization can wreck the very ecosystems upon which people depend and in which they must live. Apocalyptic visions predict the collapse of existing ecosystems and the attendant misery of those people who survive. Such visions often lead to condemnation of lifestyles held not to be in compliance with the dictates of ecological laws.

As powerful as the metaphors and concepts of ecology have been, ecology as a science has generally been rather unhelpful in providing general laws about how to delineate and manage whole ecosystems.[16] Rather, the importance of ecology has been in the vision of coexistence and codependency among the species in a community. Detailed natural histories of particular species or small groups of interacting species have also been extremely useful in understanding a limited range of interactions that go on in ecosystems.

Ecology has proven particularly unhelpful at providing insights or guidance into the dimensions of human life that most distinguish us from other species. Human beings over time have developed elaborate institutions that govern the production and distribution of biological productivity and wealth. Congruent with the institutions controlling the production and distribution of wealth are those that focus political power. Political economy is concerned with how human cultures intertwine the production and distribution of wealth with the exercise of power, or the right to make decisions that matter. Classical political economy presumed a social order composed of three classes—labor, landowners, and capitalists—and sought explanations about how these classes could and should organize and share economic production.[17]

In the twentieth century, academic institutions tended to separate political economy into two different areas of study, political science and economic science. In the former, the central concerns are the emergence and spread of philosophies and ideologies about the meaning and autonomy of an individual within the larger state or collective society. In addition, political science is concerned with the organizational structure and operation of governments, states, and political parties.

Economics, in contrast, seeks to understand how resources can be used efficiently. Typically, economists are concerned that resources such as land, water, minerals, energy, and people are deployed to produce maximum wealth or utility. Economists believe they have solved resource allocation problems when they have identified a scheme such that no other scheme exists that can enhance one person's utility without decreasing another person's. Modern economics often divides its attention between the problems individuals have in resource deployment (microeconomics) and the problems of the collective or the state (macroeconomics).

Political science and economic science had common origins in the eighteenth-century studies of philosophers like Adam Smith, who intended to forge an inquiry into the laws of political economy that would be the intellectual equivalent of Newton's studies of the universe. By the early twentieth century, the rise of democratic culture and an embrace of mathematical modeling had obscured the political dimensions of political economy to create economic science. Some scholars, such as Marx and Veblen, continued to promote the integrated study of wealth and power, but the preponderance of professional economic interest gravitated to abstract arguments, often devoid of linkages to peoples' ordinary lives.[18] Thus our language and frameworks of analysis acquired a mythology that led us to view the production and distribution of wealth as separate from the creation and exercise of authority.

Not only did the dismemberment of political economy leave us unprepared to deal with intertwined questions of wealth and power, but also both political science and economic science tended to ignore the idea that the generation of wealth depends in part upon the productivity of ecosystems. For example, agriculture allows people to channel the productivity of photosynthesis into such products as grain, which is a basis of wealth and power in virtually all human societies.

Political ecology synthesizes the concerns of ecology and political economy. Its central mission is to understand historically how people modified ecosystems and intertwined ecosystem productivities with the production and distribution of wealth and the exercise of power. Political ecology absorbs the concept of the ecosystem and emphasizes that it is the only practical source of primary production or photosynthesis. People are absolutely tied to the amount of primary production in the biosphere (the global ecosystem) because that is the sole basis of the food supply. Agriculture is one of the key concerns of political ecology because it is the most important technology with which people channel the primary productivity of ecosystems into food for survival and into the wealth and power central to human societies.

Plant Breeding and Yields

An inquiry into agriculture from the political ecological viewpoint focuses on how and why people modify and harvest ecosystems to obtain their needs, and create political economic structures to control the production and distribution of the ecosystem's productivity or yield. For most of human history, yield was always valuable and only occasionally became large enough to be considered excessive. (Generally the periods of surplus have been confined to the nineteenth and twentieth centuries.) One chronic political ecological problem to solve, therefore, was how to increase yields from the biosphere.

People who till the soil have known for millennia of two fundamentally different ways to increase the yield of the harvest. The first method is to increase the amount of land under cultivation, and the second is to increase the yield per area of land. Either way, the total yield goes up. Expansion of cultivated area was the most important way of increasing the harvest until about 1900. To be sure, history can point to a few instances in which new methods increased yields per hectare before that time. Nevertheless, from the beginning of agriculture some 10,000 years ago until 1900, the primary method of increasing the total yield was to increase the amount of land tilled.

A change of enormous importance happened in the years after 1900: farmers guided by science learned how to make each hectare of land yield more. Traces of this yield transformation were visible in the eighteenth century and before, but the most dramatic increases in yield per hectare came after 1945. Particular spots in Europe, Japan, and North America were the first locations of the transformation in yields, but ultimately the knowledge on which it was based spread to many other countries. By 1980, efforts were under way to make the knowledge available to every part of the world.

This revolution in yields was intimately connected to the factors determining land control. An individual who could successfully use the higher yielding practices was in a better position to amass wealth, with which acquisition of land might be possible. Reciprocally, control of land use was essential to using the new science-based production technologies.

The yield transformation was also one of the factors that influenced the political and military strength of nation-states. Cultures that first learned how to obtain higher yields were in a better position to control areas of land. It is thus perhaps not a coincidence that the yield-enhancing practices developed in Europe after the eighteenth century partially enabled the spread of European imperialism in the nineteenth and twentieth centuries.

New scientific and technological knowledge lay behind the transformation of yields. What were the sources of the new science and technology? Why were these new practices developed? What effects did the initial successes with yield enhancement have on subsequent efforts to increase yield yet again?

Important new technological practices in eighteenth-century Europe, spawned largely by gentlemen and farmers, were the proximate roots of the yield-enhancing practices of the twentieth century. These were the days before professional cadres of scientists, but by the early 1900s development of new agricultural technologies was largely in the province of organized, institutionally supported professionals.

A key factor in the coalescence of professional science was the close relationships among (1) the desire to develop better agricultural science, (2) the ability of a society to support a cadre of scientists, and (3) the power to allocate resources toward the research enterprise. Essentially a positive feedback loop developed in which higher yields translated into more wealth, which in turn prompted landowners and others to desire yet higher yields. The new wealth from the previous successes in turn provided the potential to support yet further research and development, and those who controlled this wealth had the power to direct its allocation to research. New practices produced a new wave of yield enhancement, which ignited the cycle again.

Plant breeders were the key people in the yield transformation because they selected the plant varieties that were genetically able to produce higher yields. Individuals from other sciences were also involved, particularly soil scientists, fertilizer chemists, hydrologists and irrigation specialists, entomologists and plant pathologists, and statisticians. Nevertheless, it was the plant breeders who more than anyone else created the conditions for the yield transformation, and it is primarily their story that needs to be understood.

What is so remarkable about the plant breeders is that they are essentially unknown by the general public. Yet plant breeders have been responsible for a radical revolution in human ecology. Larger yields after the eighteenth century increasingly enabled a higher proportion of people to forgo agricultural labor and turn to the emerging factories for work. Increased numbers of people working in factories ultimately meant a redistribution of people from the rural to the urban areas. In a very direct way, therefore, the development of higher yields must be seen as a component of the industrial revolution and the general process of urbanization, which became global in the second half of the twentieth century.

In fact, it is possible to think of increased agricultural yields, particularly of cereals, as an ability to form capital—an accumulation of goods devoted to the production of other goods. An increase in cereal yields, if it is beyond the needs of the producers for their own subsistence, can be accumulated and used to support human labor to make something besides more cereal grain. Therefore, the owner of surplus cereal grain can turn the surplus into capital and thus promote the production of many other types of goods and services.

Another way to look at the revolutionary implications of yield-enhancing technologies is to imagine life without them. First, on a dietary level, smaller supplies of cereals would mean more expensive staples and livestock products. In addition, industrial uses of grains would be less common. Lower yields would also mean that more land had to be cultivated to get the same yield. It is possible, therefore, that the earth would not now be supporting close to 6 billion people, that is, the population growth of the last 300 years would have leveled off. Finally, the need to cultivate more land, combined with fewer people working in industry, would probably mean that farmwork would be less mechanized. As a result of less mechanization, more labor would be needed in rural areas, and fewer people would live in cities. In total, the lives of each of us probably would be very different had these yield-enhancing techniques not been developed.

It is a more complex question to ask whether people would be better off without the yield-enhancing practices. What is simple to say is that our relationships with nature and with each other would be greatly different. Technologies that enhanced yields changed human political ecology, possibly forever.

Global Links: Plant Breeding and Nation-States

Britain and the United States have heavily contributed to the development of plant breeding. In the early part of the 1900s, agricultural scientists were located almost entirely in industrialized countries. By 1980 many third world countries had acquired a cadre of trained agricultural scientists, many of whom had received their advanced work in the United States, Europe, or another third world country. Plant breeders in each country worked to create and find the varieties that were suited to their locations, to the skills and aspirations of their farmers, and to the palates of local populations. Plant breeding, therefore, was a highly "site-specific science," that is, its detailed events were tied to the specific conditions where it was used. Explaining the yield transformation, therefore, requires a detailed look at specific events that are considerably less than global in scope. At the same time, it will be important to understand the links between events in different places in order to comprehend the universal features of the yield transformation.

Plant-breeding networks now facilitate the exchange of people, seeds, and ideas across national boundaries and among different crops. Industrialized countries, particularly the United States and the United Kingdom, played a fundamental role in the creation of the most important networks. How are we to understand the concerns of nations that developed the global network of plant breeders? Nation-states are the creations of Mars, and their histories are often tied to the changing tides of war. Nation-states can also be understood through their role in protecting property interests of a particular class (Marxist scholars) or by their role in promoting individual enterprise (liberal democratic theorists). Pluralists see the state as the balancer among competing groups so that all are happy enough with the compromises achieved.

Thus there are many theories about the nation-state, but plant breeders have generally ignored them. Concurrently, those who theorize about the state have usually ignored the work of plant-breeding scientists. Food supply, however, figures prominently in the strength and stability of a nation-state. Internal stability in times of peace is heavily dependent upon a safe and steady food supply, to both urban and rural people. Advent of war brings the question of food supply into critical focus. Neither armies nor urban workforces nor farmers can function to defend the nation if their food supply is interrupted, inadequate in quality or quantity, or unsafe. Targeting the enemy's food supply, a practice used more than once in the many bloody wars of history, demonstrates the strategic importance of agricultural production.

Plant breeders and other agricultural scientists became part of the strategic personnel of a modern economy in the twentieth century as they developed the ability to increase and stabilize yields per hectare. Their work was critical to assuring the food and industrial supplies of the nation. Moreover, they helped develop yet new accumulations of capital in the form of agricultural surpluses, which enabled ever more people to forsake agricultural labor. The time scale on which plant breeders work, often five to ten years to create a new variety, was disjointed from the time frame in which national security matters were settled between nations, usually in months or a few years. Nevertheless, the long-term health of the plant-breeding enterprise became one foundation of a nation's security.

Not only did plant breeders find themselves part of the modern economy, but also they became indirectly immersed in struggles over who would control land within nations and who would farm. Agriculture's story in the twentieth century is one in which landowners tended to replace human labor with capital inputs in the farm production process. The plant breeder contributed to the process of capital substituting for labor because it was the plant breeder who identified the plant varieties that did best with other capital inputs such as fertilizer, irrigation, pesticides, and machinery. A modest yield transformation could have occurred without the efforts of plant breeding, but the magnitude of what actually happened was critically dependent upon the breeder.

Farmers and other businesspeople who mastered the technical and financial aspects of the capital-intensive innovations were able to use their skill to control the land. Other farmers who were not so technically proficient often sold or lost their land. As a result, modern farms became large, quasi-industrial firms characterized by large yields per hectare and per person-hour of labor. Small farms, providing modest but dignified employment to family labor, increasingly became a relic of the past. This process has been characterized by economists as the operation of an agricultural treadmill.[19] Farmers who did not keep up with the changes in technology eventually saw their farms go out of business.

An explanation of the significance of plant-breeding science must therefore incorporate its importance for both the external and the internal political economy of the nation-state. A country's position and strength in the world depended in part on the plenty of its harvests. A farmer's position and strength in society depended in part on the magnitude of yields. Plant breeding played an important role in shaping the external and internal destinies of nations. It thereby also affected the details of human ecology: where people lived, what they did for work, and how they tapped into photosynthetic energy were all impacted by the results of plant breeding.

Plant Breeding, the Social Aspects of Knowledge, and Development

Political ecology seeks to understand how and why plant breeders modified agricultural ecosystems and thus the wealth and power of individuals and nation-states. One key to this effort is how social processes affect the development of technical and scientific knowledge. A social constructionist perspective sees specialist knowledge as one of the many artifacts that characterize a civilization or culture. It focuses on the social processes by which people identify problems, search for technical solutions, and put forth tentative answers.[20] Adoption of an answer by others indicates whether the new technical knowledge solves the problem and is the final arbiter of whether the knowledge is true.

Social processes implicit in the identification of problems and the proving of proposed new technological answers are usually the avenues by which political power enters into the issue of which technologies get developed and adopted. Those people who have power are able to argue that their identification of the problem is "correct," and they are able to guide the work of technologists and scientists toward solutions that make sense for them. Powerful individuals are also able to establish the parameters within which the verification and adoption steps are conducted.

In these ways, one component of the exercise of political power is the ability to influence what sorts of technological practices get invented and utilized. Once adopted, a new technology may increase the wealth and power of its advocates, thus giving them further abilities to influence the next round of technological development.

Plant breeding created new capital because it helped create surplus grain. Those who controlled the distribution of this grain (some farmers, grain merchants, and others) thus saw plant-breeding science as a potential route to further capital accumulation, which in turn spurred interest in further development of high-yielding varieties. In this way, the desire for capital accumulation, the fundamental motivator in capitalist societies, was harnessed to building political support for programs in plant-breeding science. New knowledge produced new and higher yielding production practices, which in turn promoted appreciation for yet further developments of plant-breeding science.[21]

A second key to understanding plant breeding comes from the work of Joseph Schumpeter and his concept of capitalism as a system of "creative destruction." Schumpeter saw that new technological processes, constantly proliferated by capitalist economies, upset the existing methods of doing things. As a result, new patterns of wealth, power, and prestige emerged to replace the old order, and the new order itself would be replaced after yet another round of innovation.[22]

A political ecological framework thus attempts to understand plant breeders as self-conscious inventors. Their views of the problems to be solved—low and often unstable yields—reflected the interests of political and economic leaders, including some farmers, particularly the largest and most technically proficient ones. Plant breeders sought new plant varieties that gave higher yields, and their work was subjected to a testing process affected by a wide range of social interests, including farmers, food industrialists, and consumers. In turn, this new knowledge fed into the yield transformation that increased capital and created Schumpeterian creative destruction. Social orders within and between nation-states changed in response to the wealth and power generated by the increased yields.

It is ironic that the word "destruction" must enter into an understanding of the work of plant breeding. After all, the science ostensibly was interested in the production of plenty and the elimination of human drudgery. If plant breeders make better plants that produce more food with less work, is that not a positive contribution? New technologies, however, invariably created the seeds for the destruction of old ways of doing things. Winners and losers emerged from Schumpeterian creative destruction. Most importantly, however, understanding the role of capital accumulation behind the science helps illuminate why the increased plenty was not necessarily channeled toward the elimination of hunger and starvation.

Knowledge of new plant varieties and how to use them, often accompanied by a panoply of other technical and social changes, was the bedrock on which the yield transformation of the twentieth century occurred. Many people who formerly were farmers found it impossible to continue in that work. This was Schumpeter's creative destruction in operation. At its foundation was the knowledge base of plant-breeding science, constantly changing through the social processes of capitalist economies.

Schumpeter envisioned creative destruction operating without heavy government involvement in the details of change. After 1945, however, several governments began consciously to promote new technologies to *develop* other countries, which was a euphemism for promoting creative destruction. Perhaps *development* was one of the most ironic concepts to enter late-twentieth-century language, and in basic ways its meaning was tied to the results of plant-breeding science and the yield transformation.

Development commonly means a society that has material plenty through urban industry and modern agriculture. Human labor in such societies is highly productive in terms of creating a great deal of wealth with relatively small inputs of labor. An industrial society, however, rests importantly on the work of the plant breeder. It was the breeder who found the plants that could be grown efficiently, that is, more harvest with lower costs per unit of harvest. Fewer people could provide all the food for a population, and many formerly rural people moved into the cities to work in industries and services.

Cultures that have not made this transformation are considered backward, traditional, or undeveloped. The historical development of plant breeding was intimately involved with efforts by people in the developed countries to spread the new technology of high-yielding plants to the less developed countries. In fact, it might be said that being developed required the adoption of technologies created by the plant breeders and other agricultural scientists. At a very fundamental level, therefore, use of plant-breeding science became synonymous with the property of being developed.

Nevertheless, considerable controversy about the use of high-yielding varieties reverberated among policy analysts of both the industrialized and less industrialized countries. At least four different critiques and assessments emerged, although the categories were not mutually exclusive.

One school of thought was developed by those connected to the actual work of agricultural modernization. It tended to celebrate the scientific triumphs, particularly as they occurred in less industrialized countries such as India, Pakistan, the Philippines, and elsewhere in the third world. Development in this school was the same as progress, and to undergo the yield transformation was a route to humanitarian salvation, prosperity, and freedom for a previously poor people. In this analysis, those who provided the technical assistance to promote the transformation acted for the good of all humanity.[23]

A second set of conclusions was a somewhat more pessimistic analysis that emanated from some who would not have disagreed with the previous analysis. Transformation of yield, in this view, may have been a technical and humanitarian achievement, but its primary function was to provide temporary relief from what was seen as the inexorable and undesirable growth of the human population. Often the term *population monster* was used to create the image of people breeding out of control and threatening to outrun their food supplies. This second image of the yield transformation had most of its intellectual and emotional roots in the political economic thought of Thomas Robert Malthus, but it also drew on elements of ecological and conservation science.[24]

Not all analysts were happy with what they saw from the yield transformation as it occurred in both industrialized and less industrialized countries. A third school of thought focused on the idea that the technology for yield transformation enabled those farmers better endowed with education, capital, or political power to outcompete their lesser endowed colleagues and thereby drive the latter out of business. This analysis saw the yield transformation not just as a technical matter based on plant breeding but as a source of social inequity and misery for small farmers. Rural sociologist Jack Kloppenburg saw a different problem emerge from plant-breeding research: the concentration of wealth through the use of legislation to protect plant varieties. Justice, economic and political stability, and the moral legitimacy of society were casualties of a Faustian bargain to get higher yields.[25]

Another branch of critical and pessimistic thought about the yield transformation focused on the environmental damage caused by the transformation of yields. In this fourth school of thought, the new varieties created by the plant breeders led to reductions in biodiversity; destruction of the soil through erosion, salinization, or compaction; increased and unhealthy dependencies upon fertilizers and pesticides; contamination or destruction of water supplies; and an inviable dependency on the use of soon-to-vanish fossil fuels. In short, this analysis criticized the technology of the yield transformation as ultimately unsustainable and therefore unwise.[26]

Each of the preceding four analyses of yield transformation is supported by an empirical body of evidence. In addition, each has support from various segments of society. Some would argue that any faults of the plant breeders' results could be mitigated by appropriate social and environmental policies. Therefore, so the argument goes, if some benefits were produced along with some faults, societies can keep the benefits while softening the harsh consequences of agricultural modernization.

Unfortunately, arguments about the social and environmental meanings of the yield transformation suffer from a limitation of view that renders each of them fatally flawed as a guide for understanding past events and for shaping the agricultural reforms of the future. This is not because no wisdom attends any of the four perspectives, but because each in its own way has critical gaps that render it inadequate. The political ecological framework for analysis helps to fill the most important gaps.

The Political Ecology of Transforming Yields

Political ecology begins with the premise that people must harvest and therefore modify the ecosystem in which they live in order to survive. Considerations of how much primary productivity (photosynthesis) must be captured, and how it should be captured, in order to support a growing population of over 5 billion people lies at the center of a political ecological analysis. Ecological science may not be able to tell us everything about how an ecosystem functions, but detailed natural histories of particular species may tell us what we and they need to thrive.

Political ecology directs our attention to the particular technologies we use in order to access natural resources to satisfy our physiological needs. This perspective reminds us that we don't have the option of forgoing technology to harvest the primary productivity of the biosphere. All we can do is understand which technologies are likely to be capable of providing access to sufficient primary productivity and perhaps some understanding of whether the use of those technologies is likely to destroy the very resource they are designed to tap or other parts of the ecosystem upon which we and other species depend.

Political ecology therefore grounds our understanding of the wealth needed to support human culture directly in the functioning of ecosystems. As such it links (1) an understanding of resources, (2) the technologies capable of accessing those resources, (3) the transformation of those resources into wealth and power, and (4) the role of capital accumulation in driving some entrepreneurs to seek new technologies in order to achieve yet higher yields. Political ecology helps focus our attention on issues of physical potential, political economic impact, and moral significance.

This book uses a political ecological perspective to explore the yield transformation in twentieth-century agriculture, particularly as it occurred after 1945. High-yielding varieties of cereals are so important as human food that it is imperative to know how these plant varieties were created and identified. It is also imperative to understand why both nations and farmers all over the world made the decision to use them.

Most of the crucial events of the story took place within a quarter century, between 1945 and 1970. Nevertheless, the roots of the change stretch easily to the late nineteenth century. More subtle traces lead back to the Neolithic period and the origins of agriculture. Thus the inquiry is a historical reconstruction of past events in science, technology, and the political economy of agriculture.

Agriculture is extremely complex, however, so this inquiry could not hope to be comprehensive for all crops and all areas. Instead, it relies on a case study approach: wheat in Mexico, the United States, India, and the United Kingdom. Other crops, especially rice, also would have been interesting and informative, but the yield transformation through plant breeding had some of its first successes in wheat. Similarly, maize could have served as the crop example, but maize has less importance in many areas as a direct human food. In addition, the genetic basis of high-yielding maize, heterosis, is still less well understood than the simpler scientific information about the genetic nature of high-yielding wheat.

The four countries that figure most prominently in the story are each present for important reasons. Important formative events in modern plant breeding took place in both the United Kingdom and the United States. Thus to understand the roots of this science, one needs to understand both these countries. Administratively, the United States provided the model, widely imitated, for organizing the work of plant breeders and the technical support system for farmers who might potentially use the new plant varieties. Chronologically, the first breakthroughs to get high-yielding varieties of wheat (outside of Japan) came in Britain, the United States, and Mexico. Events in Mexico had crucial significance for the yield transformation in India and the United Kingdom. Both Mexico and India exemplified the processes by which less industrialized countries decided to adopt the high-yielding varieties. Somewhat ironically but importantly for understanding the global dimensions of the yield transformation, the United Kingdom was one of the later arrivals to the countries that decided to adopt the high-yielding varieties. An understanding of why Britain finally embraced the science it helped create is crucial to understanding the overall reasons for the yield transformation.

One other important point linked these four countries and wheat: wheat is a major cereal crop in each of the countries. For direct human consumption, wheat has no rivals in the United States and the United Kingdom. In Mexico, maize rivals wheat as a grain for human food, and rice plays a similar role in India. Nevertheless, in each of the four countries, wheat has been a significant crop—critical for the health, prosperity, and stability of each nation. It generated a historical record that could be used to understand why each country in turn made the switch to use high-yielding varieties.

No single analytical framework can ever illuminate all facets of a complex subject. Nevertheless, political ecology can aid understanding of the world's premier industry and the earth's most important human-dominated ecosystem.

2

Wheat, People,
and Plant Breeding

Selecting improved varieties of wheat from among existing wheat plants is an ancient art that dates back thousands of years. In contrast, the deliberate generation of new varieties by controlled breeding is more recent. Wheat breeding developed from an arcane art practiced only by a few isolated individuals into a global community of professional scientists in the period from about the mid–eighteenth century to about 1925, but especially from about 1875 to 1925.

Wheat improvement, however, ultimately involved more than just finding or creating varieties with greater utility. A relationship between people and wheat developed over the millennia that increasingly left both species in a state of ever higher mutual dependency. Put another way, wheat and people coevolved in ways that left neither much ability to prosper without the other. Professional wheat breeders occupied a pivotal role in this ongoing coevolutionary process, especially after the nineteenth century. An understanding of wheat breeding thus depends upon understanding how wheat and people "grew up together."

The Wheat Plant

Wheat in everyday English designates a particular grassy plant that produces a starchy grain or seed. Most people think of wheat primarily in terms of this grain, which is used to make bread, cookies (biscuits), pastries, and pasta. Consumers easily distinguish between wheat and other grains such as rice, oats, maize, rye, and barley as they appear in manufactured products or as ready-to-consume grain in food stores.

In contrast to their savvy as consumers, most urban dwellers probably could not differentiate between these grains in the farmer's field, particularly between wheat, rye, and barley. Nor could they necessarily give a good explanation of why wheat is particularly suitable for the products in which it is used. Moreover, they probably would be unfamiliar with other uses of wheat, such as using the grain for feed or the straw for fodder and roof thatching. Finally, in all likelihood these consumers would be hard-pressed to give details about the quantities of grain that can be obtained per hectare per year or much about how yields have increased in recent decades.

In short, most consumers know and appreciate wheat but only on rather narrow and unsophisticated grounds. To understand the remarkable increases in yield that were obtained in wheat after 1940 requires delving briefly into the botanical properties of the plant and knowing how wheat came to be the single most important grain crop in the world. From a botanical point of view, three questions are most prominent. First, what is the normal life cycle of wheat? Second, what are the basic anatomical parts of the wheat plant? These first two questions are fundamental to the working tools of wheat breeders because a major part of plant breeding involved learning to manipulate the life cycle and anatomy of the plant in order to achieve objectives desired. Third, how can one most usefully distinguish the different types or varieties of plants, all of which we call by the generic term, wheat?

Wheat's life cycle and anatomy can be briefly summarized (Figures 2.1 and 2.2).[1] A wheat seed (grain) consists of a plant embryo and starchy endosperm surrounded by a protective seed coat. Under proper conditions, the seed imbibes water and initiates the sprouting process by sending out a coleorhiza, which gives rise to roots, and a coleoptile, a protective sheath that pushes above the ground and allows the first leaves contained within it to emerge into the daylight and begin photosynthesis. Until the first leaves start to photosynthesize, the young seedling is dependent upon the sugars stored in the starchy endosperm for its energy.

Growth above the ground during the first part of the plant's life consists primarily of the production of new leaves. Each leaf develops from a small ridge on the growing tip of the main stem of the young wheat plant, which for many weeks remains hidden beneath the ground. In the early stages of the plant's growth, the distance along the stem between leaves is small, and from above the ground it appears that the leaves simply emerge from a small lump of tissue (the crown) that lies just beneath the soil surface.

Events of tremendous importance to yields, however, are occurring within the crown of the young plant. The growing tip of the main stem, after four to eight leaves have emerged above ground and after producing all of the ridges that give rise to leaves, changes from vegetative growth (production of leaves) to ear and spikelet formation, the structures within which the seed (grain) of the next generation will form. Each growing tip produces about twenty spikelets on the wheat ear.

In addition to these changes in the growing tip of the main stem, the young wheat plant also begins to form tillers, or secondary stems, that emerge from the axils of the first several leaves. (Axils are the plant tissues between the main stem and its leaves.) Each tiller also develops a growing tip that produces first leaves and then ears and

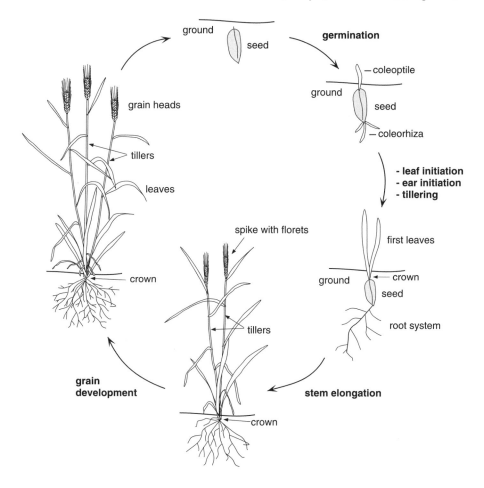

Figure 2.1 Life cycle of wheat. Line drawing by Tim F. Knight. Adapted from E. J. M Kirby and Margaret Appleyard, *Cereal Development Guide*, 2d ed. (Warwickshire, England: National Agricultural Centre, 1987), p. 4.

spikelets. Not all axils produce tillers, but the ability of wheat to form these structures is critical to obtaining high yields from the plant. Typically, a modern high-yielding variety of wheat will produce one main stem and about three tillers.

As the season progresses, the growing tips of both the main stem and the tillers move from ear and spikelet formation into floret formation. The ear of a wheat plant is called the spike, and spikelets are small structures along the ear that contain the flowers or florets of the plant. Each floret has the potential to form a new seed (grain), and each spikelet may form as many as ten florets. Seldom, however, do more than three to six florets actually mature and produce a seed.

Once the florets are well formed but still immature, the process of stem elongation begins. Elongation brings the growing tips of the main stem and of the tillers

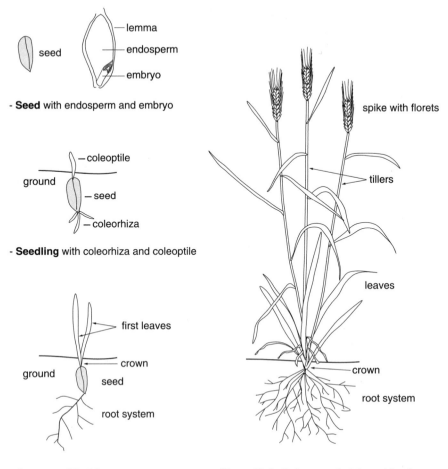

Figure 2.2   Anatomy of the wheat plant. Line drawing by Tim F. Knight. Adapted from E. J. M Kirby and Margaret Appleyard, *Cereal Development Guide*, 2d ed. (Warwickshire, England: National Agricultural Centre, 1987), pp. 4, 12, 13, 14.

above ground, to a height of 0.5 to 2.0 meters, depending upon the variety. There the florets mature, meaning the anthers release their pollen, which lands on the receptive stigmas of the female parts of the floret. Fertilization thus occurs, and the floret proceeds to ripen a new seed. If the seed is used for replanting, the life cycle starts again. If it is diverted for food or feed purposes, we call it grain.

Yields from seed to harvest are critically dependent upon the development sequence just described. Some simple arithmetic makes clear the magnitude of increase that can be obtained: one seed can typically give rise to a main stem and three tillers to create four ears of wheat. Each ear can have twenty spikelets, each with three florets that form a new seed. Thus one seed can give rise in the next generation to something like 240 seeds (4 ears × 20 spikelets/ear × 3 florets/spikelet = 240 new seeds). If the plant produces even one or two more tillers, the seed yield per plant can go over 300 seeds. To be sure, other factors, particularly soil fertility, plant spacing, temperature, water, and pest problems, can diminish these yields. But a potential for such a high return on the seeds planted is present. Total yield of grain in kilograms per hectare will also depend upon the size and weight of each grain.

In contrast to the relative simplicity of the life cycle and anatomy of the wheat plant, its classification shows a bewildering confusion and uncertainty. Some semblance of order and tidiness, however, emerges from realizing that wheat is now classified within two different but hierarchically related schemes. First, the plant is classified by botanists in ways that show how they think it is linked anatomically, genetically, and evolutionarily to other species of plants. Second, wheat is classified by agronomists in ways that distinguish critically important differences between varieties in terms of how the crop can be grown and used.

Wheat has been placed in formal classification schemes since at least classical antiquity, when Columella identified two classifications similar to what were later called the "naked" and "hulled" categories.[2] Naked wheats were those in which the seed detached easily from the ear and spikelet; hulled wheats were those in which the rachis, or backbone of the ear, broke ("shattered") and the seeds were tightly held inside chaff (the glumes of the spikelets, which enclosed the florets).

Linnaeus in the eighteenth century, in contrast to Columella, placed all wheats into one genus, *Triticum*, and identified a total of five species. Succeeding botanists of the eighteenth and nineteenth centuries continued to grapple with what were a seemingly unending series of variations by which wheat was known.[3] The number of species increased as scientific botanists became more familiar with a plant that grew in virtually all parts of the world except the very humid rainy tropics. Moreover, no scheme agreed much with the others in terms of precise names, the criteria by which names should be given, or how to relate the cultivated wheats with a number of grassy weeds that seemed to share many of wheat's characteristics.

A series of investigations in the twentieth century, however, brought a very different foundation to the question of how to organize the different varieties of wheat. At the heart of the matter was an understanding of how many chromosomes were in the nucleus of each cell of a wheat plant. Chromosomes were first identified as deeply staining bodies in the nucleus. Early in the twentieth century, Sutton and Boveri synthesized the known behavior of Mendel's inheritance factors and of chromosomes during cell division and reproduction. They argued that the factors controlling inheritance must be located on the chromosomes. Chromosomes thus became parts of the cell that were critical to the genetics and evolution of the plant. Contemporary classification schemes for wheat are based on the number of chromosomes in the somatic cells of the plants. (All cells are somatic except those giving rise to the male pollen cells and female ovule cells.)

Most authorities now agree that wheat comes in three major groups.[4] The "diploid" group have fourteen chromosomes in each somatic cell, the "tetraploid" group have twenty-eight, and the "hexaploid" have forty-two.[5] Virtually all wheat cultivated in the world today is a hexaploid wheat generally designated as *Triticum aestivum*. Substantial quantities of the tetraploid wheat, *Triticum durum*, are also grown. Only very minor quantities of other varieties are still in cultivation.

Wheat is also considered by most contemporary botanists to be related in an evolutionary fashion to a great many other grasses, many of which are of high economic value to people. T. E. Miller, of the Plant Breeding Institute of England, places wheat in the tribe Triticeae of the family Poaceae (Gramineae). The tribe contains twenty-five generally recognized genera, each of which is composed of a series of species. Some genera have only annuals, some only perennials, and some both. Within the tribe Triticeae are three genera of high importance as human food and animal feed: *Triticum*, which contains wheats; *Secale*, which contains ryes; and *Hordeum*, which contains barleys.[6] To the casual eye, in fact, wheat, rye, and barley can be easily confused. To those who grow and use these cereals directly, however, the differences between these three grains are large.

Agronomists and cereal technologists take up the classification problems of wheat where the botanists leave off.[7] Three sets of characteristics are generally of most importance in the classification of wheat by practical considerations of growing and using the grain: growth habit, hardness, and color.

Growth habit refers to the time of normal planting of wheat when it is grown in the northern temperate regions of the world. "Winter" wheats are those that are planted in the fall, grow for a short period before cold weather, remain dormant over the winter, resume growth in the spring, and ripen for harvest starting about midsummer. "Spring" wheats, in contrast, are planted in the spring, grow over the summer, and ripen in late summer to early fall. Spring wheats are generally grown in areas with severe winters that kill the overwintering plants of winter wheat varieties. Winter wheats are otherwise often preferred because they yield more than do typical spring wheats. The extra yield results primarily from the longer time they have in the ground and from the fact that winter wheats resuming growth in the spring are already established plants just waiting to take off.

Hardness refers to the texture of the starchy endosperm of the grain. In North America "soft" wheat means a wheat that when ground into flour gives large amounts of finely granulated material. In contrast, the "hard" wheats make a coarser product. Hard wheats tend to have more protein, which is thought to adhere to the starchy material in the grain and make the flour coarser than in soft wheats.[8]

Hard wheats, because of the extra protein, are good for making breads with a highly spongy texture. In fact, some baking technologists go so far as to refer to bread as "foamed gluten." Gluten is the elastic protein in wheat that is puffed up by the carbon dioxide released by yeast.[9] They are used for the leavened breads typical of the United States, Canada, and Britain. Soft wheats are good for making unspongy bread, typical of "French" bread, crackers, pastries, cakes, cookies or biscuits, and noodles typical of eastern and Southeast Asia.

Hardness is mostly an inherited trait, but environmental conditions can affect the hardness of ripened grain. Higher nitrogen fertilizer use can increase the protein content and hardness of wheat, but the effect of heredity is stronger.[10]

In Western Europe the distinction between hardness and softness refers to the differences between grains of the species *Triticum aestivum*, all of which are "soft," and *Triticum durum*, which are "hard." Millers recognize that grains of *Triticum aestivum* vary in hardness,[11] but the differences apparently are not significant enough to result in a definitive label as in North America. In North America, grains from

Triticum durum are known as macaroni or pasta wheats, reflecting the predominant uses of that variety.

Color is the third major trait for classification of wheats. Most wheats are either "red" or "white," the difference being whether or not the seed coat contains a colored, resinous material.[12] Determination of color is genetic.[13] In some areas such as the United States, different wheat-growing regions are often known by whether red or white wheats predominate.

The distinction between red and white wheats is highly visible in the harvested grain, but it tends to be of practical importance only in special circumstances. In South Asia, for example, people may have a strong preference for white wheats for making chapatis because they prefer the lighter white color of flour made from white wheats, but they will use red grains if no white wheats are available. In the United States and eastern Asia, red soft wheats are preferred to white soft wheats for producing soup thickeners. This preference derives from the higher resistance red soft wheats have to sprouting of the grain in the ear before harvest. When a grain sprouts, the starch is degraded and is less suitable for uses such as soup thickener.[14] A soup manufacturer is thus safer in buying soft red wheat than soft white wheat, which may have been damaged by sprouting in wet harvest years.

Wheat: A Global Crop

Wheat is now grown on each continent, and, in terms of its total production is one of the world's two most important cereal crops. Only rice rivals it, and maize, barley, sorghum, millets, rye, and oats come behind (Table 2.1). Unraveling the origins of wheat as a global crop involves two sorts of questions. First, what are the botanical origins of the different types of wheat? Second, what role did the evolving wheats play in the origins of agriculture as a mode of human life? Studies in archaeology, paleoethnobotany, cytogenetics, and plant biochemistry in the past forty years have been combined to suggest that the answers to these two questions are completely and inseparably intertwined.

How wheat originated as a botanical species has long occupied the thoughts of scholars, philosophers, priests, and scientists. For the Greeks, Romans, and ancient

Table 2.1   Worldwide major cereal crop production levels, 1993–94 to 1995–96

Year	Crop (million tons)			
	Wheat	Rice[a]	Coarse Grains	All Cereals
1993–94	565	529	803	1,896
1994–95[b]	528	540	884	1,952
1995–96[c]	550	545	839	1,933

[a]Paddy (grain before milling).
[b]Estimated.
[c]Forecast.

Source: Food and Agriculture Organization, *Food Outlook*, no. 5–6, May–June 1995, p. 2.

Chinese, the existence of wheat was connected to divinity. The Greeks, for example, considered wheat the gift of Demeter, goddess of the fruitful soil. Ceres was the counterpart of Demeter in Roman mythology,[15] and our English word "cereal" is derived from this ancient deity.

Science has preferred natural rather than supernatural stories for the origin of wheat. A series of studies in the twentieth century have indicated that wheat in its many varieties was the product of a complex series of hybridizations among what originally were wild grasses of southwestern Asia (Figure 2.3). Some of these hybridizations occurred most probably under conditions of cultivation, that is, after wheat had been domesticated for agriculture.

The first hybridization of importance to the origins of *Triticum aestivum* was probably a cross between two wild grasses, *Aegilops sitopsis* and *Triticum urartu*. Each of the two wild grasses was a diploid with fourteen chromosomes. The hybrid between them, *Triticum dicoccoides*, was a tetraploid with twenty-eight chromosomes.[16] Once formed, *Triticum dicoccoides* tended to "breed true," that is, pollen tended to fertilize ovules on the same plant so that each new generation would be like its parents. Such self-fertilizing plants were prominent among the plants domesticated in southwestern Asia. This trait helped keep the newly domesticated varieties from interbreeding with their weedy progenitors.[17]

Considerable uncertainty still surrounds the question of when and how *Triticum dicoccoides* became the domesticated *Triticum dicoccum*, but little doubt remains

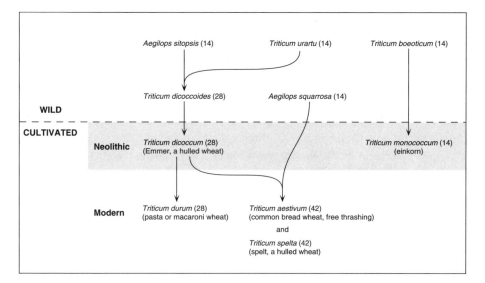

Figure 2.3   Origins of cultivated wheats from wild grasses. Numbers in parentheses refer to number of chromosomes. Line drawing by Tim F. Knight. Adapted from T. E. Miller, Systematics and Evolution, in *Wheat Breeding: Its Scientific Basis*, ed. F. G. H. Lupton (London: Chapman and Hall, 1987), p. 22.

that the cultivated variety was being farmed as early as 7800 B.C. By about 6000 B.C. it was being farmed in southeastern Europe, and by 3000 B.C. in Egypt, the Mediterranean basin, Europe, central Asia, India, and Ethiopia.[18]

Triticum dicoccum, also known as emmer, was almost certainly the dominant wheat in Neolithic farming, but at an early stage of agriculture it gave rise to what we now call *Triticum aestivum,* or common bread wheat. Cytological studies of wheat chromosomes suggest that bread wheat resulted from the hybridization of *Triticum dicoccum,* a tetraploid with twenty-eight chromosomes, with the wild diploid (fourteen chromosomes) grass *Aegilops squarrosa.* Bread wheat, as a result, is a hexaploid with forty-two chromosomes. Archaeological evidence suggests that hexaploid wheat may have been cultivated as early as 7000 B.C.,[19] certainly by 5000 B.C.[20]

Triticum dicoccum also gave rise to the other major variety of contemporary wheat, *Triticum durum,* or macaroni wheat. *Triticum durum* is a tetraploid wheat, too, but it differs from its progenitor in having a free-threshing grain with a tough, nonshattering ear, that is, the grain detaches from the ear during threshing without bringing along the rachis (backbone of the ear) or the glumes (chaff). *Triticum aestivum* is also free-threshing. The difference between free-threshing varieties that give a naked grain and those that give the hulled grains lies in one genetic trait, with the free-threshing form being dominant to its counterpart.[21]

Free-threshing was a trait that admirably suited a cereal for human use because the grain remained in the harvested ear until threshed. Ears that were not free-threshing tended to "shatter" or break apart, an adaptation that helped disperse seed in a non-domesticated plant. A plant that was nonshattering or free-threshing was dependent upon people for dispersal and propagation of the next generation. This mutation was one of the key changes in wheat that created the wheat-human codependency.[22]

During the Neolithic agricultural revolution, the dominance of emmer (*Triticum dicoccum*), a tetraploid, as the major cultivated grain was rivaled and sometimes surpassed by *Triticum monococcum,* a diploid, also known as *einkorn,* from the German meaning "one kernel." Each spikelet produces only one grain rather than the three or more that are characteristic of bread and macaroni wheats. Wild einkorn was probably collected as early as 9000 B.C. and was definitely one of the first cultivated forms of wheat. Einkorn, like emmer, spread to Europe and continues to be cultivated there in small amounts.[23]

We are particularly interested in the spread of wheat to the four countries critical to the development of high-yielding varieties in the years after 1940: England, Mexico, the United States, and India. In England (Figure 2.4) and India (Figure 2.5), wheat came several thousands of years ago and has been cultivated every year since then. In many respects both England and India are "wheat civilizations" in the sense that this cereal was a prime component of their existence as settled societies. Wheat came later to the United States and Mexico because it was an Old World crop that did not reach the New World until Columbus's voyage of 1493.

Wheat arrived in England thousands of years after Britain was colonized by people. Archaeological evidence suggests that agriculture came to Britain around 3200 B.C.,[24] about 3,000 years after the melting and receding glaciers had raised sea levels and cut the land link between Britain and continental Europe. Emmer and barley were the major cereals, but also of importance were einkorn, flax, bread (hexaploid) wheat,

Figure 2.4   Major wheat-producing areas of England. Line drawing by Tim F. Knight.
Outline of map adapted from National Geographic Society, *British Isles* (1:1,687,000)
(Washington, D.C.: National Geographic Society, 1979), 1 p. Wheat-growing areas adapted
from U.S. Department of Agriculture, *Major World Crop Areas and Climatic Profiles*
(Washington, D.C.: U.S. Department of Agriculture, 1987), Agriculture Handbook no. 664.

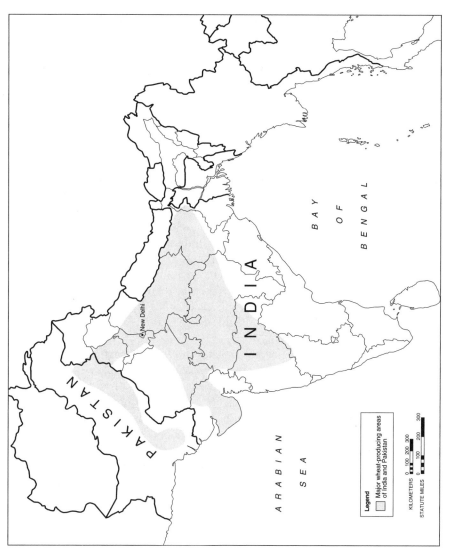

Figure 2.5 Major wheat-producing areas of India and Pakistan. Line drawing by Tim F. Knight. Outline of map adapted from National Geographic Society, *South Asia with Afghanistan and Burma* (1:6,522,000) (Washington, D.C.: National Geographic Society, 1984), 1 p. Wheat-growing areas adapted from U.S. Department of Agriculture, *Major World Crop Areas and Climatic Profiles* (Washington, D.C.: U.S. Department of Agriculture, 1987), Agriculture Handbook no. 664.

and spelt (probably *Triticum spelta*, a hexaploid wheat that has an easily shattered ear and yields hulled grain with the glumes or chaff tightly attached).[25]

Wheat arrived in India somewhat before it arrived in England, about 4000 B.C. Remnants of tools and weapons, dating to as many as 200,000 to 400,000 years ago, suggest that people from East Asia were the first to migrate into the northern parts of present-day India, while people from East Africa were the first migrants to the south. Little connection seems to have existed between the two, distinct cultures. Despite the relative nearness of India to southwestern Asia, the Neolithic agricultural revolution did not reach India until nearly 4,000 years after the first traces of wheat agriculture can be found in southwestern Asia.[26]

Regardless of why it may have taken a long time for the cultivation of wheat and barley to intrude into South Asia, one of the world's first complex and monumental societies eventually emerged along the course of the Indus River in what is now modern Pakistan. Harrapan culture, named after the major ancient city at Harrapa, thrived from about 2500 to 1600 B.C. At its peak the Indus civilization stretched into what are now the Indian states of Rajasthan and Gujarat. Wheat was its most important crop, but these people were also the first to begin using cotton for cloth, and they also relied on rice, peas, dates, mustard seeds, and sesamum. They also had domesticated many animals, including dogs, cats, camels, sheep, pigs, goats, water buffalos, zebus, elephants, and chickens.[27] To this day, wheat remains a foundation stone of the modern Indian civilization.

Human settlement of what is now the United States and Mexico (Figure 2.6) came much later than the Old World settlements of India and England. People were surely in those latter two places over 300,000 years ago, but humans probably did not reach the Americas earlier than about 40,000 years ago.[28] Wheat did not yet exist before the land bridges to North America flooded with the retreating ice and isolated the American continents from Eurasia. Therefore, the first people in the Americas had no wheat or any other Old World cereal. Complex civilizations emerged on the basis of maize and other seed crops in the Tehuacán valley in Mexico and from root crops such as manioc, sweet potato, achira, and potato in the lowland and highland areas of what is now Peru.[29]

Wheat came to the Americas with European invaders. Columbus recorded bringing wheat, beans, and chickpeas on his second voyage in 1493. The Spanish conquest of Mexico led to wheat and barley being grown around Mexico City by 1535 and exported to the West Indies. As the Spanish invasion spread northward to what is now Texas and New Mexico, wheat went along. French exploration and conquest of Louisiana also brought wheat into Texas in the eighteenth century.[30] Similarly, the invading English brought wheat with them to North America, planting wheat at Jamestown in 1607, and subsequent waves of settlers brought wheat to other European outposts.[31] Wheat now ranks as one of the foundations of New World agriculture. Only maize rivals it in terms of total production.

Wheat and People Coevolve

Many scholars have depicted wheat specifically and the Neolithic agricultural revolution in general in ecological terms. In outline, people are herbivore/carnivore crea-

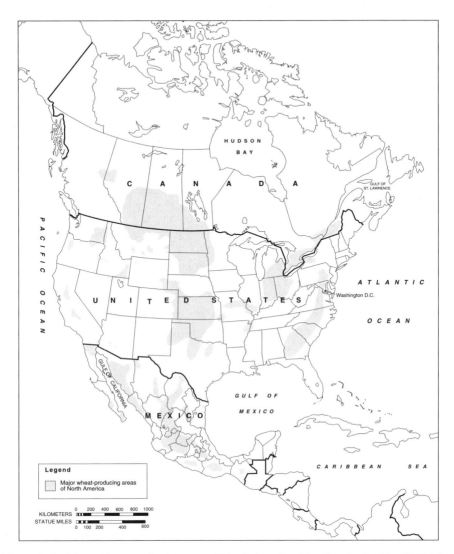

Figure 2.6 Major wheat-producing areas of North America. Line drawing by Tim F. Knight. Outline of map adapted from U.S. Department of the Interior, Geological Survey, *North America 1982* (1:10,000,000) (Reston, Va.: Geological Survey, 1982), 1 p. Wheat-growing areas adapted from U.S. Department of Agriculture, Agricultural Research Service, *Distribution of the Varieties and Classes of Wheat in the United States in 1979* (Washington, D.C.: U.S. Department of Agriculture, 1982), Statistical Bulletin no. 676, p. 3; and from U.S. Department of Agriculture, *Major World Crop Areas and Climatic Profiles* (Washington, D.C.: U.S. Department of Agriculture, 1987), Agriculture Handbook no. 664.

tures who must consume the products of photosynthesis in order to live. We do this either directly by consuming the products of plants or indirectly by consuming animals. Our consumption must include calories, proteins, fats, vitamins, and minerals. In these ecological terms, the Neolithic revolution was simply the process by which people learned to raise and eat plants and animals under controlled conditions rather than relying on what could be found growing wild. Once the technology of agriculture had been learned, people over a period of several thousand years gradually became completely dependent upon the domesticated products of agriculture rather than the less controlled produce of the land.

The Neolithic revolution, described in ecological terms, ultimately allowed people to capture more solar energy per hectare per year through the products of photosynthesis and more essential minerals that plants took in from the soil than they could reliably find through processes of hunting and gathering.[32] A major dispute, however, centers around the interpretation given to the *motivations* and *physical factors* that lay behind the efforts of our Neolithic ancestors. Did people invent and use agriculture because they saw domestication of plants and animals as a way to material and cultural progress? Did people fall by some sort of unconscious accident into a codependenct relationship with domesticated plants and animals? Or did they resort to agriculture only because they faced climatic shifts and/or population increases beyond those that could be supported by hunting and gathering?[33] In other words, did people face the necessity of learning to farm or of dying, because hunting and gathering could not garner enough resources to support the population levels of the Neolithic period?[34]

Evidence for the factors behind the Neolithic revolution remains clouded and ambiguous. What seems certain, however, is that the increased food supplies that eventually came with agriculture permitted the human population to rise far higher than it would have had people remained in hunter-gatherer cultures. Thus, whatever the reasons for turning to farming in the first place, the human species now has no choice whatever about the matter: we are desperate, and we must farm or many of us will die, rather quickly.[35] To be sure, small groups and individuals may continue to forage as hunter-gatherers, but that mode of life is not an option for the vast majority of humanity.

It is possible to argue at some length about just when the necessity to farm or die really originated, but in all likelihood it was several thousand years ago. It is the long-standing desperation implicit in our need for agriculture that powerfully weds the politics of human societies to the ecological processes of agriculture: security and survival of a group or culture came to be critically dependent upon the control of land and of the plant and animal crops that could be raised on that land. Improvement of agricultural yields very early became an event that conferred power and prestige in political terms upon farming cultures. More recently, plant breeding was a systematic effort to improve yields and therefore produce knowledge of both ecological and political importance.

The fact that plant breeding is inherently both political and ecological leads to a series of other questions that focus on wheat specifically. How did wheat as one of the first domesticated species fit into the political ecological framework of Neolithic people? How has that political ecological context of wheat changed over the past 10,000 years? What are the links between the modern science of plant breeding and earlier efforts to increase yields of wheat? And finally, what other political ecological factors before the eighteenth century strongly helped shape wheat breeding? Fragmentary archaeological and historical evidence allows only the sketchiest of answers to be given to these questions.

Domestication of wheat was complex not only because wheat was variable (diploid, tetraploid, and hexaploid varieties) but because it was not the only species that came into intimate codependency with people. Barley was domesticated simultaneously with the wheats, while rye, various pulses, and forage crops came later but still in prehistoric times. Animals were also domesticated at about the same time and place as wheat and barley, so sheep, goats, asses, camels, and horses became part of a codependency complex with people.

It is possible to ask the question, Why wheat? A variation on the same subject would be Why the changing fortunes of the different types of wheat? Some paleobotanists have argued that the hexaploid *Triticum aestivum* originated only after the tetraploid emmer moved from its origins in southwestern Asia with a Mediterranean climate to the area of what is now north Iran and southern Russia. There emmer bred with wild *Aegilops squarrosa* to produce hexaploid wheat. This new type of wheat may have been more ecologically flexible because it grew well in the Mediterranean climate of emmer (mild winters, warm and dry summers) and in the more continental climate of *Aegilops squarrosa* (cold winters, hot and humid summers).[36] Thus the hexaploid's ecological strength and its good bread-making qualities were perhaps the key to *Triticum aestivum*'s dominance in contemporary wheat growing.

Whatever the reasons people chose to grow particular varieties of wheat, it seems reasonable to assume that they would not have invested time and effort in farming these grains unless they found it worthwhile. In ecological terms this meant that people obtained more calories of energy from the harvest than they had to invest in the growing of wheat and barley. In addition, perhaps they sought wheat's high-gluten protein and its reliable yields. They may also have liked the taste and texture of wheat as a food. Regardless of why they grew wheat, the plant itself tolerated well enough the regime of temperature, water, and pest organisms that affected its growth.

Not only did people have to choose among wheat varieties; evidence from the historical period also indicates that the relative importance of the various wheats and barley have changed in the millennia since their domestication. Barley and the various wheats may have been more equal in their importance as human foods in prehistoric and early historic times. Barley withstands drier conditions, poorer soils, and some salinity.[37] Indications from late medieval and early modern Europe are that barley may have tended to yield more kilograms per hectare than wheat.[38] If anything, barley was probably the preferred grain at first. In modern times, however, wheat clearly became the premium food, and barley became used primarily for beer making and animal feed.

Within the wheats themselves, the first farmers probably emphasized emmer, the first domesticated tetraploid wheat, in preference to the diploid einkorn and the hexaploid wheats. In modern times einkorn has essentially disappeared from commercial production; emmer is a mere remnant; the newer tetraploid, *Triticum durum*, has attained a small but secure commercial niche for pasta; and the hexaploid bread wheat is commercially supreme over all parts of the wheat-growing world.

Why did such changing fortunes affect the wheats and barley so strongly? Was it a matter of ease or reliability of production of the different grains in the field? Was early barley better adapted to southwestern Asia than the early wheats? Or did people find one sort of grain more satisfying than others? If the first uses of the grains were as porridge or gruel, did the emmer wheats make a better meal than einkorn? Perhaps the hexaploid wheats were not particularly useful until people learned how to make leavened bread. Not all wheats make good bread, and barley, because it lacks gluten, is not good for bread making at all.

Whatever the reasons for the changing fortunes of the early wheat domesticates, over a period of thousands of years the fact of agricultural codependency among people, wheat, other plants, and domesticated animals had a profound effect on human ecology and politics:

- Sedentary rather than migratory life became the norm of the human condition. At first, most people were sedentary in rural areas and the occupation for most was farming. After the nineteenth century, the norm increasingly came to be sedentary in urban areas with a steadily decreasing number of people engaged directly in farming. Regardless of residence or occupation, however, the human population as a whole was almost completely dependent upon the produce of farming.
- With sedentary culture came a richer material life: at first ceramics, then metals, and in more recent years a bewildering array of new materials in an endless parade of tools, jewelry, and amusing trinkets, all of which probably would not have even been invented, or at least not widely used, without a sedentary lifestyle.
- Sedentary cultures also were the ones that made the switch from oral to literacy traditions. And with writing came different forms of learning, religion, and ultimately that peculiar form of scholarship we call science.
- Also with writing came the ability to organize larger and more hierarchical human societies with intense consequences for the distribution of political power within groups. In fact, the ability of one group of people to control another inevitably came to include authority over land and its use for agricultural purposes.
- Increased food supplies from agriculture made it possible to support more people per unit of land than was possible with most hunting and gathering cultures. Quite possibly, population pressure provided the impetus to begin agriculture, and agriculture itself became the vehicle through which more people could survive, thus increasing the necessity of continuing agriculture, with ever higher yields. Not only did the number of people increase, but the number of their domesticated animals also increased because of the new supplies of feed from agriculture.
- Increases in the number of people and their livestock in turn created incentives for expansion of agriculture into new lands not previously farmed. As agroecosystems grew in extent, other collections of species disappeared. The woodlands and meadows of southwestern Asia were replaced by fields of wheats and barley. In Southeast Asia, fields of rice grew in the place of wetlands and tropical rain forests. Mesoamericans grew maize in the place of highland forests and meadows. Thus not only was the ecology of *Homo* changed; *Homo* also changed the ecology of the globe.

Wheat breeders in the modern sense, as we shall soon see, did not appear on the scene until many thousands of years after the first crops of wheat and barley were harvested in southwestern Asia. Nevertheless, the intricate connections of wheat agriculture with the ecology and politics of human life were a major part of the context in which wheat breeding was conducted. Because of these connections, wheat-breeding science would inevitably have both ecological and political consequences.

Plant Breeding before 1900

The year 1900 provides an important but deceptively simple chronological milestone for analyzing the emergence of plant breeding science. Clearly it was an important year because three European biologists, Karl Correns, Hugo de Vries, and Erich von Tschermak, published papers that resurrected a study on hybridization in peas done over thirty years earlier by Gregor Mendel.

A casual glance at any textbook in plant breeding written after 1900 shows that Mendel's concepts now overwhelmingly provide the major framework for understanding plants and their behavior. Of particular importance were his notions of particulate factors governing inheritance, dominance and recessiveness, segregation of alleles, independent assortment of factors, and diagnosis of the existence of factors based on ratios of progeny classes in precise records of their occurrence by generation. To this Mendelian framework are appended a variety of other methods drawn from plant physiology, biometrics, soil science, plant anatomy, plant pathology, and other disciplines.

Mendel's contributions were crucial to the consolidation of what we now call "plant-breeding science." Indeed, it would be impossible to recognize the discipline in its modern form if Mendelian ideas were surgically removed from the tomes that instruct new students in the art and science of plant breeding. However, it is deceptively simplistic to focus too intensively on Mendel's "rediscovery" in 1900. Several lines of reasoning argue for caution lest we attribute overwhelming importance to Mendel.

First, a vigorous and successful group of people worked throughout the nineteenth century on the creation of new varieties by hybridization, as well as by the older method of selecting "good" cultivars from a heterogeneous mix grown in farmers' fields. Plant breeders also were active in moving seeds around the globe to test which cultivars did well in new locations. In sum, even without Mendelian concepts and methods, plant breeders saw themselves as a small but identifiable group of scientists with a mission.

The second set of reasons for moving cautiously in trying to understand the conceptual framework of plant breeding centers on what should be called the "silent" conceptual framework for plant breeding compared with the "overt" framework. Overtly, plant-breeding science is now based more than anything else on Mendelian concepts. Beneath the surface, however, is the silent or frequently unspoken concept: yield.

Yield is complex. It consists first and foremost of quantity. How much useful produce can be obtained per hectare per year from a piece of ground? Yield also must consider quality, however. To what uses can the produce be put? For wheat, is the grain useful for leavened bread, for pastries, or for cattle feed? For rice, is the grain sticky? Aromatic? For maize, are the seeds sweet or starchy? How hard is it? What color? Would you use it for tortillas, for cornbread, or for feeding chickens?

Closely related to yield quantity and quality is the concept of yield reliability. Given the climate, soil fertility, and presence of pest organisms, is it likely that the yields will be stable from year to year? A cultivar that gives yields that fluctuate greatly from one year to the next is likely to be considered unsuitable in terms of yield.

Contemporary plant-breeding scientists, when asked, are quite forward about the importance of yield to their discipline. In fact, a question about why yield is important is likely to bring the puzzled response of, "What else is there?" However, in contrast to clearly giving homage to Mendel and continually showing their direct intellectual debt to him, plant breeders are far less interested in explaining the origins of the centrality of yield in their discipline. Yet just as plant-breeding science in its current form would be unrecognizable if Mendelism were excised from its body, so, too, would the field as we know it cease to be if the presumption of yield as the primary objective were removed.

Reconstructing the roots of plant-breeding science before 1900, therefore, requires following two different threads and how they came together. First, we must be able to speak to the primacy of yields and its significance for the science. Second, we must understand the vigor, successes, and limitations of plant breeding in the first years of modern agriculture. Chapter 3 addresses when and how Mendelian ideas formed the overt conceptual framework of plant breeding and then examines the eclectic addition of other bits and pieces of scientific lore that plant breeders adopted into the heart of their knowledge.

The Primacy of Yields and Modern Agriculture

Just as the Neolithic revolution was critical to obtaining higher amounts of food per hectare per year than in hunter-gatherer societies, the scientific and capitalist revolutions of seventeenth-century Europe were key events that shaped a complex series of changes in agriculture and all other human industries. From science and capitalism came both the methods and motives for constructing new methods of wheat production. For wheat breeders, the intertwining of science and capitalism created a context in which yields were the fundamental question of their discipline.

Inquiring into the specific origins of the importance of yields to plant breeders, however, is very much like asking about the origins of modern industrial societies in general. No one person, event, country, scientific idea, or technological innovation can be identified as the "source." No one date can be advanced as "the time" after which higher yields were clearly identified as the major objective of plant breeders. Nor can improvement of yields be identified as the sole product of plant breeders rather than other sorts of agricultural improvers. Instead, the importance of yield has to be seen as a concept that emerged over a long period, at least 200 years, and in a complex array of specific contexts, "caused" by an even more intricate network of interacting factors. Indeed, the fact that physical yields of crops could be increased by plant breeders and others became an integral part of that complex of interacting factors.

Between the Neolithic agricultural revolution and A.D. 1200, farming people probably improved the yields of cereals and other crops by selecting better individuals to save for seed and by improving other farming practices. We can make only approximate estimates, however, on the yields they obtained and the magnitudes of improvements achieved. A variety of estimates suggest that, without manure or fallowing, returns for planting wheat may drop as low as three units harvested for one unit planted. Both fallowing and manuring, however, can yield substantial returns of over fifteen to one. Neolithic yields of wheat may have seldom dropped below 400 kilograms per hectare and more likely were in the area of 800 kilograms per hectare in a climate like England's.[39]

The first improvements known with more certainty in historic times involve more intensive farming practices. One of the best documented innovations was the switch from the two-field system to the three-field system, a transition that took from the eighth to the twelfth centuries in Western and central Europe. The three-field rotation system allowed an increase in crop production of 50 percent for the same amount of plowing labor invested.[40] In addition, two-thirds of the cropland used yielded a crop each year rather than just one-half. Yields of wheat from A.D. 1200 to 1700 in England under the three-field system ranged from about 500 to 1000 kilograms per hectare (446–892 pounds/acre; 7–15 bu/a).[41]

With a bit of oversimplification, therefore, we can argue that although significant changes in farming occurred between the Neolithic revolution and about 1700, these changes were not of a magnitude to move the mean yields of wheat in England above 1000 kilograms per hectare in good years. Peoples' abilities to get wheat from the land were enhanced by better plows, use of animals for traction, and learning how to use land more frequently (two years out of three instead of one year out of two). Output of wheat and other crops per hour of labor invested undoubtedly went up, especially with the use of animals for traction of tools, but the overwhelming majority of people lived in rural areas and agriculture was the basis of most peoples' lives. England, for example, had over 86 percent of its people living in settlements of fewer than 10,000 people in 1700.[42] Agriculture was also the main basis for the economies of all people all over the world. It's not as though nothing happened between the Neolithic revolution starting in about 7000 B.C. and A.D. 1700, but in many ways a person from A.D. 1700 might feel much more at home in 7000 B.C. than she or he would feel in A.D. 1900.

In barest outline, a series of events in agriculture and in other areas of human endeavor occurred between 1600 and 1900, and as a consequence the nature of human life changed fundamentally. Cause-and-effect linkages between the different events are still hotly debated by scholars, but it is easy to list some of the more prominent features of the changes:

- People learned how to get more food from a given piece of land each year. Average yields of wheat in England went from between 500 and 1000 kilograms per hectare per year in 1600 to about 1400 kilograms per hectare in 1800[43] to somewhat less than 3000 kilograms per hectare per year by 1860.[44] Particularly high rates of annual increases in yield occurred between 1660 and 1740[45] and then again between 1820 and 1860.[46]

- Rates of human population increase moved markedly upward in England after 1740.[47] The human population had been increasing on a global basis since the

Neolithic, but seldom at a rate to cause doubling of the human population in fewer than seven centuries. In many years, declines occurred, often severe. After 1740, however, England's population grew at a rate that caused doubling in just over a century. Debate surrounds the issue of whether agricultural yields kept pace with this population increase during the period 1740–1800, but in the years after 1800 the agricultural yield increases on a global basis were more than commensurate with population-level increases.

- People moved to cities and began a wide range of economically productive activities other than agriculture. In many ways it is a toss-up as to whether this resettlement pattern should be called an "industrial revolution" or an "urbanization revolution," but no doubt attends the notion that the switch from predominantly rural populations to predominantly urban ones was of fundamental importance to how people lived.

- England clearly was the paramount pioneer in this urban-industrial venture, but one could not predict this by looking at England's population in 1500 compared with other places in Europe. In that year England had 3.2 percent of its people living in towns of more than 10,000 people, but Europe as a whole had 6.1 percent urban populations. France had fourteen towns with populations over 20,000, while England had only London at that size. In 1500 both Paris and Lyon were larger than London. Nevertheless, the rate of urban growth in England jumped to higher levels after 1500, and by 1800 England was 24 percent urban compared with about 10 percent for Europe as a whole, except England.[48] England was also the clear leader in manufacturing and industry.

- Land use and land control underwent a dramatic shift from predominantly arable fields with common woods and meadows for livestock and firewood to enclosed fields for raising sheep or cereals and other crops. Significant enclosures were made before 1600, but virtually all England came under enclosure in the century beginning in 1750. Common lands essential to the subsistence living of peasants became part of estates run with technologies and practices that could produce higher yields of wheat, other crops, and livestock.

- New farm practices won acceptance over a wide area of England after 1600, and by 1800 almost all farms had moved to the methods that increased yields. Probably the most well known of these innovations involved the use of turnips to provide fodder for livestock, clover to supply nitrogen to cereal crops, and a four-field rotation system that integrated the production of livestock with turnips, clover, and cereals. The three-field system relied on fallowing for a year to restore soil fertility. Four-field systems eliminated fallowing and permitted production on all fields every year. Fertility was maintained by the clover and the manure from livestock. In addition, new machines such as seed drills and cultivators gave better seeding and weeding with less labor. In sum, the set of new technologies increased yields of wheat and other crops both per hectare and per person-hour of labor involved. That they were known in theory and practice long before they became commonly used in England demonstrates that mere knowledge of the new practices was not sufficient to promote their adoption.

- Climatic changes led to increasingly warm weather after 1700, a factor that was conducive to higher and perhaps more reliable yields in agriculture. As the "Little Ice Age" of 1550–1700 receded, wheat yields may have received a boost from the warmer temperatures.[49] Higher and more reliable yields may have created an atmosphere of trust in nature, or a sense that nature did what people wanted. Higher yields from the same amount of human labor could have supported more people, some of whom did not farm but lived in towns and cities.

- Not only were people changing their residence and ways of making a living; a small cadre of thinkers were also changing the mental constructs with which people saw the world and economic activity. Shifts in worldviews were not the exclusive preserve of philosophers from England, but the following list suggests something of the magnitude of what was involved:[50]

  — Francis Bacon of England celebrated science and experimentation as the way to control nature for human interests.
  — René Descartes from France outlined a theory of a mechanical universe and a mechanical man.
  — Thomas Hobbes of England deduced from Cartesian mechanical philosophy the need for a powerful, hierarchical central government that would keep people from killing each other.
  — John Locke of England softened the dictatorial harshness of Hobbes and argued for a democracy of property holders who owned and traded land for money.
  — Isaac Newton of England created a model of a mechanical universe whose laws could be expressed mathematically.
  — Adam Smith of Scotland outlined a theory of Newtonian-like natural laws about how economic markets work and why feudal and mercantile policies should be abandoned.

What we can now see that was created by these philosophers was an integrated ideology of capitalism and modern science. It is not that capitalism was consciously created as a part of science or vice versa, but thinking about markets and science rested upon a number of common concepts:

Natural laws govern both markets and the natural world; philosophers learn to control nature by articulating natural laws, and capitalists use these laws to forge an economy.

The material world is dead and without sacred or vital spirit; for the natural philosopher there is only atomistic matter in motion, while the capitalist sees commodities for trading.

Science is the handmaiden for controlling nature for the benefit of people and for the prosperity of industry.

The concepts of individual rights and self-interest make it laudable that the philosopher thinks and the capitalist pursues profit.

Ever-expanding knowledge allows the philosopher to predict and control an ever-expanding realm in the natural world, and the capitalist uses knowledge to bring an ever-expanding sphere of human activity into a form of commodities fit for trading and profit.

Production of agricultural goods for market created incentives for increasing production levels per hectare per year. New technologies from the industrial revolution, particularly for the making of metal tools, created possibilities for increasing agricultural yields, both per hectare and per hour of human labor invested. Markets that helped engineer the exchange of land, labor, commodities, and capital spurred on an entrepreneurial and technical spirit that could lead to greater production efficiencies in English agriculture. Higher production efficiencies, in turn, could feed into an ever-growing economy by either "freeing up" or "pushing out" labor from agriculture into the new factory system, thus creating further incentives for producing more agricultural produce from less land and less human labor.

Populations could grow, and more people would find their livelihood in the cities. More people freed from agriculture also gave the wherewithal for expansion of European systems in other parts of the world. The growing global capitalist economy provided the incentive for seizing control of land anywhere it could be found and conquered. The nation-state that built social relationships on markets, highly productive agriculture, and industrial capitalism was in a powerful position to dictate its will to people everywhere.

It was in England where all these factors interacted in a way to consolidate a system in which increased agricultural productivities could serve the needs and interests of those at the top of a new market capitalist society. And it was in England where the advantages of new modes of agricultural production shaped the objectives of a new agricultural science.

In contrast to the farming peasantry of continental Europe, India, and the Americas, English farmers were the first to face and exploit a situation that encouraged and possibly demanded a search for higher efficiencies. These efficiencies were for both increased yields per hectare per year and increased yields per hour of human labor invested. The demands for efficiency affected all crops, but the demands for cereal crops such as wheat provide a useful focus for tracing how the demands were met.

Early Traces of Plant Breeding

England provides the first traces of concerted activity by plant breeders, who tried to improve wheat yields. Thomas Andrew Knight (1759–1838) is now generally acknowledged by contemporary breeders to have been the first, in the 1790s. Later wheat breeders credited Knight with being the first to make deliberate crosses between two different wheat plants.[51]

However, the most notable successes came somewhat later with the work of John LeCouteur and Patrick Shirreff (Figure 2.7). LeCouteur's work was based on selecting individual spikes of superior individuals and sowing the seeds from each spike separately. In this sense LeCouteur was not strictly a wheat breeder because he did not cross different parents and select from their progeny. But his methods were successful in identifying the variety Talevara, which was used extensively by Shirreff.[52]

Shirreff used the selection method, but he also began crosses. Over a period of several decades, Shirreff, at his station in Mungoswells farm, Scotland, identified several important varieties. By selection he found Shirreff's Bearded Red, Shirreff's Bearded White, Pringle, and Shirreff's Squarehead. He also used Talevara for crosses and obtained King Red Chaff White, which he considered a worthy variety.[53]

Selection of pure lines and additional crosses to create and select useful hybrids continued to be the output of the pre-1900 wheat breeders. Especially successful practitioners in addition to LeCouteur and Shirreff were F. Hallett[54] (England), Henri

Figure 2.7   Patrick Shirreff. From Patrick Shirreff, *Improvement of the Cereals* (Edinburgh: William Blackwood and Sons, 1873), frontispiece.

de Vilmorin (France), Wilhelm Rimpau (Germany), Broekema (Netherlands, 1886), Hjalmer Nilsson and Herman Nilsson-Ehle (Sweden), William Saunders (Canada), William Farrer (Australia), and Liberty Hyde Bailey and W. M. Hays (United States).[55]

Successful as the breeders were before 1900, the rate of output of new varieties was low and their ability to explain how and why they performed their crosses was not entirely satisfactory. Science, as it gains control over a body of observations, moves from observation to explanation to prediction to control. Before 1900, wheat breeders were astute on observation and had schemes of explanation. They could not, however, do much to predict the results of their crosses and therefore had only minimal levels of control over the wheat plant's genetic resources. Mendel's theory about the flow of information from one generation to the next provided the basis for better prediction and thus better control. It is the subject of the next chapter.

Wheat Breeding

Coalescence of a Modern Science, 1900–1939

Darwin, Mendel, and a New Theory of Variation

Plant breeding in general and wheat breeding specifically were rudimentary activities on many grounds in the nineteenth century. Not many people engaged in the activity. Those who did were self-taught. because no formal educational programs existed in the subject. For the most part, they had only a few very modest institutional bases within which to work. Many farmers paid them little or no attention, and governments usually ignored their contributions and gave them next to no support. They had no organized way of broadly disseminating their results, which in any case were few in number.

By 1970, wheat and other plant breeders occupied a very different position within both the scientific and political economic landscapes. Many people worked as breeders.[1] They were highly trained in educational programs dedicated to the reproduction of plant breeders. Elaborate networks of institutions gave them employment. A substantial proportion of farmers cared very much what they did, and governments gave substantial, sometimes lavish, support. They had means of communicating their work that included both scientific and popular outlets. And they had substantial results to convey to farmers and the general public, some of them remarkable either for their scientific cleverness or for their broad political, economic, and ecological impacts, or both.

Another way of gaining perspective on the change in status of wheat and other plant breeders is to suggest that their absence might not have been noticed by anybody but their families had they suddenly disappeared in the nineteenth century. In contrast, the twentieth century came increasingly to depend upon the plant breeders. Cessation of wheat breeding after 1970, for example, would have put some agricultural systems in distinct danger of slow decline or even collapse and failure. In both political economic and ecological terms, an increasing portion of the global human community became absolutely dependent upon wheat breeders and other plant scientists, certainly for prosperity as we now know it and possibly for survival and security.

The transformation of wheat breeding from nearly invisible to virtually indispensable resulted from two mutually interacting events: a commercial-industrial revolution in agriculture and construction of a new science of plant breeding. The new science had its origins in a philosophy of variation and inheritance grounded in the works of Charles Darwin and Gregor Mendel. Darwin's thoughts were widely discussed after 1859, and Mendel rose to scientific prominence after 1900. The two worked independently, and it was only after both had died that their ideas joined as the foundation of a new applied science of plant breeding. Events in England and the United States were paramount, but other major contributions came from Sweden, Denmark, Germany, and elsewhere.

Nineteenth-century wheat improvers found and created new varieties with higher yields, but until Darwin's work from 1859 to 1868, variation was generally understood as the result of divine creation or the result of slow change over time of one form into another.[2] After Darwin, wheat breeders may still have been uncertain about the origins and heritability of variation, but they had an entirely new framework for seeing variation as the source of new varieties and new species of plants and animals. Darwin placed either natural selection or the actions of breeders and fanciers as the cause, over a period of time, of the production of new species or new varieties, respectively.

Darwin's provisional hypothesis of pangenesis was a valiant effort to account for how specific characteristics of a particular individual could be passed on to future generations in a way that allowed natural selection or selection of fanciers to create their new species or varieties. In pangenesis, each cell of an individual produced granules or gemmules that dispersed throughout the organism, multiplied by self-division when given proper nutriment, and ultimately produced new cells like the ones from which they were formed. The sexual elements were collections from all parts of the individual of all the different types of gemmules, and this collection gave rise to the next generation. Gemmules could also be dormant in a generation but still be passed on to subsequent generations, where they would develop.[3]

Pangenesis was a reasonable attempt by Darwin to account for a number of things he knew about inheritance and development, but for a variety of reasons it was found wanting and never attracted much support. Darwin himself seemed highly tentative about his proposal. Gregor Mendel, however, working somewhat before Darwin developed the idea of pangenesis, outlined a mechanism of inheritance that ultimately attracted support. It was Mendel's theory of the transmission of factors governing unit characters that ultimately found acceptance and elaboration among those scientists who worked to understand the origins, transmission, and significance of variation.

We will come shortly to the ways in which Darwin's and Mendel's theories entered the work of plant breeding, but it is first necessary to note that the introduction of Darwin-Mendelism occurred in England and America during a period of tumultuous economic and ecological change. Farming in both countries was undergoing a fundamental reorganization toward industrialized production for commercial markets. New varieties, new machinery, and new knowledge enabled growers to increase their output per hectare and per person-hour of labor. New lands were opened for farming in areas previously not farmed at all. New related technologies, such as railroads, made it possible to create a global network of markets for agricultural and other goods. Population increases and migration created turmoil for farmers everywhere. At the same time, shifts of people out of agriculture into urban factory work created mass markets for food. Pressures from these factors and others changed how the state regulated trade and the balance of power between the countryside and newly industrialized cities.

It is too simplistic to say that the profound changes of the nineteenth century "caused" Darwin and Mendel to propose their philosophy of variation and its transmission. At the same time, however, certain congruences and compatibilities appeared between the theories of Darwin and Mendel and the demands of the newly industrialized economies. Specifically, Darwin and Mendel brought a sense of order, prediction, and manipulation into the study of variation. These perceptions were critical to the construction of an agricultural plant science that served as the base for a complete industrialization and commercialization of agriculture. The parallels thus are useful in gaining an understanding of how plant breeding was constructed and what sorts of roles it came to play.

Industrialization and the New Science of Variations

Industrial revolution, first in England and then elsewhere, is generally depicted as a complex set of changes in technology, economics, and politics. Without doubt, political economic changes during this period were stupendous and unprecedented. What should not be overlooked, however, is that industrial revolution also resulted in a fundamental shift in human ecology. Prior to industrialization, most people lived in small rural villages or on isolated farmsteads. Virtually all members of the laboring part of the population worked directly in food production. Only a tiny minority worked outside of agriculture or lived in cities. After industrialization, an increasing portion of the population lived in cities and worked at tasks other than agricultural production.

Chapter 2 has traced some of the shifts in population to cities as industrial and commercial activities increased. It was not until the nineteenth century, however, that the political consequences of the shifts became clear. In many ways the most important symbol of change was repeal in 1846 of the British Corn Laws, a substantial tariff barrier to the free import of wheat and other grains from 1815 to 1846.[4] These Corn Laws protected English wheat growers, who could not produce grain as cheaply as it could be grown in other parts of Europe and, increasingly, in North America. For the landed interests of England, the Corn Laws were a route to preserving their traditional bastions of power and privilege in a political economy that saw urban commercial and industrial interests steadily eclipsing the old feudal power systems based on landownership in the countryside.[5]

Owners of factories and pillars of the commercial community saw the Corn Laws simply as a way to keep bread expensive to benefit landlords and farmers. They, however, wanted cheap bread in order to keep wages to their workers low. After years of bitter division, their representatives formed the Anti–Corn Law League in 1839 in Manchester. Duties came down in 1842 and, under the conditions of famine in Ireland, were phased out starting in 1846.[6]

Repeal of the Corn Laws resulted in a short-term decrease in wheat prices in England, but other factors intervened to stave off for about twenty years a permanent decline. Transport technologies were not entirely adequate to bring in large quantities of grain on a regular basis. In addition, outbreak of war in America and the Crimea served to preserve a steady domestic market for English wheat growers. In fact, repeal of the Corn Laws was followed by investment by British landowners and farmers in such practices as field drainage systems, new buildings, and fertilizers. Their grain production remained profitable, and for the period of about 1850 to 1870 England enjoyed a "golden age" of agriculture.[7]

Prosperity unraveled after 1870 as railroad and shipping technologies made it ever cheaper to import grains from abroad, especially from North America. Bad weather and livestock diseases exacerbated the situation in the late 1870s.[8] England moved from importing one-fifth of her wheat in 1841 to three-fourths by the early 1900s.[9] At the same time, rising incomes and changing tastes led to a pronounced shift in preference for bread wheats rather than barley, oats, rye, and other crops as the staple of British diets.[10] High yields of wheat in England were still possible on physical and biological grounds, but no English bread wheats could compete freely with foreign bread wheats. England became a land of grazing and dairying, not of cereal production, and labor left agriculture in ever larger numbers.[11]

Not only did repeal of the Corn Laws alter the rural landscape and economy of England; their absence also created another situation that was symbolically of highest importance to the shape and content of future farming economies and their need for particular types of scientific expertise. When wheat, the foundation of the English food supply, became simply one more commodity to be traded freely in a world market, attachment to the notion of food self-sufficiency from the nearby region was undermined. Capitalist markets would now be the arbiters for how much wheat would be produced, where, and for what price it would be sold. A farmer who wanted to survive in this market had to produce wheat for less money than the market would pay for it. Differences between production costs and selling prices became the supreme arbiter of a farmer's skill and fortune.

Repeal of the Corn Laws thus symbolized a major junction in the long transition from farming as a means of local self-sufficiency to farming as simply another commercial enterprise. Producers of basic commodities, first in England and eventually everywhere, increasingly had to come to grips with production for profit in a highly competitive global market. Yield of the right type of crop was the primary means to achieving low production costs in relation to market prices, and growers slowly came to learn of their dependence upon plant breeders and other agricultural scientists to get the best differentials possible. Higher yields at lower costs were key to economic survival.

It is worth inquiring at this juncture whether repeal of the Corn Laws was the only important factor in making yields primary. The short answer to this question is no, for reasons that emerged at several points in Chapter 2 and will continue to appear as the narrative moves on to the remarkable transformation of wheat yields between 1940 and 1980. Other factors also affected whether or not an individual grower was under pressure to maximize yields, including expansion of wheat production into new lands, expansion of the human population, and mechanization of agriculture. In fact, the political squabble in Britain over whether wheat should be freely imported from abroad was a battle that could be fought only because new wheat lands were being farmed abroad, generally by an expanded population of people of European ancestry, who used new laborsaving technology. It was this complex of factors that created the conditions for the ascendancy of yield as one of the best ways for a farmer to ensure that selling prices of wheat were greater than production costs.

Expansion of population and of the area planted in wheat were closely linked processes in the nineteenth century. From 1800 to 1900, the European population increased from about 180 million people to about 390 million, a rate of increase far in advance of any other group of people on earth.[12] This "population explosion" went hand in hand with another change: expansion of the wheat-growing lands by millions of hectares within a short period of time in North America, South America, and Australasia.

Expansion of population and of wheat area interacted to produce a complex signal to individual wheat growers about their need to maximize yields. Expansion of population meant that more people needed to eat and that more people were available to farm new lands. Extra people eating expanded the market demand for wheat, which, in turn, could be met by the production on new wheat lands. So long as the new lands were of the right size to provide for the new people, little change in wheat-growing practices or in wheat markets would occur. For two major reasons, however, the ways in which both the cultivated area and population expanded were such that severe competition fell on all wheat growers in both old and new lands.

First, only some growers were actually in a position to cultivate larger amounts of land in wheat. One or more of four conditions had to be met for an individual farmer to expand area: (1) the land was on the farm already, just not used (e.g., it might have been wooded), (2) the land was on the farm already and used for another crop that could be eliminated, (3) the farmer was on a "frontier" adjacent to land that could be appropriated easily, or (4) the farmer was willing to pick up and move a substantial distance to a new frontier. Farmers who could not meet one of these four conditions and who wanted to grow more wheat were left only with improvement of yields per hectare as a way to increase production. Even farmers who could expand, however, were not immune to pressures to increase yields.

Second, it was the European population that exploded in the 1800s, and it was the European expatriates who were most responsible for expanding the land area given to wheat during that century. Over 50 million people, the largest migration ever in human existence, left Europe to settle in what historian Alfred W. Crosby called the Neo-Europes: the temperate regions of the Americas, Australasia, and Eurasia.[13] This exodus began in a small way in the sixteenth century, but it was not until the nineteenth that it reached its peak. As a result, the North American prairies, the Argentine pampas, Australia, and, for a short period, British India became immense granaries that supplied ever-increasing amounts of wheat to a burgeoning population of Europeans and Neo-Europeans.

As a result of unequal access to new lands, some wheat producers were "left behind" on the old, settled wheat lands of Europe while others took up business on the new lands. Economic and cultural links between the Neo-Europes and Europe, however, were such that producers in the new wheat lands needed to trade with Europe. For most of the nineteenth century, and even beyond, however, the only products in abundance in the Neo-Europes were wheat and other agricultural commodities. Once transport links were cheap enough, the Neo-Europeans naturally worked to sell their goods in Europe.

If land had not been plentiful in the Neo-Europes, if transport had not been cheap enough, or if the Neo-European populations had been large enough to consume the grain they raised, then no massive amounts of wheat would ever have been sent back to Europe to compete with European farmers. Unfortunately for European farmers, these conditional assumptions were not true. As a result, the grain harvests in the Neo-Europes were sufficiently large and cheap that farmers everywhere, both in Europe and elsewhere, entered a phase of intense competition on a world grain market of limited demand compared with supplies after about 1870.

In order to better understand this competition, we now explore two of its major features: the excess of land given to wheat and the advent of laborsaving machinery. These two factors were the foundations upon which was built a receptivity for knowledge about how to increase yields per hectare. Increasing yields was a critical weapon for farmers in Europe and was the cornerstone for receptivity to the Darwin-Mendelian theory of variation and its use in an applied science of plant breeding.

Excess of Land

Evidence suggests that sometime after 1850 the world entered a period in which available technology for wheat production was sufficiently powerful to produce amounts of wheat above what was needed for food and seed on the land suitable for wheat cultivation. Disparities in wealth undoubtedly kept some people hungry and malnourished, but their starvation was a matter of distribution, not lack of natural resources to produce wheat. Economic historian Wilfred Malenbaum analyzed the changes in global wheat acreage, production, and yields between 1885 and 1939. He distinguished between the "necessary" acres needed to produce wheat for food and seed and the number of acres needed to produce the total world wheat harvest at average global yields; he called the difference "excess acreage." He found that from 1885–89 to 1899–1904, the world had an average excess acreage of 11 million acres (4.6 million hectares), or 6.9 percent excess. From 1904–09 to 1934–39, the excess grew from 14.9 million acres (6.2 million hectares) to 31.0 million acres (12.8 million hectares), an increase of from 8.6 percent to 15 percent excess.[14]

Malenbaum thus argued that the farmers of the world planted too much land to wheat in relation to the amount of land needed to provide food and seed. During the half century after 1885, the excess planted became larger, causing the price of wheat to fall absolutely and relatively to other commodities. Only the disruptions of normal farming patterns in World War I had marked effects on these trends. For a variety of reasons, farmers persisted in planting more land in wheat than was needed to satisfy the effective market demand for the cereal.

Not only was too much land being planted, but the opening of new wheat lands and the construction of railroads, especially in the United States, after 1870 led to a flood of American cereals into Europe. These new imports took a tremendous toll on European farmers who were raising wheat. Most important, it was cheaper to grow wheat in North America and ship it to Europe than it was to grow and consume the grain there. In Germany, for example, production costs for winter wheat in 1910–14 ranged from $0.90 to $1.08 per bushel. Spring wheat in North America, however, had a production cost range from $0.45 in Montana to $0.75 in the Canadian prairies. Transporting wheat from Montana to Liverpool cost only $0.15 per bushel. To make matters even worse for the European farmers, the North American wheat was almost invariably of better quality for making raised loaves of bread.[15] As a result, American and Canadian growers were highly competitive with European farmers.

Europe itself was complex in terms of how different national economies related to wheat. Some countries, such as Russia, Hungary, Bulgaria, Romania, and Yugoslavia, were wheat exporters during the period 1885–1939. Others, such as Germany, moved from being exporters before the 1880s to being importers due to competition from North American wheat. Still others, such as France, were largely self-sufficient, with occasional supplies for export. Finally, there were the British, who, after repeal of the Corn Laws, became importers on a massive scale in the last part of the nineteenth century.[16]

Governments of these countries, both importers and exporters, had to come to grips with a critical problem: How does a state govern its people and its countryside to ensure a regular supply of food at a reasonable price? This question has always been with human societies in one form or another, but it is crucial to see that the advent of a global, competitive market in wheat completely reshaped the issues that had to be faced. In Europe and the Neo-Europes, the specific question was how should a government best ensure its nation a regular supply of wheat when that crop is being grown in excess quantities and can be shipped cheaply all over the world.

Individual farmers and landowners in Europe and the Neo-Europes, in contrast, had a much smaller set of concerns, but they, too, were of critical importance in establishing the framework for decision making: How should this land be used so that the farmer or landowner will be well served this year and be in a condition to continue next year? At stake were decisions on how to exert human labor; how to deploy land, crop species, and other resources; and how to maintain an individual's power, wealth, and social prestige.

We can think of these sets of questions for farmers and landowners as the "micropolitical ecological" concerns, in contrast to the "macropolitical ecological" concerns of how a government should control its people and resources to ensure a food supply. Governments, landowners, and farmers gave different answers to these questions, according to their circumstances. Critical to their dilemmas, however, was the fact that in the half century following the repeal of Britain's Corn Laws, too much land was given to wheat while the technology for a global market emerged. Wheat growers everywhere were in deep competition with one another, and everyone was looking for new ways to produce wheat more cheaply.

Laborsaving Machinery

In wheat production, the processes leading to higher yields per hectare changed only slowly during the nineteenth century. Much more rapid was the increase in production per hour of human labor invested. An incentive to create laborsaving machinery came from the stiff competition among wheat farmers, which in turn came from the excess lands planted in wheat. Not only did new machines thus play a role in the global competitive wheat market; they also generated interest in the uses of science in agriculture. One example, the mechanical reaper, illustrates the power of the new inventions.

Before the nineteenth century, wheat was harvested with a great investment of hand labor. Sickles and scythes cut the stems so that the grains could later be threshed from the ear (Figure 3.1). In the late eighteenth century a modified scythe, called a cradle, came into use in some areas. The cradle was a frame that caught the cut stems and thus doubled the rate at which the field could be harvested, to up to four acres per day by an expert (Figure 3.2).[17] Even the cradle's effectiveness was debated, however, and agricultural improvers noted the need in America for a device that could put an animal to work in grain harvest. By 1830, eleven American patents had been issued and a number of machines from Europe had been tried.[18]

Right up until the 1830s, therefore, farming practices for wheat were based on a high labor investment to capture all of the tiny little grains of wheat. In fact, the labor investment was so high that wheat might not have been ecologically successful as a human food crop[19] except for the fact that once it was sown it required virtually no labor until it ripened. The harvest labor averaged over the total time the crop was growing was sufficiently low that wheat was considered "worth the effort" it took to get it.

Figure 3.1   Wheat harvest, scythe without cradle. U.S. Department of Agriculture.

Figure 3.2   Wheat harvest, scythe with cradle. International Harvester Company Collection, from the film *Romance of the Reaper*, 1930, State Historical Society of Wisconsin, Madison.

European expansion in the North American Neo-Europe, however, ultimately proved conducive to the development of a radically new way of harvesting wheat: the mechanical, horse-drawn reaper. Two Americans, Obed Hussey and Cyrus McCormick, in 1833 and 1834, respectively, obtained patents for machines that could harvest small grains and grass. At first the machines were of crude design, poorly made of materials that quickly wore out, and of uncertain economic return. Persistence from the inventors, promoters, and users, however, finally resulted in common adoption of the machines by the 1860s. About 7,000 harvesters were made before 1850; by 1864, nearly 100,000 per year were manufactured. Their use became general throughout the United States.[20]

With the reaper using animal power and mechanics to amplify human labor, the vast, flat, well-watered woodlands and prairies of central North America became ideal country for raising wheat. So long as Native Americans could be pushed out, the European and American immigrants saw an immense supply of land that could raise wheat. A smart farmer was one who figured out how to use the reaper rather than the sickle, scythe, or cradle and thus was able to harvest more land than his peers working with hand tools and without animal power.[21]

We can perhaps never know the exact course of the creative thoughts that led both McCormick and Hussey nearly simultaneously to develop their machines. Nevertheless, they created precisely what was needed to allow scarce human labor to work its will on the very large North American landscape. They made it possible to grow wheat far in excess of what an individual farmer's family could eat. They thereby also made it possible for North American farmers to produce wheat more cheaply than their counterparts in Europe, which in turn was part of the basis of a fierce global competition in wheat production.[22]

This, then, was the context into which the Darwin-Mendelian philosophy of variation came into existence: population was growing rapidly; Europeans were flooding the globe in the largest migration ever in human history; more land was found suitable for growing wheat, the staple of the European and Neo-European diet, than was needed; huge wheat supplies accordingly put pressures on all farmers to seek ways to produce at lower costs; and governments and growers everywhere were trying to figure out how best to make their way. The European and Neo-European world was awash with too much cereal, which hung over every farmer as a threat to financial solvency. Paradoxically, governments were threatened with famine and the ruin of their farmers if they did not establish the right policies. Farmers were faced with financial ruin and potential starvation if they could not grow a crop at a cost low enough to be sold at a profit.

What were the right moves for governments to make? Was one set of answers correct for all countries and farmers? To what extent could farmers and governments make decisions independently of one another or of their peers (farmers with farmers, governments with governments)? It is hard to judge whether people found the "right" or "best" answers. What is clear is that countries found three different responses.

First, some countries well endowed with good lands for growing wheat set a number of government policies that encouraged the development of a massive export industry. Prominent among this group were the United States, Canada, Australia, and Argentina. The United States and Canada used land and railroad policies prominently to settle their vast prairies and grow wheat. Australia emphasized railroad growth. Argentina focused on immigration policies to people the pampas.[23]

A second set of responses came from the European countries that prior to 1870 had been largely self-sufficient in wheat. They adopted tariffs to protect their growers against the more cheaply produced imports from North America, Australia, and Argentina. Included in this group were Germany, France, Italy, Spain, Portugal, Austria-Hungary, Sweden, and to some extent, Switzerland. Motives for protectionism were probably mixed between protecting farmers, preventing agrarian unrest, and keeping wheat self-sufficiency for military and strategic reasons. Germany may have been the most explicit in basing its tariff policies on the need for self-sufficiency in time of war and the need for a strong agricultural population to provide manpower for the German army.[24]

Britain essentially stood alone in Europe in not adopting protectionism. When the Corn Laws were repealed in 1846, tremendous political sentiment continued to resist their reimposition. As a result, the British became absolutely dependent for survival on wheat from other countries, and the British market for wheat became the dominant force in the world market.[25] Although the British bought wheat from many places, its special colonial and imperial relationships with Canada, India, and Australia were of special importance in guaranteeing its supplies of a vital material. Britain also had heavy investments in the United States and Argentina, which meant those countries, too, needed to sell something to the British in order to service their debts. Wheat was a prime export that served this purpose.

The third and final set of answers to the problem of wheat overproduction was on the surface paradoxical, yet it was absolutely critical to the spread of the new science of plant breeding based on Darwin-Mendelism: a number of countries created institutions dedicated to research and technological development in agriculture. Wheat was invariably a prominent object for study in these new laboratories, and the goal of research was to make it possible for their farmers to grow more wheat at lower costs so they as individuals could compete in the hostile financial environment of the global wheat market. Protectionism may have allowed farmers to survive in the enterprise of producing wheat in high-cost conditions, but science ultimately was intended to lower the costs of production.

The Beginnings of a New Science of Wheat Breeding: England and America

Two scientists, Liberty Hyde Bailey (1858–1954), an American, and Rowland Harry Biffen (1874–1949), from England, made early, fundamental contributions to a new science of plant breeding. We will trace how they adopted the new philosophy of variation from Darwin and Mendel for the study of plant improvement. They were not the only significant actors creating a new science of plant breeding, but their respective thoughts about the manipulation of plant behavior exemplify the connections between applied plant genetics and the needs of civil society.

Bailey was a major conduit for moving Darwin's ideas about variation into the mainstream of applied, agricultural science and plant breeding. He was born on a farm near South Haven, Michigan, in 1858. Bailey was interested in natural history from a very early age and read Darwin's *On the Origin of Species by Means of Natural Selection* before he was sixteen. He received his undergraduate training at Michigan State Agricultural College and then worked with Asa Gray in botany at Harvard. Gray was a leading spokesman for Darwin's views in America, and Bailey furthered his appreciation for Darwin while in Cambridge.[26]

In 1885, at the age of twenty-seven, Bailey took a position as professor of horticulture at Michigan State Agricultural College. He was insistent that the practical field of horticulture be infused with the new science of botany that had been so influenced by Darwin and others. Three years later, in 1888, Bailey was lured to Cornell by better research support to become that university's first professor of horticulture. As a horticulturalist using the science of botany, it was Bailey who coined the term "plant-breeding."[27]

Bailey was by no means the first person to attempt to identify principles about variation in plants and how they could be used to produce economically useful varieties. He was, however, one of the first to attempt a systematic synthesis of this new field of "plant-breeding." In 1891, at the age of thirty-three, he published "Philosophy of Crossing Plants," followed in 1892 by "Cross-breeding and Hybridizing."[28] In 1895 Bailey published his first book-length treatise, *Plant-Breeding, Being Five*

Lectures upon the Amelioration of Domestic Plants. This book was derived from his earlier work in 1891 and 1892 plus additional lectures he gave at the University of Pennsylvania in 1895.[29]

Bailey began his 1895 discussion with a profoundly Darwinian theme. Variation was a natural fact, and the problem was to understand its biological and evolutionary importance:

> There is no one fact connected with horticulture which so greatly interests all persons as the existence of numerous varieties of plants which seem to satisfy every need of the gardener. Whence came all this multitude of forms? . . . Whatever attempt the gardener may make at answering them is either befogged by an effort to define what a variety is, or else it consists in simply reciting how a few given varieties came to be known. But there must be some fundamental method of arriving at a conception of how the varieties of fruits and flowers and other cultivated plants have originated. *If there is no such method, then the origination of these varieties must follow no law, and the discussion of the whole subject is fruitless.*[30] (emphasis added)

Bailey then moved to a consideration of the causes of variation, which in his view consisted of fortuitous variation without known cause, the variation arising from sexual reproduction, physical factors in the environment, and the outcomes of the struggle for life. The last cause led him to ask how nature fixed or made relatively stable certain variations. At this question Bailey moved directly to link Darwinian evolution with applied plant breeding:

> "This preservation of favorable individual differences and variations, and the destruction of those which are injurious, I have called Natural Selection of the Survival of the Fittest." This is the philosophy which was propounded by Darwin, and which will carry his name to the last generation of men. It looks simple enough. . . . Yet, this simple principle of natural selection was the first explanation of the process of evolution which seemed to be capable of interpreting the complex phenomena of the forms of organic life. . . . It seems to be indisputable that natural selection is the chief force underlying the evolution of plants, and it is the only one with which the person who desires to breed plants need intimately concern himself.[31]

According to Bailey's contemporaries at Cornell, it was precisely Bailey's use of the concept of Darwinian evolution that made his teaching and research so exciting:

> The glory of the instruction in the horticultural department was centered in the course in evolution. In its day it was the most effective presentation of evolution given in Cornell University and attracted students from all colleges. By gradual and easy steps the student was led from the simple facts of variation to the most profound problems of evolution.[32]

Bailey occupied a middle ground in late-nineteenth-century biology between those who still saw a possible place for Lamarck's inheritance of acquired characteristics as a source of variation and those who accepted Weissman's dictum that germ cells (cells giving rise to pollen and ovules in plants) and somatic cells (the rest of the plant) were separate and thus precluded the permanent acquisition of an adaptation to the immediate environment. Bailey leaned primarily toward Darwin's notion that natural selection was the force molding the shape of species, but he left the door open to Lamarckian views based on his experience that "every change or variation in any organism—unless it proceeds from mere accident or mutilation—may become hereditary or be the beginning of a new variety."[33]

Bailey recounted in 1914, in a historical preface to the fifth edition of his book *Plant-Breeding*, that he still had a warm place in his heart for Darwin:

> These new investigations have taken us far from the point of view of Darwin, in which the original editions of the book were founded. I doubt whether the students receiving their instruction to-day . . . have any such feeling for a master-spirit as we had in those days when the studies of Darwin had given a new meaning to nature, when there were still a few naturalists left, and when the glow of his writings was warm in every person's work. To one coming out of a plant-growing relationship, the masterful works of Darwin had introduced order, and the forms of cultivated plants had been made worthy of serious study. . . . All these writings were fascinating to read. How to produce new forms of vegetation seized some of us with irresistible power.[34]

What Darwin did for Bailey was to take the complexity of variation in organisms away from being a distracting diversion and turn it into a focus for study. Darwin also provided a rationale, evolution by natural selection, that gave a scientist hope that natural, discoverable laws governed the function of nature's complex diversity. Bailey used Darwinian theory as the springboard for making it possible, in his words, to find a method to study the origins of the many forms of cultivated plants. Without the method, all discussion of how varieties originated become "fruitless." Armed with Darwin, Bailey was completely enthralled by an "irresistible power" to create new varieties as a plant breeder.

Not only did Bailey serve to bring Darwinian order to the study of variation; he also played a role with others in the introduction of Mendel's thoughts into plant breeding, especially in the United States. His textbook *Plant-Breeding* was apparently very popular, and it served as the vehicle for introducing Mendelism to a broad audience. The book came out in a second (1902), third (1904), fourth (1906), and fifth (1915) edition. All but the 1902 and 1904 editions had multiple printings. It was not until the third edition, however, that Bailey incorporated the thoughts of Mendel into his work. At that time (1903) he noted that the bibliography of his 1892 essay contained a reference to Mendel's papers but that he, Bailey, had merely taken them from a German book on plant crossings and had not actually read them. Bailey credited Professor Hugo de Vries of Amsterdam, William Bateson of England, and Herbert J. Webber of the U.S. Department of Agriculture as the most active in introducing Mendel to American scientists.[35]

A certain amount of confusion has surrounded the story of how Gregor Mendel's work became integrated into the mainstream of twentieth-century agricultural science. Mendel himself (1822–84) certainly did not make the connections between his work and its applications. He was an Augustinian monk, the son of a small farmer who was educated at a village school, ordained in 1847, educated at the University of Vienna in physics, math, and biology, and a teacher at a technical high school until he became abbot of the Monastery of St. Thomas in Brünn, in what is now the Czech Republic. Most of his scientific work occurred during the time he was a teacher, between 1857 and 1865, and he did no scientific work at all after he became an abbot. The paper for which he is now most noted was "Experiments in Plant Hybridization," which appeared in 1866 in the *Proceedings of the Natural History Society of Brünn*.[36]

What was confusing about Bailey's role in the integration of Mendelian thought into twentieth-century biology concerned the statement of one of Mendel's rediscoverers, Hugo de Vries. De Vries, the Dutch botanist now most remembered for his theories about evolution by mutations, wrote on two occasions that he had been led to Mendel's paper by Bailey's bibliography, but the exact publication of Bailey's that de Vries said he used was contradictory. A careful reexamination of the situation suggested that de Vries received a copy of Mendel's paper from a friend at Delft, not Bailey, probably in early 1900.[37]

Regardless of how de Vries was led to Mendel's paper of 1865, he was one of three botanists who virtually simultaneously discovered a new significance for Mendel's thoughts. Karl Correns, a professor of botany at the University of Tübingen, and Erich von Tschermak of the Hochschule für Bodenkultur in Vienna also came upon Mendel's work in 1900.[38] Of the three, only Tschermak was interested in the practical aspects of plant breeding. His work included peas, beans, wheat, barley, and rye.[39] Correns was more interested in plant physiology, while de Vries was interested in physiology and evolution. After 1900, de Vries was highly important in getting Mendel's thoughts directly to English and American wheat breeders, but Tschermak also contributed because of his contact with Swedish wheat breeders, who in turn were influential with the English.[40]

Not only did confusion surround how de Vries was led to Mendel, but uncertainty has surrounded the question of whether Mendel's paper lay relatively unnoticed for thirty-five years and, if so, why. When it became famous, what were the reasons for its transformation? A great deal of historical literature has dealt with the "rediscovery" of Mendel, and some have suggested that in fact Mendel was not rediscovered in the sense that his work was never noticed. Instead, Mendel's work reentered twentieth-century biology because it was relevant to an entirely new and different set of questions than those to which it had been directed.

Mendel at mid-nineteenth century very likely was interested in understanding how species evolve through hybridization. As such, his paper was consistent with a substantial amount of other investigation at the time on this question. In contrast, when Mendel's work was cited in 1900, the debate in evolution had shifted to two questions: (1) Could Darwin's mechanism of natural selection work if only very small variations were present? (2) What might replace Darwin's faltering theory of pangenesis as a mechanism of heredity? De Vries was adamant that the answer to the first question was no, and he viewed Mendel's observations on the stability of transmission of large character differences as supporting evidence for his theories of evolution by "saltation," that is, the importance to evolution of the inheritance of large character differences.[41]

Once Mendel's thoughts were taken into the crucible of the debate over de Vries and his saltation theories of evolution, the significance of Mendel's work was entirely changed. It ended up caught between what historian William Provine has called two warring camps of English biologists who were searching for ways to prove or disprove Darwin's mechanism of evolution by natural selection working on almost imperceptibly small differences among individuals. On one side of the debate were the biometricians Karl Pearson (1857–1936) and W. F. R. Weldon (1860–1906), both of whom adhered to Darwin's notion of evolution by natural selection acting on very small variations. Both were inspired to use statistical methods by Francis Galton (1822–1911), and they claimed him as their intellectual progenitor. Galton formulated, and Pearson revised completely, what Pearson called Galton's law of ancestral heredity, a statistical approach to inheritance that studied the means of measured characteristics of interbreeding populations and not the particular outcomes of crosses between specific parents.[42]

On the other side was William Bateson (1861–1926) from Cambridge University. Bateson, like Thomas H. Huxley (1825–95) and Galton, had become convinced that Darwin's theory of natural selection acting on small differences was inviable. Only if large differences arose by mutation could one expect to see the evolutionary changes envisioned by Darwin. Their reasoning was based on Galton's earlier formulation of the idea of regression, in which offspring tended to have means of characteristics more like the population than like their parents. In other words, differences between parents and the rest of the population tended to be lost in the next generation. Thus it was not possible to select for small, continuous variations in order to create new species.

When they met for the first time in 1899, Bateson and de Vries were both advocates of evolution by selection for large, discontinuous variations. When de Vries in the following year brought Mendel's work to light, Bateson was delighted and became the major champion of Mendelism in England.[43] Bateson and de Vries both were influential in taking Mendel to America. They both attended the International Conference on Plant Breeding and Hybridization in New York City in 1902. According to conference participant Liberty Hyde Bailey, Bateson's book *Mendel's Principles of Heredity: A Defense* (1902) was the thing to read for all work in plant breeding.[44] De Vries also lectured extensively in the United States in 1906 and promoted his ideas of inheritance and evolution.[45] Bateson worked to create a community of researchers at Cambridge who shared his enthusiasm for the new experimental way of studying inheritance and evolution. One of his colleagues who immediately saw a way to use Mendel's ideas was Rowland Biffen.

Biffen, born in 1874, entered Cambridge University in 1893, where he studied for the natural science tripos. He did so well that he obtained a studentship to study fungi in 1896. A trip to explore for sources of rubber (1897–98) led him to see the importance of botany for industrialized economies that increasingly ran on tires. Accordingly, he switched to economic botany. In 1899 he became a lecturer in botany in the university's new department of agriculture. Biffen was an early enthusiast for Bateson's crusade for Mendelism, and he began to gather a collection of wheats and barleys for studies in variation and inheritance. He outlined his research program by 1903 and published his first extensive results on wheat in 1905.[46]

What makes Biffen's work more interesting than the squabbling of Bateson, Pearson, and Welden is that Biffen deemphasized the controversy over the mechanisms of evolution. This is not to say that Biffen may not have been interested in evolution or hostile to evolutionary theory, but he gave no evidence that he thought Darwinian theory was the critical issue at stake in Mendelian studies.

An anecdote recounted years later by the American botanist Edgar Anderson gives some personal dimensions to Biffen's lack of engagement with the burning evolutionary debates of the times. It seems that a young Russian wheat breeder, N. I. Vavilov (1887–1943(?)), came to Cambridge to study wheat with Biffen. Soon, however, Vavilov switched to studying with Bateson because of Bateson's encouragement to study comparable (homologous) variations in different cereals. This work was preliminary to Vavilov's later theory that the greatest variability of an organism could be found near the origin of the species,[47] but long before he became famous in his own right he was asked why he left study with Biffen for Bateson. Vavilov is said to have replied in his thick Russian accent: "Ahh, yes. Beeffeen. Beeffeen yess, he is a gude man, a fery gude man, but you see, he has no phee-low-so-*phe*."[48]

Perhaps it is fairer to say that Biffen was interested in other philosophical issues than the ones that so engulfed his colleagues at Cambridge. Biffen had a vision for the importance of Mendelian genetics that, when properly considered, was astoundingly radical in its implications for the use of the English landscape, the role of agriculture in the British economy, and the place of the British state in governing the food supply of the British people.

Consider the introductory paragraphs to Biffen's first article reporting the results of his researches on Mendelian characteristics in wheat. He began not with Darwin but with a recitation of the fate of the declining English agricultural economy during the previous thirty years:

> We [England] can grow on the average over 30 bushels to the acre where the United States grow 14, Russia 10, and the Argentine 7. Yet the acreage under wheat in this country has fallen from three and a-half million acres in 1876 to one and a-half million in 1903, and we now grow approximately only one-fifth of the wheat we consume. Further than this there is good evidence to show that the quality of the grain now grown is inferior to that of twenty years ago. It has been sacrificed to yield, and many of the better class varieties . . . have been more or less driven out of the field by [new] varieties . . . which are capable of giving slightly larger crops of grain and straw. These inferior varieties have now to compete with wheat imported from Canada, the United States, Russia and other countries. The seriousness of the position becomes evident when one finds English wheat selling at 28s. 6d. a quarter when Manitoba Hard is selling at 35s.
>
> . . . the miller tells us that English wheat . . . is lacking in "strength." . . . The flour of English-grown wheat, alone, will not produce a loaf which is marketable under present conditions. . . .
>
> Since the opening up of the wheat-growing districts of the United States and Canada . . . the milling trade has to a large extent found its way to the ports. The millers so situated grind the strong wheat brought direct to their mills by sea. . . . The whole question then pivots on the strength of the grain we can produce.[49]

Biffen then went on to report the results of his crosses in wheat in which he studied the inheritance patterns of over a dozen characteristics with at least two forms of each characteristic. He found that Mendel's laws fit the behavior of characters that were morphological, histological, and constitutional in nature. For Biffen, the Mendelian scheme moved the art of predicting the outcomes of plant hybridization experiments away from being a complete game of chance to being an exercise in order. Whereas before 1900 little existed other than intuition to guide breeding experiments, Mendel's laws gave a method for how to plan and study such crosses.[50]

Bateson was perhaps more direct in stating the profound change that Mendelian genetics gave the practical breeder: "Though, as naturalists, we are not directly concerned with the applications of science, we must perceive that in no region of knowledge is research more likely to increase man's power over nature."[51]

Biffen pursued his investigations in a way that reflected the optimism of Bateson, but he used an entirely different perspective and language: he used Mendelism to seek a new role for English wheat in the bread of Britain. Superficially, that may not sound like much, but a fuller perspective would see Biffen's work as an effort to reverse the half-century decline and decay that had beset the English agricultural economy since repeal of the Corn Laws in 1846. He saw no reason why England should not produce high yields of strong wheat. He saw no reason why England should not produce a substantial amount of its own wheat rather than rely on imports.

In essence, Biffen was countering the results of what the industrial revolution and the opening of the North American prairies had done to English agriculture, but he was using the tool of science rather than new Corn Laws to make British agriculture more rational and more efficient. In a very real sense he was a prophet for bringing the industrial revolution into agriculture. By 1910 he had released his first successful new variety of wheat, Little Joss, which was resistant to yellow rust. He had determined in 1905 that susceptibility to yellow rust was a Mendelian character,[52] and Little Joss was the first refined product of that basic research.

In 1912 Biffen became the first director of the Plant Breeding Institute, funded partially by Cambridge University and partially by a government grant. He also became professor of agricultural botany at Cambridge in 1911, was elected fellow of the Royal Society in 1914, and was knighted in 1925.[53] His dream of an England capable of raising large amounts of strong wheat was not to be realized for many decades, but by his actions Biffen showed he had a sweeping "phee-low-so-*phee*." Moreover, Biffen's philosophy of remaking British agriculture has continued to dominate English agricultural science for decades, right up to the present.

Like Biffen in England, Liberty Hyde Bailey in America also became a powerful spokesperson for a philosophy of agriculture, and agricultural science was seen by him to play a pivotal role. Bailey's scientific career began in 1883 when he went to work for Asa Gray;[54] it thus overlapped substantially the fundamental transformation of the American economy from agrarian to industrial. It was during his life that the United States moved from being a predominantly rural and agricultural country to one in which most people lived in cities and worked in industry or services. Business firms became integrated units controlling as much as possible of raw material supplies, manufacturing, transport, marketing, and finance.[55] In a similar vein, agriculture changed from being a mixture of subsistence and market-oriented activities to being almost entirely market-oriented.

Transforming agriculture to a fully capitalist, entrepreneurial activity was as wrenching for the American countryside as it was in England. American agriculture had strong commercial attributes from the beginnings of European settlement in North America. Until 1860, however, local rural communities had significant

capacity for self-sufficiency in food and many other materials, except in the planta-
tion economies of the Deep South. Skills suitable for farming under such conditions
required little in the way of financial management or scientific and technical know-
how, other than what could be learned by young people as they grew up on farms.

After the American Civil War (1860–65), the nature of farming began to change
toward more commercial, specialized operations in which entrepreneurial skills and
business acumen were increasingly important. The fruits of technical invention,
particularly machinery, also became more important. What one had to know to thrive
in market-oriented agriculture was very different from what was necessary to prosper
as a largely subsistence farmer.

During the period 1870–1900, American agriculture became for many a life of
pain and suffering. Some distress was caused by drought, but many problems had
their origins in how societies organized their use of land and in the structure of the
economy. Opening the new lands, the resultant low prices, battles over the control
exercised by railroads and grain elevators, the effects of an almost continuous defla-
tionary policy by the U.S. government, and a shortage of credit were in many ways
symptomatic of the unceasing movement of capitalist markets and capitalist social
relationships into rural America.[56] Some farmers learned how to work the rules of
the new agriculture and prospered. But some farmers did not, and their productivity
and prosperity remained low.

A variety of reform strategies emerged, each seeking a way to bring order, tranquility, and prosperity to the countryside. Fraternal and cooperative movements like
the Grange spread, especially in the new lands in the Great Plains.[57] Political back-
lash against the crop-lien system and the railroad monopolies, the "Populist Revolt,"
swept many into partisan politics in the South and the Midwest.[58] New business
methods brought forth another stream of reformers who labored to teach all farmers
the new, rationalized accounting methods of capitalism.[59] Scientists, like Bailey,
preached the benefits of scientific farming.

All the ferment in rural America continually bubbled in American politics. Resolution of the turmoil raised by the fiery Populists and others came after 1900, but it
was partly in the shape of higher farm prices and partly in a growing dominance of
scientifically progressive agriculture. Ultimately, the rationalism of the scientists and
the industrial commercialism of agricultural publishers and various agrarian philosophers who voiced their views through the country life movement articulated the new
essence of American agriculture: it would be business-oriented and based on scientific knowledge and technical expertise.

President Theodore Roosevelt created the Commission on Country Life in 1908
and at its helm put Liberty Hyde Bailey as chair. In some ways Bailey was an agrarian romantic, but as chair of the commission he was a forceful voice for progressive,
scientific rationalism that would improve country living by consolidating the scientific revolution in American agriculture.[60] Bailey himself noted that the closing of
the frontier mandated a transition in American agriculture from ceaseless exploitation of new lands to scientifically intensive management of a stable land base.[61]

Complexity of motive, however, permeated the country life movement, which
reflected the many dimensions of the context in which wheat breeding emerged. For
some, promotion of rural reform was a romantic nostalgia for an imagined bygone

day in which yeoman farmers were the economic, political, and moral backbone of the American Republic. For others, reform was a matter of hardheaded business sense by urban capitalists and industrialists who recognized that industrial economies had to have an efficient, highly productive agriculture in order to support manufacturing economies. Yet a third group saw reform as an antidote to their fear of urban industrial immigrants and a search for American native values in the rural areas. In many ways, remaking the rural economy was parallel to the ideology of progressivism and a penchant for rationalism in all facets of life.[62] The country life movement encompassed all these complex motivations, and together they served as a powerful foundation for support of a scientific enterprise that aimed to rationalize the wheat plant and the methods of growing it.

By chairing the Commission on Country Life, Bailey was transformed from a disciple of Darwin and Mendel among the agricultural biologists to a policy adviser on reform of rural America. Bailey saw three fundamental conclusions in the final report of the commission: (1) that society needed to have a great deal of factual information about rural areas in order to create site-specific plans for reform and improvement; (2) that the agricultural colleges must have the capacity to provide the best scientific knowledge to every farmer; and (3) that the nation needed an ongoing series of conferences on rural life at the local, state, national, and perhaps international levels.[63] Each of the three was in some way followed. The first was influential in creating disciplines of rural sociology and agricultural economics. The second was important in promoting the Smith-Lever Act of 1914, which provided extension services in every county. The third helped inspire numerous country life conferences after 1909.[64]

Bailey fully developed his philosophy of agriculture within the modern economy in his book *The Country-Life Movement in the United States*, published in 1911. Here Bailey spelled out quite explicitly that American agriculture had to change to an industrial model. He went so far as to distinguish the "country mind" and the "town mind," which had very different and antagonistic ways of approaching the problems of life. Country folk were isolated and individualistic, while town minds saw the solution to problems through cooperation and collective action. As a result, the city could sit like a parasite on the countryside and determine its fate and the disposition of the wealth and people produced in rural areas. In order to be perfected, civilization had to forge organic links between city and countryside. Broadly trained people working in the countryside and collaboration of city and country people were essential to making the needed adjustments.[65]

And what was the mission to be accomplished? It is here that Bailey was most explicit. He saw the whole basis of civilization changing toward an industrial order with trade over wide areas. To be sure, he saw a growing brotherhood among men and the need for industrial societies to promote human well-being as well as financial gain. But above all the new industrial order was to complete the physical conquest of the earth. The rural people had a key role in that it was every farmer's prime duty "to conquer his farm" in an unending struggle with better tools and methods. Bailey wanted the countryside and urban areas evenly matched, because for him the real battle was a rural and urban coalition working together to bring an industrial order to the entire world.[66]

Two characteristics stood out in Bailey's rural philosophy. He had a nostalgic sentimentalism for the economic and moral virtues of country living.[67] He also had a profound belief in the value and power of education at all levels to mold human life and economic enterprise.[68]

Unfortunately, a profound and irresolvable contradiction lay between Bailey's two major beliefs. The romantic in him wanted rural people to find prosperity and happiness in the countryside through agriculture as a living and a way of life. Yet prosperity in this enterprise, under the conditions of America's burgeoning industrial capitalism, required efficient production methods, which in turn required sophisticated education in the science and business of agriculture. As farmers turned through education to more sophisticated methods of running their farm businesses, they found they had to become materialistic, to adopt varieties and growing methods that produced the most at the least cost, to mechanize and thus reduce labor costs, and to become fully integrated into the manufacturing, marketing, and finance worlds of modern America. To keep the beauty of Bailey's agrarian dream was to eschew the fruits of education in science and business methods. To adopt the products of research and education was to leave behind the myth of an idyllic rural utopia.

In the decades after 1900, Bailey's promotion of scientific education for farmers had far more impact on American farming than did his admiration of a simple and wholesome agrarian society. Bailey explicitly realized that the transformation of agriculture to an industrial and scientific mode was simultaneously a transformation of society and of peoples' lives and the ways they worked: "It has been necessary to eliminate much of the old farm method in order to clear the way for the new. We have also been undergoing a process of assorting the people, to determine who will make the farmers that we need; this process is not yet completed."[69]

In the face of the wealth and comfort created by industrial economies, little attraction or capacity remained for staying in the countryside and facing hard work for little material reward.[70] Unless one became a progressive, educated, technologically sophisticated farmer, agriculture in the twentieth century was not a very good way to live. Bailey—as dean of agriculture at Cornell, proponent of scientific means of plant breeding, and apostle of Darwin-Mendelian natural laws governing plant variation— did far more to shape the emergent entrepreneurial agriculture than did Bailey, the sometime romantic agrarian philosopher.

Bailey may not have wanted to take the message of a radical industrialism into agriculture and have it contribute to the transformation of agriculture into a completely capitalist enterprise. Indeed he called for keeping a holy attitude toward the earth.[71] He maintained a spirituality that contrasted with the materialism and "irresistible power" with which he approached the making of new plant varieties that provided ever more efficient and higher yields in agriculture. His successors tended to lose sight of Bailey's more spiritual attitudes, but they never lost sight of the fact that he used Darwin-Mendelism to open up "a new meaning to nature," an "order" to the myriad variations then known among cultivated plants, and made those crop species "worthy of serious study." Bailey saw the need for "conquest" of earth, but his spiritual dimension modified his quest for a new industrial order into a plea for wise conservation and use of natural resources. Still, he was out to remake agriculture, civilization, and the earth.

Bailey had as sweeping and radical a philosophy in America as Biffen did in England, even though the specific circumstances under which the two men worked were quite different. England had no opening of new lands to throw its agriculture into turmoil, but the new lands in North America and elsewhere had a profound effect on England. England had no waves of immigrants pouring into its borders; instead, English people were part of the wave of immigrants to hit the United States. England had no sentimentality for a bygone era of yeoman farmers as backbone of English democracy; nostalgia for rural England was nostalgia for feudalism, and democracy came through the struggles of the urban industrial class. In spite of these differences, Bailey and Biffen both came to the same conclusions: agriculture, indeed civilization, would be better if rationalized and industrialized through science and technology. Scientific plant breeding with its intellectual roots in Darwin-Mendelism was their contribution to the promise of tomorrow.

Plant Breeding Coalesces into a Formal Science

Contemporary plant breeding rests on a foundation of Mendelian genetics.[72] On the surface, therefore, it appears that the community of plant breeders was in some way "receptive" to Mendelian ideas. But what does receptivity to a new idea or doctrine mean? Is it possible to get a more precise notion of how individual scientists and scientific communities received new principles and worked with them to create a body of knowledge that had credibility and prestige within that group? Most important, does such insight explain why certain forms of knowledge prevailed against others?

In the years immediately after 1900, a Darwin-Mendelian theory of variation was developed and propagated within scientific institutions created to bring scientific rationalism to agriculture. As will be developed in chapter 4, many of them were created to promote commercial efficiency in the face of the deluge of new production that flowed from the newly developed lands of the Neo-Europes. Exactly what the scientific questions were, however, involved a tangled complex of issues that can be seen from at least two vantage points: first, all scientists have questions or problems upon which they formulate their immediate research objectives, and their results in some sense must be understood in the context of those queries.

Second, results of investigations frequently end up being interpreted or found useful in contexts rather different from the concerns that motivated the work. With hindsight it is frequently possible to see the importance of research in ways that were not even imaginable before the work was done. In some ways, therefore, it is unfair to original investigators to interpret their contributions from the present because our hindsight makes so abundantly clear what their unknowns really were, at least from our point of view. Looking from the vantage point of the present, therefore, can bring an unfortunate sense of smugness about the "mistakes" made by an earlier generation of people. Nevertheless, when we see the list of questions that we can clearly ask and answer, but which original investigators could not, we can see best the impact of Darwin-Mendelism on the work of plant breeders after 1900.

Just what were the questions and uncertainties that a plant breeder in 1899 (pre-Mendel) faced and for which the answers given by 1939 were vastly different and, from our vantage point in the present, show the confusions and uncertainties of plant-breeding efforts before Mendel? The following questions highlight the important issues, and the answers briefly recapitulate the results of extensive study and reorganization of the knowledge of plant breeding in the forty-year period from 1899 to 1939.

QUESTION 1: No two individual organisms are identical, but is variation governed by a set of principles, or is it chaos?

ANSWER: By the 1890s, Darwin's work clearly gave shape to a philosophy that viewed variation not just as a random, chaotic event unguided by any principle or natural law. *On the Origin of Species* (1859) provided the framework to see variation as essential to the process of evolution by natural selection. *The Variation of Animals and Plants under Domestication* (1868) rounded out Darwin's immense appreciation of variation by reviewing all the important examples of it he knew from domesticated species. Cataloging variation in this way gave hope that the subject could be studied and understood. Liberty Hyde Bailey clearly drew his inspiration from Darwin in his efforts to codify the methods of plant breeding in the 1890s.[73]

Despite Darwin's great contribution of giving the plant breeders faith that variation could be mastered and controlled, the theory of evolution by natural selection was not particularly helpful in the design of methods for breeding. For reasons noted in the following, Darwin's name began to drop from the cited references and from the explicit discussion of practical plant breeding after about 1920—not because he was no longer believed but because he was no longer needed as an authority figure.

QUESTION 2: What is the relationship between variations among individuals and the classification terminology that botanists and plant breeders developed in order to identify and think about the organisms with which they worked: type, variety, subspecies, elementary species, species, genus, and so forth?

ANSWER: This question has at its heart the mire surrounding efforts to understand what a "species" was, an issue that continues to bedevil biologists even today. From the time of Linnaeus in the eighteenth century until well into the twentieth century, most classification schemes were what Ernst Mayr calls "essentialist" and "typological," that is, they emphasized that a species was characterized by an unchanging essence and that a type specimen existed that showed this essence, usually through morphology. For Linnaeus, species were real, sharply separated from each other, and constant. A countercurrent to the essentialist and typological scheme was a "nominalist" concept of species, in which the claim was made that only individuals were real and all taxonomic groupings of individuals were, in the final analysis, arbitrary.[74] Nominalist classification schemes were popular among botanists of the nineteenth century, and Liberty Hyde Bailey worked within this framework.[75]

Practical plant breeders before 1900 were not hindered by the different philosophies of classification. Instead, they made their crosses using parents that were interfertile and tried to find the progeny they wanted. Then they hoped these progeny would breed true so that the breeder could "fix the type." Major problems, however, came from new types that "ran out" over a period of several subsequent generations or types that reverted to earlier ancestral forms ("atavism"). Whatever a species was, "types" seemed to be an unstable category and their properties were not considered reliably heritable.[76]

Instability of progeny potentially linked plant breeding to the deeply philosophical question of what was a species. If a species had material reality based on an "essence," usually manifested by a morphological character, then passing on traits to the next generation was simply a matter of transmitting the essence. In this framework, reversion or instability in a progeny class would be troublesome because either the "essence" was paradoxically plastic or the essence of the species lay elsewhere than in the unstable character. Similarly, if species were mere arbitrary constructs without material reality, and only the individual organism had reality, then presumably some sort of rules had to govern how that individual passed on traits to offspring. Otherwise, it was hard to explain how the offspring looked anything like the parents. In both frameworks, therefore, the researcher seemed left standing in a quagmire in terms of how to understand what a species was and from there how to build a workable research program in plant breeding. This uncertainty may be why Bailey, who was a nominalist, introduced the whole subject of plant breeding with an extensive discussion of the nature of variation and the concept of species in his 1895 book.[77]

Establishing a working concept for "species," therefore, was a problem for plant breeders who were attempting to elucidate the laws of inheritance with Darwin-Mendelism. It may be that de Vries achieved substantial influence between 1900 and 1910 among plant breeders precisely because he linked a concept of species with ontological standing both to Mendel's scheme for inheritance and to observations made by practical breeders. De Vries proposed "elementary species" as the building blocks of a "species." Elementary species were individuals that possessed definitive traits, easily noted compared with the individuals without the trait.[78] Applied breeders could plan their work as crosses between elementary species and interpret the results with Mendel's rules. Ironically, de Vries himself lost interest in Mendelian inheritance, because he saw it as of only occasional importance.[79] His 1907 book, *Plant-Breeding*, does not even mention Mendel in the index, although it mentions other people liberally.[80] Nevertheless, de Vries exercised substantial influence among American scientific plant breeders such as Bailey, Edward Murray East, and Mark Alfred Carleton in the first decade after the reappearance of Mendel's ideas.[81]

Despite a considerable amount of debate before 1920 about the concept of species and the best terminology to describe it, the issue disappeared from the foundation texts of plant breeding by the 1940s. Plant breeders after about 1920 also dropped explicit reference to Darwin, de Vries, and evolution — not because these issues had been resolved but because plant breeders no longer saw them as critical. Getting useful new varieties was a matter of crossing parents with useful characteristics and no longer seemed to have much relationship to the quasi-metaphysical debates over the ontological reality of species.[82]

QUESTION 3: How do you distinguish and interpret variations that are large and conspicuous rather than smaller and more subtle? How do you distinguish and interpret variations that are heritable and those that result from environmental factors?

ANSWER: This complex question reflects the fact that scientific plant breeding had its historical origins in two different problems: What was the role in evolution of small rather than large differences among two or more individuals? And what was the role in the development of an organism of variation caused by hereditary factors rather than differences in the "environment" in which hereditary factors were expressed? Although we can now ask these questions with some clarity, much of the difficult work in plant breeding and genetics during the period 1900–1920 consisted of sorting through the conceptual complexities involved, interpreting a wide variety of empirical evidence, and developing new methodologies. Statistical methods and field-testing techniques were new tools that were of special importance to scientific plant breeding.

The story begins with wheat improvers of the nineteenth century, for example, LeCouteur and Shirreff. Both made their fame by selecting marked differences among individual plants in fields with much variation. These early workers, however, found only a few new varieties. With the intensification of competition in the new global agricultural market, however, a new problem began to take on increasing significance: What strategy is best for a breeder who wants systematically to find a series of ever-improving new varieties?

In essence, this was the problem Liberty Hyde Bailey set for himself in the 1890s when he began to compile his thoughts on plant breeding, which he considered the first "sustained attempt to account for the evolution of all garden forms" by making "very brief statements of some of the underlying principles of the amelioration of plants."[83] Simultaneously with his own writing, Bailey was involved with a series of editing efforts to bring the best of scientific thinking to a wide variety of popular works on agriculture and horticulture.[84] Unraveling the complexities of small differences and heritable variation, therefore, was part of the larger project of rationalizing agriculture.

Mendelian genetics, as noted earlier, began by playing a role in the debate about whether evolution proceeded by the selection for small differences or the emergence of new forms by an instantaneous jump or mutation. This was the debate among Pearson and Weldon on one side and Bateson and de Vries on the other. Pearson and Weldon argued that small, continuous differences were heritable and provided the raw material for evolution. Bateson and de Vries argued that small differences were not sufficiently heritable to provide the basis of evolutionary change; only large differences were reliably inherited and thus were the material selection acted upon. Biffen provided evidence about the reliable heritability of large, discontinuous differences in wheat, which Bateson used in his defense of Mendelian genetics.

Methodologically the two camps were sharply separated. Pearson, Weldon and the other biometricians measured the means, standard deviations, and correlation coefficients of populations of parents and offspring. The Mendelians, in contrast, identified parents with interesting and well-developed character differences, crossed them, and followed the numbers of progeny in distinct character classes in subsequent generations. They had no need for the statistical measurements of the biometricians, and the biometricians saw no need for counting numbers of progeny in conceptually distinct classes. As Edward Murray East said in 1907, "Galton's law is statistical and deals with averages; Mendel's law is physiological and deals with individuals."[85]

The biometrical methodology did not provide an explanatory framework for the hereditary behavior of differences that were large and discontinuous. Similarly, the Mendelian methodology tended to relegate the small, continuous variations to the uninvestigated and uninteresting category of "fluctuations." Neither biometricians nor Mendelians were searching for how to distinguish environmental from heritable variation, because both were chasing a different objective: the raw material of evolution.

Although the initial research programs in Mendelian genetics after 1900 did not immediately set out to distinguish between genetic and environmental variation, writers from Darwin and before knew that both types were always present. Consider two brief remarks from Darwin's *Variation of Animals and Plants under Domestication*:

> The wonder, indeed, in all cases is not that any character should be transmitted, but that the power of inheritance should ever fail.[86]

> These several considerations alone render it probable that variability of every kind is directly or indirectly caused by changed conditions of life.[87]

Although Darwin knew both inheritance and environment were important sources of variation, he also knew, perhaps painfully, that he had no firm way to distinguish whether a particular variation was due to inheritance or environment.[88] Neither experimental nor biogeographic methods of studying variation could claim to distinguish the two sources of variation. Bailey, even in the last edition of his book published in 1915, could go no further than Darwin in distinguishing the two causes of variation.

Bailey's book *Plant-Breeding* (plus his numerous other publications) provided important syntheses of the effort to rationalize agriculture and agricultural research. Nevertheless, *Plant-Breeding* was essentially conceived in the decade before Mendel's work became important, and Bailey did only a little to make use of and synthesize the four conceptual developments that ultimately created a methodology for distinguishing heritable from environmentally caused variation: Johannsen's work on pure lines, Nilsson-Ehle's work on multiple-factor inheritance, adoption of biometrical techniques for plant breeding, and development of field-testing techniques. Bailey and his coauthor on the fifth and final edition, Arthur W. Gilbert, cited no literature after 1912, which made the book somewhat obsolete by the time it came out in 1915, with reprintings in 1916, 1917, and 1920.

Other, younger authors, however, soon prepared the broad, synthetic textbooks that indicated plant breeding had reached a new level of maturity and consensus. Ernest Brown Babcock and Roy Elwood Clausen, both geneticists at the University of California, released *Genetics in Relation to Agriculture* in 1918, with a second edition in 1927. By 1918, however, Babcock and Clausen had a very clear sense that separating variation into heritable and environmental categories was the "fundamental distinction" to be made in understanding variation.[89] Because textbooks are written for didactic purposes rather than rigorous proofs, they do not neatly assemble all elements of the methods for making the distinctions between the two classes of variation. Through the book, however, the key elements emerge as the products of work on pure lines, multiple factors, biometrics, and field-plot techniques.

Pure-line theory came from the work of Wilhelm Johannsen (1857–1927) of Copenhagen, who in 1903 published a paper in which he reported his investigations on selecting for change in a continuously varying character, the size of seeds (beans) in self-fertilizing *Phaseolus vulgaris*. Johannsen was seeking to gather experimental evidence on whether evolution proceeded by mutations making large jumps or whether it was more likely to come from selection acting on small, continuous variations. He was, in other words, working on the problem framed by de Vries.[90]

What Johannsen found was that, within pure lines started by one seed of the self-fertilized plants he used, he could not further select for true-breeding larger seeds instead of true-breeding smaller seeds. After numerous generations in which he continuously picked large seeds and small seeds, within each line, he found that both large and small seeds still gave a range of seed size in the next generation that was just about identical. Johannsen concluded that within a pure line, variation was the product of the environment working on the internal, fixed, genetic properties of the pure line.[91]

Johannsen's interpretation of his results went straight into the battle between de Vries and Bateson, on the one hand, who were arguing for evolution through the origins of substantial mutations, and Pearson and Weldon, on the other hand, who were arguing for evolution through natural selection acting on small, continuous variations. Although Johannsen may not have intended to provide a definitive and formative experiment for practical plant breeders, Babcock and Clausen used his work as part of the argument that variation was either heritable or environmental. By the 1930s, Johannsen's work had been appropriated to articulate explicitly a principle that some variation was due to heritable differences and some was due to environmental differences that are not heritable.[92] Thus some small, continuous variations were explainable not as differences in genetic constitution but as differences in the environmental conditions in which the individuals had been reared.

Johannsen's work, however, did not mean that *all* small, continuous variations were the result of environmental differences. Mendel's work focused on character traits of markedly different form (e.g., tall and short stems or green and yellow seeds) that were easy to distinguish and gave no intermediate forms (e.g., of medium-length stems or of yellow-green seeds). Stark differences between the traits Mendel studied, incorporated into de Vries's search for an evolutionary scheme based on big differences ("mutations"), steered the early investigations into distinguishing between the big, discontinuous variations that behaved in "Mendelian fashion" and the small, continuous variations whose behavior was better studied with biometrical methods.

Was it possible, however, that continuous variations also could be attributed to Mendelian factors? Did any material differences exist between the continuous and discontinuous variations? Herman Nilsson-Ehle of Sweden conducted a series of studies on variations in oats and wheat between 1900 and 1908. His work at the Svalof plant-breeding station involved crosses between what appeared to be good, discontinuous Mendelian characters that would yield a typical 3:1 segregation pattern in the F_2. What he obtained, however, were, in different crosses, ratios he interpreted as 15:1 and 63:1 as well as some with 3:1. Nilsson-Ehle interpreted his results with the explanation that some of what appeared to be simple Mendelian characters were in fact traits governed by one, two, or three independent Mendelian factors. Those governed by one gave ratios of 3:1, those governed by two gave ratios of 15:1, and those governed by three gave ratios of 63:1. In other words, multiple factors could govern the same character.[93]

Nilsson-Ehle went on to calculate that if each factor had two forms and neither was dominant to the other, then the offspring in a cross could appear as 3^N different forms, where N is the number of factors involved. If ten factors governed a character and each factor had two forms that were not dominant to one another, then nearly 60,000 different combinations could appear, and the differences between them would appear as a continuous variable with little difference between individuals. Nilsson-Ehle recognized that his results suggested a way to reconcile the differences between the continuous and discontinuous variations: both types were due to Mendelian factors, and it was just a question of how many independent factors were affecting the same character in the individual. Edward M. East independently came to the same conclusions in the United States shortly after Nilsson-Ehle completed his work in Sweden. In addition, Thomas Hunt Morgan, also in the United States, found Mendelian factors in 1912 that caused very small but heritable differences in *Drosophila melanogaster*.[94]

Johannsen's and Nilsson-Ehle's works together demonstrated that both environmental and genetic factors could be responsible for small, continuous variations among individuals. The problem still remained, however, of how to tell the difference between the two causes for a specific variable feature. Distinguishing the cause of variation was perhaps the single most important problem for plant breeders: if they could not tell the difference between genetic and environmental differences, they could not recommend either a variety or a set of farm practices with any confidence or authority. Without a reliable, practical method of analysis, plant breeding could contribute little more than Darwin and Bailey's philosophy of variation. In fact, no rationalization of agriculture could occur through varietal selection or fertilizing and other soil manipulation techniques unless the conceptual distinctions between heredity and environment could be clarified and reliable tests developed to identify the source of specific variations.

A series of studies, many of them in the United States and Britain, in the first fifteen years of the twentieth century provided the nucleus of a practical methodology for distinguishing on a reliable, cheap basis whether a particular variation was genetic or environmental. The heart of the matter lay in adapting biometrical techniques and field-plot study methods to practical plant genetics and physiology. Babcock and Clausen were able to provide an introduction to these subjects by 1918, and over the following two decades both statistical and field techniques were refined considerably.[95]

Conclusions reached by Johanssen, Nilsson-Ehle, East, Morgan, and others wended their way into the standard lore of applied plant breeding in the years after 1910. Over a period of time, they provided a new foundation for a philosophy of variation: that variation came in at least two forms, heritable and environmental; that heritable differences between individuals were caused by Mendelian factors, ultimately named "genes" by Johannsen; and that heritable differences could be distinguished from environmental factors with appropriate field experiments analyzed by statistical methods. Johannsen also contributed the two terms now so familiar to all who study genetics, *phenotype* and *genotype*, to distinguish the superficial appearances of an individual, which was the manifestation of genetic and environmental forces, from the genetic constitution, which could be identified only by breeding experiments.[96]

QUESTION 4: Where in the organism are the Mendelian factors, and does their location make much difference to the plant breeder?

ANSWER: Very soon after Mendel's work entered the mainstream of biology in 1900, a number of speculative thinkers noted the similarity of the behavior of Mendel's postulated factors and the behavior of chromosomes during the formation of gametes or germ cells (sperm and eggs in most plants and animals). Walter Sutton (1877–1916) in the United States and Theodor Boveri (1862–1915) in Germany were the first to note that Mendel's segregation of factors to form pure germ cells was like the separation of homologous chromosomes in gamete formation. As a result, they independently postulated that Mendel's factors were located on the chromosomes.[97] Other than similarities in behavior, however, neither Sutton nor Boveri had any real evidence that linked the hereditary factors with chromosomes.[98]

Through a separate line of work by Edmund B. Wilson and Thomas Hunt Morgan, both at Columbia University, chromosomes were implicated in the determination of sex in embryos in a wide range of organisms. Morgan had a long-term interest in sex determination, as well as an avid interest in the model for evolution promoted by de Vries, with whom Morgan was a good friend. Efforts to find evidence for de Vriesian evolution eventually prompted Morgan in 1908 to begin work with the tiny fruit fly, *Drosophila melanogaster*, to see if he could find de Vriesian–type mutations that would be useful in studying evolution on an experimental basis.[99]

Morgan's search ultimately led him to a set of conclusions that, when compared with his position in the early 1900s, was a startling reversal for him and garnered him a Nobel Prize in 1933. Fruit flies proved to be a useful organism for genetic studies because they had easily examined chromosomes, numerous character differences, and rapid generation times, and they were cheaply raised in large numbers. Morgan utilized all these traits and had experimental evidence by 1910 that "Mendelian factors" were real in the sense that they must be made of biological material in cells and that the factors were located on the chromosomes. In his earlier work Morgan had expressed severe skepticism that any experimental evidence supported either of these conclusions, but ultimately he provided the empirical studies needed to justify the claims.[100]

Morgan had little interest in the implications of his work for plant breeding, but his results were quickly absorbed by both botanists and zoologists, who concluded that the chromosome theory of inheritance had wide validity among many organisms. Work by Harriet Creighton and Barbara McClintock in 1931 demonstrated simultaneous crossing-over between chromosomes and crossing-over between Mendelian factors in maize, thus providing the final proof of the chromosome theory of inheritance, and in a plant species to bolster the theory's universal applicability.[101] Moreover, the chromosome theory of inheritance united two fields of biology that had previously existed as separate lines of research: cytology and genetics.[102] Within twenty years, this union was the foundation for a number of conclusions in wheat breeding that were of extraordinary importance.

First, Tetsu Sakamura in Japan studied wheat chromosomes and postulated that wheat had a basic number of chromosomes that was seven, that is, its ploidy was seven.[103] Different kinds of wheat could be classified based on whether they were diploid with fourteen chromosomes ($2 \times 7 = 14$), tetraploids with twenty-eight ($4 \times 7 = 28$), or hexaploid with forty-two ($6 \times 7 = 42$). Tetraploid and hexaploid wheats are called "polyploid species," and understanding their polyploid nature allowed a series of studies, detailed in chapter 2, about the evolutionary origins of wheat. Once the evolutionary origins were understood, it became feasible to develop methods of crossing cultivated wheats with their ancestral, weedy grasses and thereby introduce valuable traits such as disease resistance from the ancestors to the modern wheats.[104]

Beyond understanding the evolutionary origins of wheat and devising a classification scheme, the chromosome theory of inheritance was critical to understanding many basic phenomena, such as those now called "linkage" (i.e., when the genes controlling two different characters are located on the same chromosome and thus do not assort independently) and "aneuploidy" (i.e., when the number of chromosomes present in the individual is not the usual number, which usually causes severe abnormality). By the early 1940s, one standard textbook for advanced students in plant breeding stressed the critical dependence of plant breeding on the chromosome theory of inheritance as follows:

> A knowledge of the chromosome basis of heredity is essential to the breeder. The characters of a plant are the end result of the interaction of genes, carried in the chromosomes, under particular environmental conditions. . . . The linear arrangement of genes in the chromosome has been generally accepted, and the division of the gene in mitosis and its segregation in meiosis have furnished the mechanism for the transfer of the unit of inheritance, the gene, from cell to cell. A knowledge of the number and nature of the chromosomes in each crop plant and their behavior in cell division is fundamental to the study of plant breeding.[105]

QUESTION 5: How does one construct a systematic program of plant breeding for the improvement of a crop species?

ANSWER: This question was, of course, not strictly a matter of science. Deciding that a systematic improvement of a crop species was needed was as much or more a political economic decision as it was a scientific one. Hence getting some understanding of how systematic breeding programs came to be constructed requires attention to both social and scientific dimensions. We turn here to the scientific problems and take up in Chapter 4 more details about the political economic dimensions.

First, it is important to note that crop improvement circa 1900 fell into several broadly recognized categories. In 1916 Mark Alfred Carleton, in charge of cereal investigations for the U.S. Department of Agriculture, prepared a treatise, *The Small Grains*. Carleton saw the work of the scientifically trained breeder encompassing three major strategies for improvement of wheat and other cereals: introduction of new varieties from other places to see how they perform, selection of better individuals from highly heterogeneous mixed populations that characterized most varieties then grown, and hybridization or the sexual crossing of two different varieties with subsequent selection of the best progeny as a new variety. In shorthand, these three strategies were referred to as introduction, selection, and hybridization, respectively.[106] Introduction and selection were both ancient in their origins, and both were extensively practiced in the nineteenth century, although only occasionally on a systematic basis.

For wheat, one of the first, major, *systematic* efforts of improvement came from Sweden and was based on selection. Unfortunately, the work began before Johannsen's notion of a pure line was formulated, and disappointment followed the failure of continued selection to effect desirable shifts in the characters of a wheat plant.

Hjalmer Nilsson began his selection work in Sweden in 1890 with what de Vries called the "German method," in which the breeder took a bunch of ears from a number of plants in a wheat field and used the mixture of seeds as stock for the next generation. The ears selected seemed appropriate for the characteristics desired, for example, higher yields or stiffer straw, but many individual plants served as sources for these ears. Nilsson worked, therefore, on a theory that selection could mold a plant in a highly elastic manner. Within a few years he gave up this strategy because it was not yielding stable new collections of plants with the desired characteristics. Instead he went to a method that involved selecting ears from an individual plant that seemed to have the desired characteristic. The seed from this one plant was the source of subsequent generations, and Nilsson found that starting with just one plant for a parent could ultimately yield a uniform variety that had the desired characteristic.[107]

What Nilsson discovered in his systematic improvement scheme was that plants were not limitless in their plasticity. It was Johannsen's theory of pure lines that finally clarified the reasons why breeders must start with one plant rather than many and why they must pay close attention to the parentage of the progeny selected. Breeders had to cross different parents if they wanted to extend the limits of plasticity. As Mendel's theories were developed after 1900, hybridization came increasingly into use along with introductions and selection.

Being able to delineate three major strategies for plant improvement, as Carleton did, was necessary to the confidence of breeders that their exertions would be rewarded with new crop varieties. Mere identification of strategies, however, was not sufficient to give a method for proceeding with their work. Breeders also needed specific tactics to guide them. Perhaps the best way to see the sizable changes in how biological theory guided action is to compare the general guidelines drawn from Liberty Hyde Bailey's pre-Mendelian treatment of the subject in 1895 with the advanced textbook of 1942 by University of Minnesota breeders Herbert Kendall Hayes and Forrest Rhinehart Immer.[108]

Bailey

1. Avoid seeking features that are antagonistic or foreign to the species; for example, don't try to get a tuber-bearing plant like a potato to bear fruits.
2. Faster results in hybridizing plants will come from species that are more variable than from those that are less variable.
3. Breed for one trait at a time.
4. Don't attempt to get contradictory attributes in a plant; for example, if you select for larger tomatoes, don't also try to select for higher numbers of fruits per plant.
5. When selecting seeds, look at the whole plant, not just one branch or one part of the plant.
6. Plants may look alike but give offspring of very different sorts.
7. The less an individual varies from the norm of the species, the more likely it is that its variation will be heritable; that is, wild aberrations are generally unstable.
8. Crossing two plants is only the beginning, not the end, of successful plant breeding.
9. If you cross two parents to get a new variety, use the preceding rules to select each parent and have each parent be strong in the characteristics you wish to combine.
10. Have firmly in mind what you are seeking before you begin crossing plants.
11. Seek to make your plants vary in the direction you have in mind, either by crossing or by altering the conditions under which they grow. (Bailey, in contrast to most of the breeders who took up Mendel's work, was a Lamarkian who believed acquired characteristics could be inherited.)
12. When seeking new varieties of perennials, look for bud varieties or sports that can be propagated asexually.
13. All permanent progress lies in continued selection.
14. Even when a new variety is successfully achieved, it must be kept up to standard by constant selection; that is, "there is no real stability in the forms of plant life."
15. The best progress will come as a result of "the best cultivation and the most intelligent selection and change of seed."

Hayes and Immer

1. Genetic and cytogenetic principles are the fundamental principles of plant breeding.
2. Know thoroughly the mode of reproduction of the crop you are working on, for example, whether it reproduces asexually or sexually and, if the latter, whether it is self-fertilized or cross-fertilized.
3. Self-pollinated crops can be improved, very marginally, by removing clearly aberrant types or by selecting good seeds, from an already successful commercial variety, for seed multiplication. (This rule follows the results of Johannsen's work on the limits of hereditary influences in pure lines.)
4. Cross-pollinated crops can be improved, sometimes substantially, by allowing natural selection to operate over a period of years, for example, by letting climate eliminate those individuals not suited to the area.
5. Self-pollinated crops can be improved substantially by selecting from a variety that is quite heterogeneous those individuals that are desirable and propagating these individuals as new varieties.
6. Cross-pollinated crops can also be improved substantially by selecting individuals of high promise.

7. Self-pollinated crops can be improved substantially by combining in one individual two or more traits from pure-line parents that had these traits separately.

8. Cross-pollinated crops can also be improved substantially by combining in one individual two or more traits that were previously in separate parents, but the seed from these individuals may not or almost surely will not breed true, in contrast to self-pollinated crops.

9. Statistical tests are needed to distinguish reliably the quality of the new varieties produced by any of the methods of plant improvement.

10. Testing new varieties in the field requires picking test plots similar to the field used by farmers and in using techniques to eliminate variation caused by differences in soil types or climate.

Several differences between the two summaries of principles for plant breeding are striking. First, Bailey could not even speak of "genetic principles." The word "genetics" had not been coined, and Mendel's work was not well known and certainly was not understood in a context of breeding experiments. Similarly, Bailey's sixth point was merely a statement summarizing many experiences in plant breeding, but why and how parents might produce offspring of different types had no order, rationale, or explanation. By 1942 Hayes and Immer could simply assert as their first principle that the breeder needed to follow "genetic principles," that is, the scheme of analyzing parents and offspring derived from Mendel.

Second, Hayes and Immer very carefully separate their mode of studying plant breeding into categories based on whether a plant is self-fertilized or cross-fertilized. Bailey knew the different habits of plants, but the critical importance of knowing how a plant reproduced simply was not clear. Bailey's context of understanding "self-fertilization" was derived from Darwin's observations that most flowers functioned to promote pollination of flowers on one individual by pollen from another individual plant. Darwin generalized what was then known about self-fertilization by noting that it tended to weaken the offspring.[109] In fact, Bailey used "close-fertilization" for what we now call self-fertilization and "cross-fertilization" for pollination of one individual by another. "Hybridization" for Bailey was a cross between two different species.[110] Both terminology and the conceptual links between reproductive modes and genetics had changed mightily by 1942.

Finally, plant breeding by 1942 had incorporated a full complement of statistical and field-testing techniques that Bailey could not have known and that provided a way of separating heritable variation from that caused by unknown causes or the environment. Plant breeders worked on the detection and manipulation of the heritable variations, while their intellectual siblings, the soil scientists, worked on the latter. Bailey made no such clear separation. In addition, because he was one of the remaining Lamarkians at the turn of the century, he still sought to propagate the environmentally induced variations (his eleventh principle).

Plant breeding thus moved a great distance in the first four decades of the twentieth century. It began its scientific debut as an offshoot of the study of variation and evolution. Its roots were in the studies of variation conducted in the eighteenth and nineteenth centuries by people who called themselves philosophers or scientists (like Knight, Darwin, de Vries, Bateson, and Bailey) and by people who considered themselves practical men (like LeCouteur, Shirreff, and Hallett). Mendel's ideas were at first treated as abstract formalisms without a material base, but Morgan and his colleagues gave them a material foundation in the cell. With time, the burning concern with evolution, along with explicit reference to Darwin and de Vries, passed out of the pages of plant-breeding tomes. Only Mendel and those who directly added to the ability to manipulate plants were left. Plant breeding was firmly an applied science with the announced aim of producing as many new varieties as possible that would have commercial utility.

Plant breeding by 1939 was first and foremost a compendium of knowledge, a science and art form practiced by trained experts. In many ways, the experts wanted simply to call themselves scientists who were doing their work and who served the public welfare by making better plant varieties. Concurrently with the development of plant-breeding theory, however, governments became more and more active in the support of institutions that hired and supported the plant breeders. In addition, in some cases large but private commercial interests became interested in the products of plant breeding—better seeds—as a source of profit.

Involvement of public and private agencies, spread over many countries, tightly linked the outcome of plant-breeding science to affairs of the state and to power politics. In various ways, the considerations of political economy also shaped the very development of plant-breeding knowledge. Chapter 4 will describe how the state and other groups became involved in scientific wheat breeding in several countries. With an understanding of state involvement, we will be in a better position to pursue an understanding of how states came to use plant-breeding science as part of their devices to govern earth and its people.

Plant Breeding in Its Institutional and Political Economic Setting, 1900–1940

Geneticists such as Liberty Hyde Bailey and Rowland Harry Biffen were prominent leaders in the new science of plant breeding. By 1940 they and their successors had constructed an elaborate body of theory and methods and had acquired a working collection of plant germ plasm. Plant breeding was an ongoing enterprise in a few countries, and production of such crops as wheat and maize already showed the commercial importance of the science.

As noted in chapter 3, the promotion of plant breeding and other agricultural science was part of the industrial revolution and stemmed from (1) the repeal of the Corn Laws in Britain, (2) the development of international markets for wheat and other grains, (3) the population growth and emigration of European peoples to many other parts of earth, (4) the subsequent expansion of land in agriculture, and (5) the increasing mechanization of agriculture. Each of these interlinked factors reflected that the growing of wheat and other grains was increasingly a commercial matter and enhanced the role of science.

Conditions conducive to the use of plant breeding also spawned organizations to train plant breeders, support their research, and provide easy access to their results. Agriculture, in turn, came increasingly to depend on a steady stream of new varieties from plant-breeding research.

Institutional development in plant breeding did not occur in neat synchrony with the conceptual and methodological breakthroughs outlined in chapter 3. Instead, the growth of organizations depended upon (1) the promotion of the science by scientists, (2) the stresses imposed on farmers by market competition in agricultural products, (3) the perils nations faced in war, and (4) in the case of India, efforts by the British to alleviate famine so that India would remain profitable and governable. Both the organizational infrastructure created by 1940 and the conceptual developments were critically important to the subsequent transformation of wheat yields that occurred between 1940 and 1970. This chapter explores the institutional developments to 1940.

Prelude to 1900

Britain, America, and India each started to organize agricultural science before 1900 by forming scientific societies, agricultural improvement associations, private and public experiment stations, educational institutions, and government ministries to promote science. These various institutions created networks of people, all concerned with the promotion of a more productive and efficient agriculture.

It is fruitless to search for an event that clearly separates early forms of prescientific agriculture from later forms that incorporated science. Suffice it to say that before 1800, only a few tangible examples of agricultural improvement based on systematic, abstract principles of knowledge existed, and only a small number of professional scientific investigators engaged in systematic research.

Members of the Royal Society in Britain, for example, had professed an interest in turning science to the advantage of agriculture since the mid–seventeenth century, and several voluntary associations for the promotion of efficient agriculture formed in eighteenth-century England.[1] The Royal Botanic Gardens (Kew Gardens), founded in 1841, served as a place for botanical study and the movement of new plant species and varieties within the British Empire.[2]

Although English settlement of America was concerned first with conquest and expansion of landholdings, Americans soon developed similar societies to promote the use of science in agriculture. Experimental gardens were established in South Carolina, Pennsylvania, and Georgia long before the Revolutionary War, and in 1742 the American Philosophical Society began its promotion of scientific agriculture.[3]

Transfer of British modes of science to India was much slower than to America, most likely because India was treated as a source of revenue and trade, not a place for English settlement. Nevertheless, botanical gardens at Samalkot in the south and Calcutta in the east were resources for study, new introductions, and experimentation in the late eighteenth century. The British East India Company was interested in the potential of revenue enhancement of such study, but the work was done primarily by European, not Indian, scholars.[4]

Continuing growth of science, commerce, industry, and population led to a quickening of the pace of agricultural science in the nineteenth century. During this period the organizational foundations emerged for professional, systematic investigations in agricultural science.

A faith in science for progress and prosperity was an important feature in the organization of the Agricultural Society of England in 1838, which became the Royal Agricultural Society of England (RASE) in 1840. Major landowners and agricultural publicists were the leaders, and RASE specifically provided a strategy for English agriculturalists who were economically depressed but believed that perpetuation of the Corn Laws was not the way to build a strong rural England.[5] "Practice with Science," the motto of RASE, was the objective to restore prosperity to English agriculture.[6]

RASE was a network of commercial agriculturalists and scholars that provided, through its exhibitions and journal, an outlet for new scientific knowledge. Of more importance, however, was the invention of what is now called the agricultural experiment station, which can conduct a research program over a period of years. John Bennet Lawes (1814–1900), born at his country manor, Rothamsted, was an uncertain student with an avid interest in the new science of chemistry. Between 1836 and 1842 he conducted a series of studies on how to make bone powder into a reliable fertilizer. His development of superphosphate warranted a patent that enabled him to start a profitable fertilizer business. It also enabled him to finance and conduct, with Joseph Henry Gilbert (1817–1901), a long series of soil fertility experiments on wheat and other crops at Rothamsted. Primarily, Lawes and Gilbert were trying to understand the origins of nitrogen in plants.[7] The Rothamsted Station became a model for systematic research in agriculture in Great Britain, the United States, and India.

Formal governmental structures for agricultural science were slower in coming to Britain. The Board for the Encouragement of Agriculture and Internal Improvement, founded in 1793, was not able to establish itself as an indispensable part of government and disbanded in 1822. The RASE took over some of its efforts to promote agriculture on a voluntary basis in 1838, and the government had no further official duties in agricultural science until the Cattle Plague Department was created in 1865 to control rinderpest.[8]

Sharp depression of Britain's agriculture in the 1870s, caused by bad weather and the continued pressure of free trade in grains and meat, led to new calls for a section of government to promote agricultural interests. A new Board of Agriculture was created in 1889, and its president consolidated scattered government work on animal diseases, control of insects, information collection, and agricultural research and education. Britain's agriculture, in spite of the board's presence, continued its decline. Land previously given to cereal production was changed to livestock and dairy farming. Free trade in low-priced grains meant imports continued to dominate Britain's food supply until 1916.[9] No significant infusions of money for research came until 1910.[10]

Although German states and universities began to support agricultural research in the early nineteenth century, Americans were among the first to unite government with agricultural science in a comprehensive way. Teaching was the job of the land-grant university system, which was created by Congress in 1862 and 1890.[11] Research was institutionalized by the Hatch Act of 1888, which funded an agricultural experiment station at every land-grant university. The U.S. Department of Agriculture (USDA) was created in 1862, became a part of the cabinet in 1889, and developed an extensive capacity for research after 1897.[12] Cooperative agreements linked university and federal researchers.

The American system was completed with the creation of the Cooperative Extension Service in 1914, which tied the federal and state governments together in a scheme to promote change through science in every county of the nation. Rothamsted may have inspired the notion that systematic long-term research was useful, but it was the American system of education, research, and extension that provided a model for government support of systematic, sustained agricultural research.

Despite experimental gardens and the founding of the Royal Agri-Horticultural Society in Calcutta in 1820, British India remained virtually devoid of systematic agricultural research during the nineteenth century. India before 1857 was the province of a private company, and its management sought revenue for its owners. The revenue came from taxes on agricultural land, which was the major part of the economy of India. Moreover, the British tended to set revenue expectations based on an estimation of what the land should produce, not on actual production. Such a land-tax system indicated that the British had expectations about how the land *must* be managed: with technology sufficient to produce at least enough product to pay the land revenue.

After the rebellion of 1857, control of India passed to the British government, and, in the United Provinces (now Uttar Pradesh), revenue policies were altered so as to encourage the favor and support of the landlord castes. Land revenue remained important, but other commercial activities served to reduce its relative importance in financing the Government of India.[13] Movement away from strict reliance on land revenue may have reduced incentives for the Government of India to support agricultural science so as to maximize production and hence potential tax receipts.

Two sources nevertheless promoted the introduction of agricultural science into India in the last third of the nineteenth century, but both were denied by London. First, the Cotton Supply Association of Manchester, speaking for the cotton textile industry, implored the British government to start departments of agriculture in each province of India in order to increase the supplies and decrease the costs of raw cotton. Second, the earl of Mayo, governor-general and viceroy from 1869 to 1872, saw great potential for agricultural science to increase the productivity of Indian agriculture. Mayo, with the assistance of Allan Octavian Hume of the Indian Civil Service, proposed to London that the Government of India start a Department of Agriculture to promote the use of science. London permitted only a Department of Revenue, Agriculture, and Commerce to be formed, but it never functioned to promote agricultural science. Military control remained the primary purpose of the Government of India.[14]

No change occurred in London's attitude toward India until a particularly horrific famine in 1876–78 affected 60 million people, killing over 5 million. A famine commission report in 1880 recommended both central government and provincial government departments of agriculture, and the provincial departments were created.[15] Incorporation of agricultural expertise in the central government was delayed until 1892, however, and then only a chemistry staff was added, for research and teaching. Finally, in 1897, the Government of India appointed an inspector general of agriculture, who provided the structure for bringing agricultural science systematically into the central government. No further appointments of scientific staff were made, however, until after 1900.[16]

At the close of the nineteenth century, only the United States had a substantial agricultural research capacity that included national and state activities. Britain had the rudiments of agricultural research and education, but they were not well supported and lacked effective ways of reaching most farmers. In turn, the British showed at best a desultory interest in promoting agricultural research in India. In both Britain and India, the beginnings of scientific infrastructure were present, but conditions had not yet persuaded London to develop that nascent system.

Developing the Systems, 1900–1914

Momentous changes in the organization of agricultural science appeared between 1900 and 1914. In India, the precipitating causes for reform were efforts by the British-controlled Government of India to continue its hold on the subcontinent. Reform in Britain came from efforts to end the stranglehold of the landed aristocracy on Britain's politics and to revitalize its depressed rural economy. In the United States, reform consisted primarily of an entrenchment of agricultural science as the major method for the management of land and the rural economy.

In all three countries, these reforms established the case that the modern state required scientifically based agriculture just as much as it needed control of its borders, criminal behavior, and currency. For the remainder of the twentieth century, the question of state-sponsored research virtually ceased to be an item of political debate. This period, therefore, was formative to the many changes that came to wheat production in the years after 1940.

George Nathaniel Curzon (1859–1925), baron of Kedleston, served as British viceroy in India from 1899 to 1905, the period in which, at Curzon's insistence, the Government of India embraced the task of conducting agricultural research in a systematic and ongoing manner.[17] Curzon came to India on the heels of a collapse of the value of the Indian rupee in relation to the pound, a severe depression stemming from the rupee's collapse and the disappearance of European investment, an outbreak of bubonic plague that caused many deaths and nearly closed the port of Bombay in 1896, and severe drought in 1897.

Further failure of the 1899 monsoon rains, coupled with the earlier disasters and a continual heavy drain on crop reserves by the revenue collection of the Government of India, created the famine of 1899–1900. Given these calamities, the population of India actually declined in number in the period 1895–1905, a sharp change from the period 1872–1895, which saw an annual increase of about 2 million people per year.[18]

Despite a surface tranquility caused by lack of a strong, coordinated resistance movement, Curzon knew the Government of India was not doing well, and he feared for the future ability of Britain to continue its domination of the Indian subcontinent. He feared the ultimate rebellion that would flow from the racist attitudes of the Europeans, which blocked Indians from securing more than low-level positions in the army and the civil service. He also knew that unpunished crimes committed by Europeans and corruption in the police were time bombs waiting to explode under British rule. Curzon was also a progressive who believed that human improvement would come with education, science, and industry, and he foresaw a future India that was strong economically and industrially. For Curzon, therefore, extensive reforms on many fronts were the hallmark of his rule in India.[19]

Curzon's comprehensive reform agenda included agriculture. He promoted irrigation development, extended the Indian railway network and improved its management, appointed an inspector general of agriculture for the Government of India, and, most important for long-term changes, collaborated with the provincial government of Bengal to create the Imperial Agricultural Research Institute (IARI) at Pusa in 1905. In the latter task, he obtained an important donation of £30,000 from Henry Phipps of Chicago, a friend of his wife's family.[20]

Curzon also arranged to have a fund of 2 million rupees devoted annually to agricultural education, demonstration, and research in the provinces. Six colleges of agriculture were initiated, and each major province acquired a director of agriculture. The research envisioned by Curzon in the provinces was not considered necessary by London, however, so the major research efforts were confined to the new IARI at Pusa.[21]

Despite the inability of London to see a need for a comprehensive center-provincial system of agricultural research, the establishment of the IARI was the first in a long series of events that had significant effects on India's agriculture. Within two decades, substantial change came to Indian wheat production. In the long term, the IARI was the center of activity for India's embrace of plant breeding as a major vehicle of reform in wheat production and other agricultural enterprises.

Albert Howard (1873–1947), a young botanist of thirty-two, accepted an appointment as imperial economic botanist at the IARI in Pusa in 1905. He had obtained a first-class natural science tripos at Cambridge in 1898, a bachelor's degree in 1899, and a master's degree in 1902. Howard grew up on a farm in Shropshire before heading to Cambridge, from which he went first to the British Colonial Service in the Caribbean. In 1903 he returned to England at the Wye College of Agriculture. In 1905 Howard married Gabrielle L. C. Matthaei, who had also completed a first-class natural science tripos and a master's degree at Cambridge and begun her scientific career with work on transpiration and respiration in plants.[22] Although Albert left Cambridge in 1899, just before the excitement in 1900 caused by de Vries and Bateson over the work of Mendel, he and Gabrielle were fully in touch with the new methods of plant breeding promoted by de Vries, Bateson, and Rowland Biffen.

From 1905 to 1924, the Howards worked closely as a team in India on an immense range of problems germane to agricultural botany.[23] Perhaps their most notable work was on the wheats of India, a topic for research that they received immediately upon arrival. Wheat was such an important topic for the Government of India that some consideration had initially been given to having Albert Howard work only on wheat, a decision that was rejected in favor of the wider duties he received. Within four years the Howards published a joint monograph, *Wheat in India: Its Production, Varieties and Improvement*.[24]

Although the Howards were not the first to study wheat in India, earlier work had been sporadic, uncoordinated, and done before the reintroduction of Mendelian thought. Much of their early work was devoted, therefore, to a thorough survey of the varieties and conditions of cultivation then known across the vast lands and conditions of British India. They also began a series of breeding experiments, assisted by Abdul Rahman Khan[25] and based on the methods pioneered by Biffen[26] at Cambridge. From this work the Howards achieved wide respect in the scientific community and helped establish the reputation of the new IARI.[27]

The Howards themselves noted that IARI came into being because of the advocacy of Lord Curzon, the viceroy.[28] The complexities of Curzon's reform agenda make it difficult to explain his motivations, which brought modern plant breeding to India. In all likelihood, however, Curzon understood that modern agricultural science promised enhanced yields and better prospects for commerce, which in turn could enhance the economic and political strength of the British Empire. The Howards, botanists par excellence, probably seldom thought about imperial security planning.

It is ironic but telling that the British government fostered an invigoration of agricultural science in India before it turned to its own land and agricultural economy. Possibly the widely shared assumption that Britain was to manufacture and the empire was to raise crops was the basis for this situation. Nevertheless, with the coming of a Liberal government in 1906, Britain, too, entered a phase of intense reform[29] that by the 1940s had immense implications for agricultural science and agricultural production. Significantly, a major part of the reforms in Britain was based on a struggle for control of land and its uses. From these efforts came the basic institutions in which a few key researchers fully integrated the new Mendelian-based methods of plant breeding into British agriculture.

Creation of a comprehensive system of agricultural research was part of the larger reform that changed Britain from a feudal, rural, agrarian economy to a capitalist, urban, manufacturing country. Prior to the industrial, commercial, and scientific revolutions, the primary basis for wealth, power, and prestige in Britain was ownership of land. By 1900 ownership of industrial facilities and commerce had come to provide a new basis for wealth and power. Similarly, training in the learned professions had provided a vehicle for advancing in society. Enlargement of the franchise brought many new occupations into political life, but the Parliament, both the Commons and especially the Lords, was still heavily dominated by the remnants of the wealthy landlord class.[30]

This landed aristocracy was broken permanently in the years after 1900. Working-class fermentation over issues of wages, working conditions, and control of wealth was a threat to power and privilege in all of industrialized Europe and in America. In Britain, the firmly entrenched system of free trade in food and reliance on imported grain had also left the farmers in a state of near-constant financial hardship since the 1870s[31] and farm laborers with a sense that they fared considerably less well than their urban counterparts.[32] With the cities seething and the countryside in dismal decay, the Liberal Party, in conjunction with allies close to labor interests, sought a broad reform in British politics after their 1906 victory.

Although the first two years of Liberal government brought little change, matters began to heat up after 1907 when the Liberals started to use the budget as an engine for social reform. A change in Liberal leadership brought Herbert Henry Asquith (1852–1928) to the prime ministership and David Lloyd George (1863–1945) to be chancellor of the exchequer in 1908. They launched reform through the budget in a way that ultimately ended the landed aristocracy's control over Parliament through their domination of the House of Lords.[33] In addition, they created conditions that led to a massive sale of estates by the landed nobility to middle-class farmers[34] and began the construction of a comprehensive system for state-supported agricultural research. British agriculture was thereby reformed over the next forty years. Farmers gained incentives to seek out methods of high-yielding agriculture, which were provided by the newly enhanced research laboratories.

Introduction of the budget bill in April 1909 was the first step in this long line of changes. Major points of contention in the budget were imposition of a steeper graduation in income taxes, a capital gains tax on land sales, and a tax on new leases of land.[35] These taxes struck at the heart of land as a special form of wealth and at the class of people who owned large amounts of it. Inconspicuously riding on the backs of these political giants were provisions to spend money to develop rural roads and fund agricultural research.[36]

The Development and Road Improvement Funds Act (1909) began state funding for agricultural research on a systematic basis. Lloyd George proposed that the Development Fund have £2.5 million to last for five years. Expenditures from the fund were directed by the Development Commission of seven unpaid volunteers and one full-time salaried employee. At £500,000 per year for expenditure, the Development Fund eclipsed the previous levels of support for agricultural education of less than £20,000 and for agricultural research of less than £500.[37] Earlier funding by the Board of Agriculture and Fisheries was quickly eclipsed by the Development Fund. Expenditures made by universities and private trusts for agricultural teaching and research, such as at Rothamsted, were expanded, and for the first time all parts of Britain saw a more coordinated approach to agricultural science.[38]

One of the first beneficiaries of development funds was the School of Agriculture at Cambridge University. This part of the university had begun its work informally in the 1890s when the county councils were beginning to offer technical education in agriculture, and it needed a supply of well-trained teachers. Instruction began in 1893, a diploma in agriculture was initiated in 1894, and some large donations led the university to accept complete financial responsibility for the Department of Agriculture in 1899. Rowland Biffen was appointed lecturer in agricultural botany that same year. The department began to flourish, partly because it found an outlet for its graduates not only in Britain's counties but also in the empire.[39] Albert Howard, one of its first products, went immediately into the Colonial Service and then spent most of his later career in the Indian Civil Service. The department became the School of Agriculture in 1910.

Biffen's first theoretical successes came by 1904. His first major practical success was in 1910 when he released a new wheat variety, Little Joss, resistant to yellow rust (*Puccinia striiformis*) and suitable for light as well as heavier, wet fen soils (see Chapter 3). Partly as a result of Biffen's successes, the Development Commission recommended £18,000 be given to the university to establish the Plant Breeding Institute (PBI) in 1912, with Biffen as its first director.[40]

PBI thus joined the earlier stations—Rothamsted Station (1843), the Long Ashton Research Station (1903), and the John Innes Horticultural Institute (1910)—as part of a growing network of experiment stations in Britain. The PBI was one of fourteen institutes and research centers founded in 1911–13 by the Development Commission.[41] The institutional importance of the new Mendelian genetics in this network was suggested by the fact that the first director of the John Innes Horticultural Institute was William Bateson, while the PBI was directed by Biffen—two of the major pioneers in England in this type of research.

John Fryer and E. John Russel gave credit to Alfred Daniel Hall (1864–1942) for the scientific leadership on the Development Commission in establishing the network of research stations in Britain. Hall was trained in science and had a long career of promoting scientific farming in Britain.[42] It was David Lloyd George as Liberal chancellor of the exchequer, however, who was the major leader in the political crusade to reform rural Britain. In his first presentation to Parliament of the 1909 budget, which started the Development Commission, Lloyd George made it clear that he had a magnificent vision for his complex reform package:

> This is a War Budget. It is for raising money to wage implacable warfare against poverty and squalidness. I cannot help hoping and believing that before this generation has passed away, we shall have advanced a great step towards that good time when poverty, and the wretchedness and human degradation which always follows in its camp, will be as remote to the people of this country as the wolves which once infested its forests.[43]

A. Whitney Griswold argued that a fundamental difference separated British and American efforts at agrarian reform, a difference that provides useful insights into what Lloyd George was trying to do. For Griswold, the British people, of all classes, never developed an identity between democracy and the small family farm.[44] For many centuries British rural life was characterized by a tripartite division into owner-landlords, tenant-farmers, and landless-labor. The lords owned and ran the nation, the farmers ran the parish as the lords' tenants, and the laboring class did what they were told.

Lloyd George was obsessed with the land, its ownership, and its importance for remaking Britain into a nation that empowered and benefited all its members, not just the tiny aristocracy that owned most of it. One song of the Liberals, "Song of the Land," indicated that land and its ownership was a cornerstone of their vision for a new Britain:

> The land! The land! 'Twas God that made the land.
> The land! The land! The ground on which we stand.
> Why should we be beggars with the ballot in our hand?
> God gave the land for the people![45]

Lloyd George was also enthusiastic about agricultural science as one of the vehicles for carrying out his program for social reform. For centuries science had carried with it the promise of bringing nature under human control and thus enhancing human prosperity. But science alone could not transcend the barriers to prosperity created by lack of ownership of land or lack of access to the fruits of knowledge because of poor schooling. Lloyd George believed science was essential for his program of promoting more equality in Britain. In this respect he was very much like his American counterparts, even though the social history of landownership and its relationship to democracy was so different in the two countries.

Despite the differences, the United States, too, went through important reforms between 1900 and 1914. Earlier discussion (Chapter 3) outlined the ideological change involved in turning American agriculture into a business rather than a romantically idealized "way of life." Liberty Hyde Bailey, plant breeder and member of the Commission on Country Life, was a leading figure in articulating the philosophy by which the rationalized scientific agriculture would be run as a sound business.

Bailey's eloquent philosophy of agriculture was complemented by a series of political, organizational, and administrative developments that affected the federal and state governments' capacities to conduct research in the United States. The most significant changes occurred during President Theodore Roosevelt's administration (1901–9). In contrast to the situation in Britain and India, however, the changes in the United States did not flow from catastrophic threats to governmental authority, farmer distress, or conscious efforts by one class to supplant the power of another. Instead, American reforms occurred in a period of recovery from years of agricultural depression and a wave of strikes in the 1890s.

In the election of 1896, William McKinley and the Republicans decisively beat William Jennings Bryan, candidate of the Democrats and of the Populists. Prosperity for entrepreneurs was thus enshrined as paramount in politics, rather than the interests of the working class and technologically unsophisticated farmers.

Farm incomes rose after 1900, effectively quashing the Populist revolt of a decade earlier, and urban-industrial unrest also quieted with economic recovery. After the assassination of McKinley in 1901 and the rise of Roosevelt to the presidency, the relative domestic tranquility permitted a flowering of reform efforts built upon rationalization, science, and efficiency.[46] In industry and in agriculture, a loose coalition of scientists, entrepreneurs, publicists, and their allied elected officials consolidated the role of science and efficiency as the arbiter of how to proceed.[47] The ideology of the Progressive Era was the hallmark of the Roosevelt administration, and it brought changes in the form of irrigation and reclamation of arid lands, conservation of forest resources, control of fraud in the marketing of food and of agricultural chemicals, regulation of railroads and trusts, and increased control of the movement of animal and plant pests.[48]

Amid this enthusiasm for science, several specific developments were critical to the fortunes of plant breeding. First, in 1901, the Bureau of Plant Industry (BPI) in the USDA came into being, part of the reorganization begun in 1897 when Secretary James Wilson began to expand the scientific capacities of the department. The BPI became a center for plant studies, including genetics and plant breeding.[49]

Second, in 1906 the Adams Act substantially increased the support of the agricultural experiment stations in each state. Moreover, funds from this act were specifically directed to be spent on original research, not for administration, general maintenance, or publications.[50] Perhaps the most telling indication of the increase in capacity for agricultural research was the growth in personnel, expenditures, and buildings during the period 1906–14. In 1906 the agricultural experiment stations had 950 administrative and research workers, a number that nearly doubled, to 1,852, in eight years. Support grew from $2 million to $5 million, and the physical plant grew from $347,000 in 1906 compared to $1 million in 1914.[51]

Plant breeding had a well-supported home in what was starting to become the sprawling USDA land-grant university complex of research and education. More significantly, in America as well as in Britain and India, plant breeding and the allied agricultural sciences had become an established feature of government policy. Without any articulate opposition, the support of scientific work had become a feature of what a government needed in order to conduct business in the twentieth century.

War, National Security, and Agricultural Science, 1914–1918

The First World War served both to destroy existing patterns of agricultural research and to force creative change, or at least to force discussions about creative change. At the outbreak of hostilities in August 1914, the United States was the only country to have a comprehensive agricultural research system, and it was shortly thereafter significantly expanded by passage of the Smith-Lever Act, which created the extension service. Britain and India both had small and relatively young institutions for agricultural research, with a minuscule number of research workers compared with the thousands engaged in the United States.

Agriculture in each of these three countries was affected differently by the conflict, but two major developments were especially important for postwar attitudes toward agricultural production and science. First, food production as an element of war became fully integrated into government planning efforts. Second, the capacity for agricultural science became essential to build the food production system and thus the military and economic strength of a nation. Britain and America were immediately affected by these developments. India was affected only in modest ways until three decades later, after winning independence from Britain.

Britain confronted military weakness from inadequate food production from its own soils. The country began the war with no strategic stockpiles of food and a reliance on imports for over 60 percent of the calories needed. In the decade before the outbreak of the war, considerable political debate swirled around the vulnerability created by this dependence upon imports, whether the military and economic power of the empire were adequate for national security, and whether some sort of protective tariff (i.e., a revival of the Corn Laws) should be established to protect domestic grain growers from severe foreign competition.[52] Nevertheless, the only specific steps to improve British agricultural production during this prewar period were the grants made by the Development Commission for research institutions.

Increased prices, which encouraged increased domestic production of cereals and potatoes, plus continued imports of food through 1916 kept the British food system relatively intact. Unrestricted submarine warfare by Germany in 1916 and especially in 1917 and after, however, threatened Britain's security. Wheat supplies dropped in May 1917 to fewer than seven weeks' supply, a dangerously low level, but increased production of ships and improved methods of protecting them from submarines reversed the trends. By September, after a new domestic harvest, Britain had a six-month supply.[53] A disastrous decline in North American grain production in 1916, in contrast to the huge harvests of 1915, however, sent another signal that reliance on foreign sources for food staples created national security risks.[54]

Lloyd George became prime minister in late 1916 in a complex series of political realignments prompted by the stalled British war effort.[55] He regarded production and distribution of food as essential to British success in the war, and he appointed Rowland Prothero (later Lord Ernle) as president of the Board of Agriculture and Fisheries with a seat in the cabinet. Prothero developed a series of strong measures that included guaranteed prices for farm products and wages for farm labor, measures to ensure that every farm was worked as efficiently as possible for production of calories, and government-controlled distribution and rationing of food.[56]

One of Prothero's key officials was Thomas H. Middleton (1863–1943), who had served as the second chair of the School of Agriculture at Cambridge from 1899 to 1902. Although Middleton's strengths probably lay more in teaching and administration than in research,[57] it is interesting to note that he was at Cambridge during the first years of Mendel's rediscovery.

Middleton's administrative leadership in the Food Production Department of the Board of Agriculture was important to the campaign in 1917 and 1918 to substantially increase British food production. In Middleton's view, the submarine warfare campaign in 1917 forced the British government to seek an increase in grain production by plowing up grasslands devoted to production of livestock.[58] The Food Production Department also sought higher production by utilizing the War Agricultural Committees, which had been organized in 1915 in each county.[59] These local people organized the supply of agricultural labor and promoted the use of fertilizers, machines, and other supplies to get the highest production possible.[60]

In addition to these voluntary efforts, the Board of Agriculture, through the Defense of the Realm Consolidation Act (1914), had the power to enter private land not considered in top production, evict the current operator, and provide for new management. The local committees, renamed War Agricultural Executive Committees in 1917, were responsible for implementing the board's control of farm land use. Coercion was supplemented with an incentive through the Corn Production Act (1917), which provided a guaranteed minimum price for wheat, a minimum wage for agricultural laborers, and an injunction to keep landlords from raising land rents.[61]

Largely due to increased areas planted, wheat and potato production rose in 1918 to 64 percent and 40 percent more, respectively, than the average production levels in 1904–13.[62] New technology, however, did not raise the average yields per acre during the war years. Historian Margaret Barnett has concluded, even more pessimistically, that the campaign to grow more food in Britain may have raised the rate of self-sufficiency in calories from 38 percent by only 1 percent. In addition, she has raised serious questions about whether the Food Production Department's programs could have maintained the increased production levels in 1919 and beyond.[63]

Despite, therefore, what might have been modest accomplishments that came only late in the war, British agriculture was changed by the events of that conflict. State intervention in the use of private land and the primacy of technically efficient production had been accepted by portions of all political parties, at least for the duration of the war. Some agricultural interests within the United Kingdom wanted to reconstruct agriculture so as to create continuing self-sufficiency after the war,[64] but, as discussed later, such a reconstruction was not to happen in the immediate postwar period.

It is interesting to note that Germany, in contrast to Britain, began the war with official statistics indicating near self-sufficiency in food. For years Germany had maintained tariff barriers against foreign grains in order to protect and nurture domestic agricultural production.[65] Mistakes made during the war, according to the postwar analysis of the British Food Production Department, probably contributed to substantial shortages of food in Germany by 1917 and imbalances in needed dietary ingredients even before that time. These defects in the German food supply probably contributed to the country's eventual surrender.[66] Blockade of imports plus the general collapse of domestic production severely reduced German food supplies and led to severe malnutrition and hunger-induced disease in the postwar period.[67]

India, in contrast to Britain, did not undergo any fundamental agricultural changes during the First World War. Instead, the Government of India, beholden to London, rallied some support for the British war effort from portions of the Indian middle class. India served as a source of manpower for the armed forces, which served in France, East Africa, the Suez Canal zone, and the Persian Gulf.[68] Rewards for military service included the allocation of thousands of acres of newly irrigated lands in the Punjab, a largesse that was used partly for horse breeding for the army and increased agricultural production but which under British rule had little effect on development of an independent economy.[69] Exports of Bengal jute and imports of textiles and kerosene were disrupted by the war, and these trade problems adversely affected millions of Indians, especially the poor.[70]

Other than these types of indirect effects on Indian agriculture, the First World War served mainly to promote more activity in nationalist and independence movements. The Ghadar movement based in North America attempted to raise support for a violent overthrow of British rule, and the Home Rule League in India agitated for more control of India by Indians. It was the latter movement that ultimately breathed vitality into the Indian National Congress.[71] Perhaps the most significant feature of the war from this point of view was the demonstration in Europe that Britain was not omnipotent.[72] Sowing the seeds for national independence had profound effects on Indian institutions for agricultural science, but not until after 1947. During the war itself, however, the programs started by the Howards at IARI in Pusa remained the major features of wheat breeding in India.

Just as India's remoteness from the battlefields of Europe provided some insulation from the direct effects of war, so too was America partially insulated from the conflict, at least from 1914 to 1917. Disruption of European agriculture led to a substantial expansion of export opportunities for American wheat and other produce. Nevertheless, the United States finally entered the war on 6 April 1917, on the side of Britain, and direct participation in the war had immediate and far-reaching effects on the conduct of American agricultural production and science. New organizations, new patterns of behavior, and the indelible impacts of war on individuals all left their traces and scars in ways that profoundly influenced the developments of high-yielding agriculture after 1945.

Americans in 1918 were deeply affected by the recently ended war, although in ways quite different from those experienced by the British. America was a food-surplus, exporting country, while Britain faced food-deficits and relied on imports.

In the absence of invasion and conquest, therefore, Americans were never threatened with starvation.

Despite this fundamental difference, after entering the conflict the United States moved swiftly (Food Control Act of 10 August 1917) to establish central control over the production, trading, and distribution of foodstuffs, partly to channel needed food supplies to the Allied powers in Europe and partly to avoid disastrous fluctuations in prices and supplies on the domestic market.[73] In addition, USDA mobilized scientific resources, such as the newly created extension services, in an effort to maximize production.[74] It was from this scientific program that the most far-reaching effects came in the field of wheat breeding.

The story begins with the outbreak of the war in 1914, which was followed by a nearly immediate rise in wheat prices.[75] Because of the continued expectations of favorable prices, American wheat farmers increased their planted acres for the 1915 crop from 54.7 to 61.6 million acres. Yields in 1915 were the highest on record for the United States, nearly 17 bushels per acre, and the total crop was 1.026 billion bushels, a record and 66 percent more than needed for domestic consumption. Unfortunately for the American farmer, yields elsewhere in the world were also uniformly good, and prices dropped to levels below those in the fall of 1914. As a result, American growers cut their planted acres for the 1916 crop to 56.9 million.[76]

Lower acreage plus a drop of average yields to about 11.2 bushels per acre led to a production drop of 38 percent in 1916, to about 636 million bushels.[77] Poor weather and an outbreak of black or stem rust (*Puccinia graminis tritici*) contributed to the lower yields. Although many farmers had little to sell, the price of wheat rose dramatically, from $1.06 in June 1916 to $3.40 in May 1917.[78]

Although the United States still exported 246 million bushels of wheat in 1916,[79] the crop for that year provided a surplus estimated at the time of about 20 million bushels.[80] Only the large 1915 crop kept supplies moving to Europe. Acreage for 1917 remained about the same as in 1916, and severe winter kill in parts of the wheat belt kept total production at about the same level, 637 million bushels.[81] It was this second year of low yields, combined with the devastating losses of ships to German submarines, that created the conditions for a rapid price rise of wheat in the months after the United States entered the war in April. America still exported over 200 million bushels of wheat in 1917, but this was possible only because of the last supplies remaining from 1915 and a decrease in domestic consumption.[82]

After American entry into the war, all government agencies moved to increase food production in order to support the Allied powers. USDA, for example, doubled the number of extension agents between 1917 and 1918. Of more immediate importance to farmers, prices and marketing moved to full central control by the U.S. Food Administration, headed by Herbert Hoover.[83] Thus was the United States able to play a preeminent role in supplying its own and its allies' larders during the last eighteen months of the war.

American participation in the war lasted only eighteen months and had major effects on only one cropping season (1917–18). Therefore, the efforts to mobilize scientific resources to increase production probably had only minimal effects on the outcome of the First World War. Nevertheless, the war ultimately had far-reaching consequences for American institutions of agricultural science.

The disastrous stem rust epidemic of 1916 stimulated federal and state scientists to organize a massive campaign to control the disease. In the process, those involved created a coalition of scientists, farmers, and industrial and political leaders. Expenses were high, and for a period this effort to control the rust represented the largest single expenditure made for crop improvement research. Symbolically, the stem rust campaign became a moral crusade that united science, farmers, and the state in a effort to increase yields in the name of national security and industrial well-being. Efforts to create high-yielding varieties of wheat later used aspects of the stem rust control campaign as a model of how to proceed. Given the magnitude and novelty of this campaign, it was hardly a coincidence that a prime leader of later efforts to improve agricultural production obtained his first and formative experience by leading the campaign against stem rust.

Elvin Charles Stakman (1885–1979) (Figure 4.1) was just about a month shy of his thirty-second birthday when the United States entered the war in April 1917. Born in Algoma, Wisconsin, Stakman was by then an associate professor of plant pathology at the University of Minnesota, where he had received his education and had taught since 1909. Already he had a reputation for intelligence, hard work, energy, commitment, and a love for argument and persuasion. His 1913 doctoral thesis broke new ground in understanding the ecological and genetic relationships between plant pathogens and their host plants. Although it was not explicit, Stakman's thesis work was compatible with ideas of Mendelian genetics.[84]

In early 1917, before the declaration of war, Stakman was invited to Fargo, North Dakota, to address the Tri-State Grain and Stock Growers' Convention on stem rust and its control. In collaboration with his colleagues Henry L. Bolley of North Dakota Agricultural College and Mark A. Carleton of the Bureau of Plant Industry, USDA, Stakman decided to emphasize breeding wheat for resistance to the rust, studies on the origins of rust epidemics, and the eradication of the barberry plant, at least in the local neighborhood of wheat fields. Barberry (*Berberis vulgare*) had long been known to be an alternate host of stem rust, so eliminating it from the vicinity of wheat was reliably considered to help prevent the spread of the disease to wheat.[85]

As a result of Stakman's talk, North Dakota moved immediately to pass a state law providing for the eradication of barberry. However, it was only after the outbreak of the war and Stakman's appointment to the War Emergency Board of Plant Pathology that the next developments occurred. As the board toured the country to promote the use of plant pathology as a means of increasing food production, Stakman continued to argue for the importance of barberry eradication on a very large scale. He and other plant pathologists met in Chicago in 1918, and this meeting endorsed a full-scale, national campaign as a wartime effort.[86]

Bolly and Stakman left immediately from Chicago for Washington, D.C., where they similarly convinced the Bureau of Plant Industry and then the secretary of agriculture, David Houston. Houston quickly approved the effort and put $150,000 per year toward it, an extraordinarily large allocation for the time. On 21 February 1918 the bureau formally asked the University of Minnesota to release Stakman for full-time work as director of the barberry eradication campaign, a request that was granted by March.[87] Stakman was already well known to bureau scientists through his work as a part-time collaborator since 1914, and he had been offered a job in Washington

Figure 4.1   Elvin Charles Stakman in a wheat field at Chapingo Agricultural Station, Chapingo, Mexico, 1953. Courtesy Rockefeller Archive Center, North Tarrytown, New York.

at the bureau during a temporary assignment there in 1919–14. He served as full-time director until May 1919.[88]

To get a better sense of just how novel and ambitious the barberry eradication campaign was, it is important to recall that plant pathology itself was a very young field of study. When Stakman started graduate work, only the University of Minnesota and Cornell University had programs.[89] In addition, Stakman had to start an organization that ultimately spanned thirteen states,[90] each of which had to pass its own eradication law; organize its state department of agriculture, farmers, and agricultural university; and develop cooperative arrangements with the bureau. Opportunities for conflict among these players were plentiful. Stakman was also an early advocate for the active involvement of private industry, mostly from the millers. He believed that their participation was important to promote an expensive operation.[91] The fact that he obtained the collaboration of the industry people is indicative that they had come to see steady supplies of high-quality grain as essential to their own operations.

Stakman's forceful evangelism for agricultural science was remembered by E. M. Freeman, his dissertation adviser and department head at the University of Minnesota:

> The most vivid picture which I carry of the early days of the campaign took place . . . at the Minneapolis Club, where [F. M.] Crosby [of General Mills] had assembled the business heads from every kind of business or other activity centered in the Twin Cities[:] Presidents of the largest banks, transcontinental railroad presidents, heads of milling and large mercantile establishments, governors of several States, legislators, judges. . . . Crosby . . . called immediately on Stakman . . . [who] jumped up and with that semi-belligerent air of eagerness to enter a fray of discussion and wits, strode solidly down the center aisle. Before he reached the speaker's platform . . . he began his speech and . . . continued . . . for an hour a veritable verbal barrage that held the intense interest of every man in the audience. He told simply and effectively the story of the rust of wheat and the role of barberry, the complicated problem of barberry eradication, and the need of support not only from public agencies and the farmers but from business men and the urban centers.[92]

Stakman's initiative and energy were successful in launching and continuing the barberry eradication campaign. Although he relinquished the directorship of the effort in 1919, he toured Europe in 1920 to gather firsthand evidence of the success of barberry eradication there. Upon his return home, he was even more convinced of the campaign's necessity.[93] Some of his disciples, such as F. M. Crosby of General Mills, continued to lobby for funds for eradication, even to President Hoover in 1929.[94]

In that same year the Bureau of Plant Industry estimated that over two-thirds of its funds went for barberry eradication, clearly making it a major focus of activities.[95] Eradication efforts continued until the 1970s, when federal efforts ended, due at least partially to the paucity of barberry plants remaining and to the development of genetically resistant varieties of wheat.[96]

After 1919 Stakman continued to serve as an adviser to the campaign, but he resumed his research on stem rust. He had, in 1914, identified the concept of physiological races in stem rust, an idea that secured his reputation and to this day continues as a major concept in plant pathology. Stakman also obtained evidence, after 1918, that stem rust epidemics spread from Mexico up through the southern American states to the upper Midwest and Canada each summer.[97]

Barberry eradication to control stem rust was a campaign that had few direct precedents, but it fit well into the creation of a national rationalized food production and distribution system. Despite the fervor of what has been called the Progressive Era, by 1914, or even by 1917, the United States had done little to rationalize its agricultural system other than to create the educational-research-extension services of the USDA and the land-grant colleges. Farming was still primarily an unregulated business in which many small and a few large entrepreneurs made their way as best they could. To be sure, the country life movement had called for new standards of efficiency and profitability as the way to the future, but many American farmers were still just beginning to embrace the promise of new technology and science.

Many forces acted quickly after April 1917, however, to change the relationships among the farmer, the state, and modern science. Loss of labor because of conscription for the armed forces was an immediate blow that disrupted farm operations and spurred considerable interest in mechanization. The wartime responsibility of feeding all of America and a great deal of Europe required the state to signal that it wanted, indeed insisted upon, maximum production from American land. Although American entry into the war came after little could be done to improve the 1917 wheat harvest, by early 1918 the USDA was moving resolutely to ensure that stem rust would never again take a major tax from the farmers' wheat crops. Stakman and his collaborators brought all wheat farmers in the spring wheat belt into full embrace with modern plant pathology.

As it turned out, the war effort never benefited much from the barberry eradication campaign because Germany and Austria-Hungary collapsed in the fall of 1918. Nevertheless, the United States had invented a new institution for control of land use by the state. Even in the absence of wartime emergency, the American state would not sit idly by while a controllable fungus wreaked havoc on the harvest.

Twenty-three years later, when Stakman was fifty-five, he was named by the Rockefeller Foundation to head the survey commission to Mexico, an exercise that led to the development of high-yielding varieties of wheat, or the green revolution. Even though the barberry eradication campaign was very different from the foundation's work in Mexico, they were joined by a common theme: the unrelenting, systematic planning of land use by experts and policy makers to extract maximum agricultural production through the use of modern science, to build a strong, industrial state.

Given the immense geographic scope of his organizational and research work, it is not surprising that Stakman could later imagine remaking the agricultural systems of Mexico. Indeed, his work on barberry eradication and stem rust, and his later work for the Rockefeller Foundation in Mexico all had the trappings of a moral crusade to remake agriculture and govern the use of land through a complete partnership among the farmer, the state, modern industry, and modern science.

American wheat, soldiers, and guns, possibly in that order of importance, may be said to have tipped the balance in favor of Britain and the other Allied governments in their conflict with Germany and Austria-Hungary. Both Britain and the United States, in different ways and for different periods of time, marshaled agricultural land and people into service for the military strength of the country during the war.

At the end of hostilities, however, both countries had readjustments to make. The British faced the issue of whether they should permanently improve their abilities to feed themselves, a mission that could be accomplished only with an invigorated agricultural science. The British also faced the continuing and now more difficult problem of Indian pacification. Any "success" they would have in such a venture would surely also depend on the invigoration of the Indian economy, which in turn meant invigoration of agriculture through science. The period between the two world wars was thus filled with further developments of the institutions of agricultural science.

Developments between the Two World Wars, 1918–1939

Events during the First World War brought farmers, industry, the state, and science into new types of relationships, but the collaboration born from wartime emergency did not long survive the end of hostilities. In both the United States and Britain, the government backed away from guaranteed prices for staple grains in 1921, which put farmers back on the untender mercies of competitive markets, even though some producers welcomed freedom from regulation.[98]

Agricultural science fared somewhat better than farmers in each of the three countries. In Britain, lack of state interest in maintaining more food self-sufficiency again dominated, but agricultural science found legitimization in a new Agricultural Research Council. The gathering of European war clouds, however, rekindled serious government interest in self-sufficient agricultural production, a transformation that was to prove stable for many years. India remained a producer of raw materials for imperial Britain, but the extensive inquiry of a royal commission pointed the way toward a vision of more science in agriculture, a vision unfulfilled until after independence. In The United States the system of rational industrial agriculture came to dominate the landscape and was of direct importance in the development of high-yielding agriculture. Each of these threads is traced in the following.

United Kingdom

British agricultural scientists at the end of the war were anxious to resume the growth they had experienced from the Development Commission in the few years before 1914. They saw an opportunity unlike any they had seen before because of the high priority given to domestic production as a result of the wartime emergency. Alfred Daniel Hall, who had designed the agricultural research system for the Development Commission, in 1918 sought to mobilize the advice of all the directors of the research institutes on what he hoped would be a thorough reconstruction of British agricultural science once the war was over:

> The [previous] scale of operations . . . was limited by pre-war conceptions of the sums which it might reasonably be expected that the Development Fund could spare, and there was in the background also a certain superior limit set by conceptions of what sums it was likely that Parliament would be willing to provide. We now live under another sky. Not only has the need for a liberal expenditure on industrial research been recognized and practically unlimited sums been made available for the purpose, but the paramount and vital importance of increased home production of food places the claims of agricultural research on a pinnacle which overtops all other demands.[99]

Hall clearly had a vision for the use of the land in Britain to provide something closer to food self-sufficiency. He also fit well with the Liberals like Lloyd George because he was a fervent supporter of land nationalization. Hall wanted Britain's agriculture to be organized into large farms on which the best scientific practices would predominate.[100]

Rowland Biffen, director of the Plant Breeding Institute at Cambridge University, may or may not have shared Hall's vision of a more egalitarian Britain, but he certainly agreed with Hall's hopes for a more scientifically productive agriculture. He responded to Hall's request on a number of fronts, including research philosophy, research priorities, and the need for a more stable and predictable career pathway for researchers beginning their work.[101] It was on this latter theme of developing a career ladder for agricultural researchers that the next developments were based.

Rowland Prothero, who became Lloyd George's president of the Board of Agriculture when the board joined the wartime cabinet, turned his attention to enhancement of the board's research capability near the end of 1918. He asked for £2.5 million over the following five years, a figure the Treasury reduced to £2 million or £400,000 per year, a considerable increase over the prewar expenditures of £142,000 per year.[102]

The Board of Agriculture and Fisheries continued this theme in the fall of 1919 by asking the Development Commission for £90,000 per year to establish a "quasi-service of research workers." New funds were not forthcoming above the £2 million already approved, but the commission did accept the notion of a uniformly graded salary scale for staff researchers of the board, to be comparable to salaries in colleges, research institutions, and elsewhere in the government service.[103] Thus was a step made to establishing a career in the civil service for agricultural scientists.

Also in late 1919, Hall, by then the secretary to the Board of Agriculture and Fisheries, moved to upgrade the technical advising received by the board on agricultural research. He reconstituted the Technical Committee so that it had two standing subcommittees, one on research composed of directors of institutes and one on advising (extension) composed of advisory officers and some members of institutes. The new research council held four meetings by February 1921, with discussions focused mostly on the progress of work, the graded salary scheme for scientists, and the need to coordinate research programs.[104]

For nine years after 1921, however, the research subcommittee did little, possibly reflecting a sense in the Conservative (1922–24, 1924–29) and Labour (1924) governments that research was not a high-priority issue for either party. From 1922 to 1924, neither Conservative prime minister (Bonar Law and Stanley Baldwin) was likely to have been impressed by the need for research. Law generally disliked all government activity, and Baldwin believed more in the spiritual values of rural life as an escape from the crassness of urban commercialism. Prime Minister Ramsey MacDonald's Labour government was in office for only nine months in 1924, so it had little time to initiate reforms.

Baldwin's return to the prime ministership in 1924 created another five years of relatively low interest in agricultural science. Such discussions as there were on agriculture during his second tenure were oriented to the question of tariff protections, credit provision for land purchases, marketing reform to standardize products, emergence of domestic agriculture as a market for British industry, and debates about the privileges that were to be granted to the import of empire-produced farm products. All of the Conservative initiatives in agriculture pointed toward a growing vision of agriculture as a business based in science, but the formation of the second Labour government in 1929 put the Conservatives in opposition.[105]

Prime Minister Ramsey MacDonald's second Labour government held only 288 seats (47 percent), so collaboration with the 59 Liberals (10 percent) was needed to keep the government in power.[106] Unlike its Conservative predecessors, MacDonald's government was unambivalent about its vision of agriculture: it was to be based on the tools of modern science and run with businesslike efficiency. Both Labour and the Liberals were free-trade advocates, so neither party wanted tariff barriers to protect farmers from foreign competition. Higher prices for urban consumers were the inevitable result of protectionism, and MacDonald's government was based primarily on urban votes. Enhancing the capacity of agricultural science, reform of marketing, and encouragement of larger scale farming were thus the keys to the new government's agricultural vision.[107]

Although the Conservatives might eventually have moved toward an enhancement of agricultural science, too, MacDonald's government took the first major steps since Lloyd George's move to establish the Development Commission in 1909. Within four months of coming to power in May 1929, he had appointed the Committee on Agricultural Research Organisation, part of the Economic Advisory Council. The committee reported on its recommendations on 29 April 1930, and by 28 July 1930 the Labour-Liberal government had invented a new institution: the Agricultural Research Council (ARC).[108] A royal charter made the ARC an independent agency in 1931, and its first members included Daniel Hall and Thomas Middleton,[109] both of whom had played prominent roles in the Development Commission and in the Ministry for Agriculture and Fisheries to build momentum for scientific agriculture.

Lord Richard Cavendish, chairman of the Development Commission, became chairman of the ARC as well. Staff from the Development Commission served as staff for the ARC and the Committee of the Privy Council for Agricultural Research.[110] Thus transition of responsibility for agricultural research was a cooperative devolution of responsibilities from the Development Commission.

One of the immediate consequences of the ARC's formation was that the previous advisory council to the Ministry of Agriculture and Fisheries, composed of the directors of the research institutes funded by the Development Commission, were no longer the primary advisers to the ministry. They would meet only as an annual conference of directors.[111]

Both the language and the structure of the new ARC indicated that agricultural science had legitimacy at the highest levels of government. Medical and industrial research councils had been formed in 1913 and 1915, respectively. From the beginning, they had reported directly to the Privy Council, which suggested they were seen as vital to all aspects of government, at the highest level.[112]

A major battle in establishing the ARC involved deciding whether it should report directly to the Ministry of Agriculture and Fisheries and the Department of Agriculture of Scotland. Both the ministry and the department argued strenuously, in May 1930, that the ARC should report to them, primarily on the grounds that they were responsible for virtually all agricultural research, even though it was sponsored by grants from the Development Commission.[113] Counterarguments were that an independent ARC was more likely to attract scientists of the highest caliber and that research free from bureaucratic intrusion of the departments would be more productive.[114] A compromise was reached in June, with an arrangement for the ARC and the departments to consult extensively. Reporting, however, would be to the Committee of the Privy Council for Agricultural Research.

The ARC's structure, responsibility, and reporting lines are of more than casual interest, because the arguments over how they should be arranged reflected the exercise of political muscle and over the next several decades influenced how agricultural science in Britain would proceed. The most plausible explanation for the conflict was that at the first level it was a battle waged by scientists for financial support under conditions of autonomy. In a word, the scientists advocated and wanted the enhanced support of the state, but without the fettering ties of the departmental bureaucracy in Whitehall. Ministers, for their part, were unhappy giving up the control that an independent ARC represented.

That the scientists were successful suggests that MacDonald's government bought the argument that better scientists, working under conditions of relative autonomy, would be more productive in meeting the government's goals for British agriculture. Those goals were to make farmers more productive so as to relieve their financial distress, but to do so without resorting to tariffs on imported food. Urban politicians essentially sought an arrangement in which their primary constituents, urban consumers, would not be harmed yet the discontent of the countryside would be answered. Scientists, for their part, achieved relative autonomy, because an ARC heavily oriented to scientific appointments would be preferable to supervision by the nonscientists of Whitehall.[115]

In the debates on how to enhance and structure agricultural research in the 1920s and early 1930s, one particular argument disappeared, despite the fact that it had effectively driven thinking about agriculture during the war period of 1914–18: food self-sufficiency as a means of national security in wartime. Given the precarious nature of Britain's food supply during 1917, it is in retrospect difficult to understand why the succeeding Conservative and Labour governments were both indifferent to the issue. Price supports and control of agricultural practices ended after the war, and the British habit of dependency on foreign imports again predominated.

Several voices made complaints and called for a continuation of the massive reforms begun during the First World War. Daniel Hall, who had designed the research institutes for the Development Commission and then become secretary to the Ministry of Agriculture and Fisheries, was one of the most vocal advocates of a new program for British farming. In 1916, in the midst of the Great War, he laid out a plan that called for state ownership of the land, reorganization of farms to be large enough for laborsaving machinery, programs to support more food (i.e., wheat and potatoes) production rather than livestock products, and the adoption of the most modern scientific and technical practices possible.[116]

Hall, of course, had spent his entire career in agricultural science, first as a researcher and teacher and later as an administrator. Perhaps, therefore, he remained unheard because he seemed to be pleading for more jobs and support for his peers in science. When the war ended in 1918, most British people involved in policy matters probably thought that never again would such a horrible and dangerous war erupt in Europe. The empire was still together, the Royal Navy continued as one of the preeminent navies of the world, and British prosperity and safety may have been seen as more attainable by careful nurturing of the empire and dominions rather than reverting to the agricultural production practices of the 1870s and before.

Furthermore, agricultural science, despite some new and useful developments such as Biffen's Little Jost and Yeoman wheats, could not yet provide production costs as low as those of foreign competitors in North America and elsewhere. Hall was faced with arguing that science *might* create an agriculture that would be competitive and that would protect the nation in case of war, *if* such an unexpected thing should happen. Perhaps it was remarkable that he had any support outside a narrow group of security-conscious "food-firsters" or technocrats who couldn't tolerate the notion of continued inefficiency in British agriculture.

Yet Hall was not completely alone. Another prominent spokesperson was Christopher Addison (1869–1951), who by 1939 had become Lord Addison of Stallingborough. Addison, the minister of agriculture and fisheries from 1930 to 1931, agreed to the final compromises that led to the ARC. Although Addison was raised on a farm, he was trained as a physician, his first career. Addison showed an early interest in politics that brought him as a member of Parliament in the Liberal Party. During and after the war he served as a minister both in munitions and in Health and Housing in Lloyd George's government, but he resigned in 1921 in a falling-out with the prime minister and then lost his seat in the elections of 1922. He was always on the left of the Liberal Party, and he switched to Labour in 1923, from which he reentered Parliament in 1929. During his time out of office, Addison reformulated Labour's policy on agriculture, especially with his 1929 work, *The Nation and Its Food*. This laid the groundwork for the establishment of marketing boards, through which farmers could fix higher prices for their goods. After 1945, his policy for Labour formed the basis for Labour's further reform of British agriculture. He was made parliamentary secretary to the minister of agriculture, Noel Buxton, in 1929. At Buxton's resignation, MacDonald named Addison to be minister of agriculture and fisheries in 1930.[117]

Although Addison was not kept in the government when MacDonald's government had a falling-out with the Liberals in 1931 and had to reconvene as a National government with Conservative support, he maintained a strong interest in British agriculture. In 1939 he published *A Policy for British Agriculture*, which, like Hall, put forth a strong plea for major reform. Addison's first concern was that people working in agriculture should not be consigned to poverty. Beyond these welfare concerns, however, he made a strong plea that the world was in troubled times and that national security in war depended upon a much revitalized agriculture. Moreover, he felt that with coordinated planning from government and state ownership of the land, Britain's farmers could produce much more high-quality produce for the home market.

Although Hall, Addison, and their allies could not convince the Labour, National, or Conservative governments of the 1930s to embrace radical reform, memories of the famine scare of the Great War were not completely absent. With the rise of Hitler in Germany, Stanley Baldwin's Conservative government in 1936 took steps to reorganize the War Agricultural Executive Committees. As in the First World War, these were bodies of farmers in each county who, under the general supervision of the Ministry of Agriculture and Fisheries, would ensure that every farm in the area was run with the best technology possible. Farmers who could not or would not adopt scientific practices could be replaced. Activation of the committees came immediately after the British entered the Second World War in September 1939.

In the two decades between the world wars, therefore, British governments for the most part tried not to alter their agricultural policies any more than necessary from the benign neglect of laissez-faire. Historian Jonathan Brown argues that all governments seemed most interested in promoting Britain's industries, keeping bread cheap for the urban working class, and keeping Britain the industrial workshop of an empire that supplied raw materials and food.[118]

Perhaps the only significant difference in agricultural policy between the Labour-Liberal, Conservative, and National governments was one of style. Labour, as represented by Christopher Addison, believed that the state should play a prominent role in organizing agriculture, perhaps drawing its models from the military and feeling free to use coercion. Conservatives, represented by Addison's successor, Walter Elliot, wanted to use the state to help farmers organize rational production and marketing schemes that vertically integrated production with processing and marketing.[119] No party saw a farm economy with larger production to be an important goal.

A few reformers sounded the alarm that all was not well, but only modest changes were made in the institutions of agriculture. Science in the form of the Agricultural Research Council received a prestigious place in the halls of power, but very little money came with the honor. Moreover, the council's first priority for research was on animal diseases and their control, not on crops like wheat and potatoes to increase the food self-sufficiency of Britain. Ambitious plans envisioned by reformers such as Hall and Addison lay stagnant and unheeded. Britain continued to eat imported food, and her own farming industry continued in a very depressed state. Only Germany's invasion of Poland forced a change of course.

India

Changes in Indian institutions of agricultural science between the two world wars were largely derivative of the arguments and concerns that originated in London, on the one hand, and the developing strength of the nationalist freedom struggle in India, on the other hand. Perhaps the best explanation for why Britain sponsored reform in agricultural science in India is that Stanley Baldwin's Conservative government (1924–29) needed to demonstrate action to counter the growing challenge from Mohandas K. Gandhi's Indian National Congress and others who would have India free.

Although the congress had its origins in the late nineteenth century, until after the end of the First World War it was a tiny, ineffective movement of some wealthy and well-educated Indians. British solicitation of Indian collaboration in the war, however, created a set of expectations in India that loyalty in time of war would win concessions and some sort of political self-governance. Within a year of the armistice, Lloyd George's coalition government managed to convince the Indian nationalists that nothing was to be gained by further cooperation with London. Most horrific in their impact on Indian consciousness were the Rowlatt Acts and the massacre of about 400 Indians by General Dyer at Jallianwala Bagh in Amritsar, Punjab.

Although Gandhi had arrived in India in 1915, it was not until the Rowlatt Acts were proposed in 1919 that he emerged into political struggle at the national level. These bills, designed to stop terrorist strikes at British authority, served powerfully to unite Indians of all political persuasions in the belief that the British could not be trusted. Gandhi called for non-cooperation with British authority to begin on 6 April 1920. Demonstrations were held across India, including Amritsar. Brigadier R. E. H. Dyer, military commander of the area, had previously forbidden gatherings of any type. When thousands came to participate in a Hindu festival, he ordered his troops to fire, leading to the massacre at Jallianwala Bagh.[120]

Further fuel was added to the fire in 1920 when the British made it clear that they intended to dismantle the Ottoman Empire, Germany's ally, as a part of the spoils of the war. India was far away from this seizure of Arab lands, but the Muslims of British India accepted the caliph of Turkey as the spiritual leader of Islam. Destruction of the Ottoman Empire thus became an insult to Muslims in India, which pushed them into noncooperation with the British. Gandhi joined their efforts and soon thereafter brought the Indian National Congress to agree to noncooperation, which was formally launched on 1 August 1920. Indians were called on to stop the use of British textiles, schools, law courts, councils, titles, and honors; to resign from government service; to pay no taxes; to preserve peace among Hindus and Muslims; and to renounce "untouchability" and violence.[121]

Gandhi's noncooperation movement ended in 1922 because of an outbreak of violence against Indian constables and, perhaps, because Gandhi believed the movement had spent its enthusiasm and needed to regroup. Gandhi personally moved out of direct confrontation to British authority and into contemplation and working with villages on issues of education and sanitation. British authorities then arrested him and sent him to prison for his promotion of the noncooperation movement.[122] Yet a significant hurdle had been passed, from which British ability to rule never really recovered. Millions of Indians had been mobilized and showed they could resist the most powerful imperial force on earth.[123] To be sure, Britain held on to India for another quarter century, but doing so required a continuous stream of concessions and reforms designed to thwart the rising passion for freedom.

A new Government of India Act (1921) created a two-chambered Parliament, each with a majority of elected members. In 1923, Indian nationals who wished to join the Indian Civil Service were permitted to take the examination in New Delhi instead of making the long trek to London. Also, Indians were accepted for officer training in the army for the first time. Establishment of a Tariff Board provided some fiscal autonomy for Indians working within the Government of India.[124]

It was into this atmosphere of reform out of desperation to keep power that Stanley Baldwin's government came after the Conservatives' election win in 1925. Baldwin's viceroy in India, Edward Wood (Lord Irwin), was a committed Christian reformer[125] who, somewhat like Lord Curzon, viceroy from 1899 to 1905, saw a need to promote prosperity in Indian agriculture. And like Curzon, the new viceroy sought that prosperity through agricultural science.

In 1926 he appointed the Royal Commission on Agriculture, charged to study how the Government of India might organize agricultural research and education.[126]

As chair was Victor Alexander John Hope (1887–1952), second marquess of Linlithgow, from Scotland, and among the members was Sir Thomas Middleton. An extraordinary itinerary occupied them from October 1926 through December 1927, as they interviewed 395 witnesses from all parts of India and in Britain, at a cost of over 1.3 million rupees.[127]

The Commission's *Report* in 1928 was most unusual in two respects. First, its fifteen volumes were undoubtedly the most thorough inquiry into agricultural research that any British commission had ever made, either for the United Kingdom or for any part of the British Empire. Perhaps the sheer length of the document reflected the fact that the British government itself had no ready solution to the question of how to organize agricultural science, either for the United Kingdom or for any part of the empire.

The second and related feature is that the *Report* recommended that a truly national system of research be created between the Imperial Agricultural Research Institute (IARI) at Pusa and the provincial departments of agriculture. A newly created Imperial Agricultural Research Council (IARC) would

> promote, guide and co-ordinate agricultural research throughout India. It would not exercise any administrative control over the Imperial or provincial research institutions. It would be a body to which the Imperial and provincial departments of agriculture could look for guidance in all matters connected with research and to which such research programmes as they might choose would be submitted for criticism and approval. Our object in proposing that such a body should be constituted is to provide provincial governments with an organisation embracing the whole research activities of the country, veterinary as well as agricultural, in which they can feel that they have a real and lively interest. . . . The Council should be entrusted with the administration of a non-lapsing fund of Rs. 50 lakhs [5 million rupees] to which additions should be made from time to time as financial conditions permit.[128]

What is interesting about this proposal is that it seems to have stemmed directly from Daniel Hall, who by the time of the Royal Commission's work had left the Ministry of Agriculture and was director of the John Innes Institute, where he had succeeded William Bateson. In his testimony to the commission during its tour of England, Hall stated directly that the Government of India needed a Council of Agricultural Research, primarily to create the conditions that would reduce the proportion of the Indian population in agriculture from 60 percent to 10–20 percent.[129]

As he had for Britain, Hall advocated for India a scientific and technically efficient agriculture. His desire that the proportion of the population engaged in agriculture should decline was also implicitly an advocacy that India should mechanize its agriculture and that the people should move into an industrial economy. Creation of a council to guide agricultural development was a major vehicle by which this transformation would occur. Hall was not clear, however, on exactly how India was supposed to become an industrial power when its "function" within the empire was to purchase manufactured goods from Britain. It seems unlikely that Baldwin's government, representing as it did the strength of the industrial capitalists of Britain, would have been enthusiastic about creating a competitive industrial infrastructure in India.

Despite the likely incompatibility of Hall's vision for Indian agriculture and the political realities of Tory Britain, his proposal for a council resulted in the viceroy's creation of the Imperial Agricultural Research Council in 1929, two years before the formal chartering of the Agricultural Research Council in Britain. India thereby had a mechanism at the level of the central government for coordinating and guiding the course of agricultural research in India. Ultimately this reform was not enough to stem the tide of nationalist sentiment. In fact, the Indians who served on the council may have had their appetites for self-governance whetted by the experience.

After 1947, independent India kept the council with merely a name change, to the Indian Council for Agricultural Research (ICAR). ICAR was a prominent focus of efforts to create the green revolution. For scientists like Sir Daniel Hall and Sir Thomas Middleton, it just may be that their ability to create a coordinating council in India gave them an impetus and legitimacy to do the same thing in the United Kingdom.

The United States

In contrast to Britain and India, the United States had a comprehensive agricultural science establishment by the end of the First World War. For this reason, developments in the United States between the world wars were considerably more minor.

As was the case in Britain and India, prices received by American farmers dropped precipitously after 1919, rebounded briefly in 1925, and then dropped steadily until 1931.[130] Thus American farmers were in a state of economic depression throughout much of the interwar period. Unlike Britain, however, the political force of the farming population was sufficiently strong to force continual debate in Congress about how to provide relief to farmers. As scientific research was already provided, most of the proposed reform ideas focused on schemes to get prices for basic commodities to a higher level.

Reforms at the periphery of American agriculture had some impact on the institutions of agricultural science, particularly after President Franklin D. Roosevelt's government came to power in 1932. Reclamation schemes to develop irrigation; the Tennessee Valley Authority, which brought flood control, irrigation, and cheap power and fertilizer; and the Rural Electrification Administration, which brought electricity to American farms, all encouraged the further commercialization and intensification of agriculture. In addition, the price supports provided by the Agricultural Adjustment Acts of 1933 and 1938 provided a guaranteed, stable price regime under which farmers could plan their efforts to adopt more technologically complex practices.[131]

Thus despite the absence of defining new scientific institutes, America, too, just like Britain and India, underwent changes that further solidified the role of agricultural science in the national political economy. In all three countries, the advent of more scientific inputs to agriculture created conditions under which it became increasingly difficult to farm without those inputs. Farmers, agricultural scientists, the industrial economy that depended upon a productive agriculture for inputs and markets, and the power brokers in government and industry all came to be increasingly intertwined. Political power was increasingly reflected in the ways the people on the land used the landscape.

5

The Rockefeller Foundation in Mexico

The New International Politics of Plant Breeding, 1941–1945

Events during World War I and in the years between the two world wars demonstrated that agricultural production was essential for the security of individual nations. No country could afford to neglect its food supply if it wished to maintain its status as a major military power. In addition, pressures from technically sophisticated farmers and industrialists, both interested in efficient agricultural production, solidified the use of scientific research in reforming the agricultural economy. Underlying the drive for both military power and efficient agricultural production was a powerful vision of the nation-state as an industrial economy in which all natural resources, including agriculture, were marshaled by the rational control of modern science. Both people and nature were subservient to the imperatives of power and rationalism in the new scheme of things.

What was largely missing from the pre-1939 vision, however, was a sense of how nations might interact to address issues of industrialization and agricultural modernization. By 1939 industrial states like the United Kingdom and the United States developed a sense of how individually they should manage their industrial and agricultural resources, and the British government certainly had a clear sense of how the Indian economy should be controlled. Outside of the realms of direct imperialism, however, industrial countries had only vague notions about how to use scientific and economic policy to foster their aims internationally. Furthermore, no country had any profound sense, incorporated into policy, that rich and powerful countries should assist the poor countries to achieve a better standard of living for humanitarian reasons.

Aside from imperialism, therefore, in 1939 no analytical framework existed to see how agricultural science and technology and modernization of agriculture fit into the overall scheme of international relations and power. Perhaps the only exception to this situation was a small program of the Rockefeller Foundation in China. In 1924 the International Education Board of the Rockefeller Foundation began to assist the University of Nanking with wheat improvement, economic issues, and other projects.[1] In addition, during the 1920s, the foundation supported medical reform in China.

Despite the fact that China was the second-largest recipient of Rockefeller Foundation funds (after the United States), the different projects were not coordinated. Selskar M. Gunn, vice president of the foundation, traveled in China in 1931 and recommended a substantial increase in foundation activities there. Most important, Gunn articulated a vision that the foundation should structure its program to raise the educational, social, and economic standards of rural China. Gunn's report was adopted in 1934.[2] China thus became the foundation's first large, coordinated foreign effort for rural reform based on medical and agricultural projects. The Chinese programs emphasized technical reform in the midst of substantial social unrest, about which the foundation was silent.[3]

The outbreak of war in Asia and Europe in the 1930s led to a series of changes in the philanthropic programs of the Rockefeller Foundation. The foundation's support of scientific research in both China and Europe came to an end. Until 1945, the foundation had to create a new program of giving that was centered in the Americas. From these new endeavors created by wartime exigencies came new concepts of how agricultural science could be a tool for strategic planning. Within a period of ten years after 1939, assistance in agricultural science came to be seen by the developed countries as a comprehensive need for all less industrialized countries. Moreover, efforts by industrialized states to help modernize agriculture in less industrialized states came to be seen as a critical component of political relations between nations.

Rockefeller Foundation work in China and Mexico helped officers of both the foundation and the U.S. government understand agricultural development as a part of international politics and relations. The foundation's Mexican Agricultural Program (MAP), especially, was a critical event in the transformation of agricultural science from a tool merely for industrial modernization into a device for power relationships between nations. In order to appreciate the catalytic effect of the MAP, it is first necessary to understand the circumstances and context in which it originated.[4]

Formation of the Mexican Agricultural Program

In 1963, the Rockefeller Foundation reorganized its giving programs. One of its major areas was to be "Toward the Conquest of Hunger," which in turn was closely linked to two companion areas, "The Population Problem," and "Strengthening Emerging Centers of Learning." This reorganization strongly reflected the foundation's sense of success with the Mexican Agricultural Program (MAP), begun in 1943.[5] The foundation thus gave a clear statement that had it found its interactions with Mexico to be positive, a basis for future action.

Despite this positive judgment, the foundation's statement about hunger in 1963 reflected little sense of Mexico as a place with a history:

> About half of the human beings on earth have an inadequate diet, and millions live constantly on the edge of starvation, despite the fact that an overabundance of food is being produced in a few technologically advanced countries. A world which possesses the knowledge and methods to confront the demands of hunger must accelerate its efforts to increase the production and improve the distribution of food supplies.[6]

A brief recapitulation of Mexican history will shed light on the situation that the MAP faced when it began its work in 1943.

"In the beginning there was malnutrition. . . . "[7] With this stark description, Gustavo Esteva characterizes the food situation that existed in Mexico starting with the conquest by Spain in 1519. But, as Estava continues, the prevalence of malnutrition was only one of the many dimensions of what European conquest meant for the Indian peoples who lived in Mexico before the arrival of Cortés.

Put bluntly, European conquest was an unmitigated environmental disaster for the Indian civilization of Mexico. In a process that lasted for nearly 400 years, the new elite systematically appropriated the land of Mexico and all its attendant natural resources. As a result of this theft, with its concomitant enslavement and pauperization of the people, the Indian population dropped from a high that has been estimated at somewhere between 7 and 25 million to a low of about 1 million after 100 years of Spanish rule.

Independence from Spain in the early nineteenth century did little to alleviate the poverty and oppression of the Indian majority in Mexico. An elite of European descent continued its iron grip on the natural resource base of the country. By the end of the century, during the dictatorship of Porfirio Díaz (1876–1911), a tiny minority of about 1 percent of the population controlled 90 percent of the land. A mere 8,000 of the largest haciendas held 113 million hectares of land, about three-fifths of Mexico's total land area. Ninety percent of the people were completely landless and at the mercy of the landholders for "opportunities" to labor for subsistence survival.[8]

The Díaz government favored the production of export crops in place of staples as a way to earn foreign exchange. Consequently, malnutrition and poverty were intensified in his period of rule.[9] Rebellion over the gross inequities erupted in 1910 into a revolution that dragged on for seven years of intensely bloody warfare. Death, destruction of buildings and railroads, and disruption of economic activity were extremely widespread. Some place the death toll in the millions.[10] Bitterness at the oppression and poverty of the Díaz period elevated land reform and alleviation of hunger to among the highest priorities of the revolution.

With victory, the revolutionaries made the first efforts to redistribute Mexico's land to the millions of people without access to this most fundamental resource in an agrarian economy. In halting steps, between 1917 and 1934, 7.7 million hectares were distributed to 783,000 people. Impatience with the pace of reform was an important factor in the rise to power of Lázaro Cárdenas, Mexico's president from 1934 to 1940. During his administration, between 18 and 20 million hectares were distributed to about 0.75 million people. This was an estimated 65 percent of all land distributed between 1917 and 1940. By 1940, 13 percent of Mexican land had been redistributed to small holders.[11]

Cárdenas was an idealist dedicated to the egalitarian spirit of the Mexican Revolution.[12] When he took office in 1934, he quickly moved on agrarian reform by redistributing land and breaking up large estates, including some lands owned by Americans. In 1938 he seized oil properties belonging to American, British, and Dutch concerns, including lands held by Standard Oil, which was the basis of the Rockefeller family's fortune.[13]

His reform actions brought Cárdenas into direct conflict with the United States. During his entire tenure as president, much of the correspondence between the American embassy and the State Department in Washington concerned claims of Americans who demanded compensation for seized property.[14] Secretary of State Cordell Hull asked for daily reports from Ambassador Josephus Daniels on the situation surrounding the seized oil properties.[15]

Conflict between the Mexican and American governments about seized property, however, was simply one dimension of the overall ferment in which the Cárdenas reforms occurred. Europe seethed with fascist and communist reformers, and these conflicts had a direct spillover into Mexico. Cárdenas made clear that his government supported the Republican forces in the Spanish civil war, and Mexico became a source of relief for these people. Cárdenas's support for the Republicans put him at odds with conservative elements of Mexican politics, who were more in sympathy with the fascists. Moreover, Mexico granted asylum to Leon Trotsky, which aroused the anger not only of Trotsky's enemies in the Soviet Union but also of the organized communists of Mexico allied with the Soviets.[16]

Conflicts within Mexico were as much a concern as disputes over oil and other seized property, because the Roosevelt administration wanted neither a socialist nor a fascist country on its southern border.[17] Although Roosevelt had pledged the Good Neighbor Policy as a way to reform the interventionist history of U.S.–Latin American relations, some private citizens put pressure on the American government to take action against Mexico if solutions suitable to American interests were not forthcoming.[18] To its credit, the United States does not appear to have acted on these suggestions. Instead, negotiations continued through the efforts of Ambassador Daniels.

Despite the dramatic initiatives made by Cárdenas in land reform and seizure of oil properties, his government also pursued avenues of change that were very different in character and, ultimately, had as much or more effect on the agriculture of Mexico as did the land reforms and empowerment of the *ejidos* (land held in common by local groups). First, Cárdenas continued the development of large-scale irrigation works, an effort that had its roots in the prerevolutionary period of the Díaz government.[19] Second, Cárdenas weathered threatened and attempted coups against his government from 1934 to 1940, but his sentiment was to pick a successor who was less a fiery leader and more a conciliator with conservative elements within Mexico and with the United States. General Manuel Ávila Camacho became president in 1940, over the strong challenges by General Francisco Mugica, who was more in line with the radicalism of the Cárdenas reforms, and General Juan Andreu Almazan, who had the support of Catholic and conservative elements.[20]

In his campaign, Ávila Camacho stressed the need to secure title to landownership more than the need for further reforms. He also favored moves toward small proprietorships rather than the communal ownership of the *ejidatarios*. Ávila Camacho also stressed the need for industrialization and modernization rather than further radical social reforms. He also wanted foreign investment for Mexico and felt it was essential to reach a settlement with the Americans.[21] The United States responded in kind, and President Roosevelt designated Vice President Elect Henry A. Wallace to be "Ambassador Extraordinary and Plenipotentiary" at the inauguration of Ávila Camacho in 1940.[22] This recognition of Ávila Camacho as the new head of the government of Mexico may have been a key element in quieting a remaining claim to the presidency by General Almazan.[23] All indications are that the Roosevelt administration felt it could work well with the new president at a time when Europe and eastern Asia were already at war and the United States wanted first and foremost to secure its base in North America.

Concern about security in a dangerous world was therefore the primary framework within which the Rockefeller Foundation worked to create an agricultural program in Mexico. Whatever the meaning of the foundation's formal purpose of furthering the well-being of humankind, it certainly could not counter the efforts of the U.S. government to create a working relationship with Mexico, despite the mutual antagonisms between them. More important, given the foundation's desire to provide leadership in reshaping modern life through science and technology, it would be most likely to support the development of modern technology in Mexico, as advocated by Ávila Camacho.

Creation of Rockefeller's Mexican Agricultural Program

Proposals for the Rockefeller Foundation's involvement in Mexico began as early as 1933 with discussions between the foundation's regional director for public health, John A. Ferrell, and Josephus Daniels, the American ambassador to Mexico.[24] Little came of these early talks, however. It was not until Vice President Elect Wallace's trip to President Ávila Camacho's inauguration that serious movement began.

After Wallace's return to the United States, a number of conversations among Wallace, Nelson Rockefeller, Ferrell, and Daniels led to a meeting of Wallace, Ferrell, and Raymond B. Fosdick, president of the Rockefeller Foundation.[25] Fosdick recounts that Wallace remarked on the benefits that could come to Mexico if anyone could improve the productivity of corn (maize) and beans, the staples of the Mexican diet. Fosdick was also probably seeking new philanthropic activities for the foundation because the war had ended the foundation's major programs in Europe and China.[26]

Upon his return to New York, Fosdick consulted with Warren Weaver, director of the natural science program at the foundation.[27] Weaver, whose technical background was in mathematics and the physical sciences, remembered that at first he did not know how to respond to Wallace's suggestion. Nevertheless, the natural science staff met with Albert R. Mann, who had formerly been dean of agriculture at Cornell University and who had directed the foundation's work in Chinese agriculture. They proposed to send three experts to Mexico: Paul C. Mangelsdorf, a geneticist and plant breeder from Harvard University; Richard Bradfield, a soils specialist from Cornell; and Elvin C. Stakman, a plant pathologist from the University of Minnesota.[28]

The survey team toured extensively in Mexico in the summer of 1941. Its lengthy report was in some ways sensitive to the cultural and social diversity of Mexican agriculture, but it was quite specific in its main recommendation: the Rockefeller Foundation could best assist the improvement of Mexican agriculture by establishing a four-man commission in or near Mexico City to advise the Mexican Department of Agriculture. In priority ranking, the four men should be (1) an agronomist/soil scientist, (2) a plant breeder, (3) a plant pathologist/entomologist, and (4) an animal husbandman.[29]

Once the experts' report was in, the foundation negotiated with the Mexican government for an invitation to provide assistance in developing Mexican agriculture. Invitation in hand, the foundation selected J. George Harrar as director of the program and launched activities in February 1943.[30] In 1943 Harrar was joined by Edwin J. Wellhausen, a maize geneticist; in 1944, Norman E. Borlaug, a plant pathologist and plant breeder, joined the MAP.[31] Both Harrar and Borlaug had completed their doctoral degrees with Stakman at the University of Minnesota.[32]

Before practical work could begin, the foundation had to reach an agreement with the Mexican government about the subjects of research and the organizational structure within which foundation staff would work. Negotiations on both began in early February 1943, when Harrar, Stakman, and Henry M. Miller Jr. (the latter a permanent staff member of the foundation who had been involved in the MAP from its inception) arrived in Mexico City. Negotiations were concluded within a week, and the foundation signed a memorandum of understanding with the Mexican government.[33]

Two specific research problems were identified by the Mexican government for research: wheat rust and improvement of maize varieties. Ironically, these two activities were different from what was expected by Stakman, the chief adviser to the foundation on the MAP. Stakman felt that since Mexican scientists were already trying to obtain improved maize varieties, it would be inadvisable for the foundation to start a parallel investigation. Offended feelings might be a problem. Instead, he anticipated that Harrar and his staff would advise on maize work. For wheat, Stakman noted that the secretary of agriculture, Marte Gomez, felt that wheat rust was the most important single problem. Stakman did not agree with this assessment, but he recognized that Gomez reflected the desires of President Ávila Camacho, who wanted increased wheat area and production. Stakman therefore agreed that control of wheat rust should be the beginning of a general project for wheat improvement and expansion of acreage.[34] For reasons that will be developed in chapter 8, the inclusion of wheat rust as a high-priority research topic had an important influence on the research program that led to the green revolution.

Organizational structure was of equal or more importance to the foundation staff. Essentially, the foundation wanted to hire U.S. scientists who would pursue research on topics mutually agreeable to the secretary of agriculture for the Mexican government. Foundation funds would pay for salaries, equipment, library resources, travel expenses, and office expenses. Mexico was asked to pay for land, greenhouses, manual labor, and technical assistants. Foundation officers would select, from the technical assistants and others, Mexican nationals to go on for advanced training in the United States and elsewhere. Research would be the primary mission, with demonstration or extension work done only as time and resources permitted.[35]

The memorandum of understanding, signed in Mexico on 10 February 1943, included wheat rust and maize improvement as top-priority research items.[36] Harrar had already moved to Mexico, so the wheat rust work began immediately and he hoped to have the maize work under way by 1 July 1943.[37] The Office of Special Studies (OSS)[38] was thus established as a semiautonomous research unit directly within the Mexican Department of Agriculture. Until its transformation in 1966 to the International Center for Wheat and Maize Improvement (CIMMYT),[39] the OSS was the research center that stimulated a major transformation of Mexican agriculture. Harrar and subsequent leaders at the OSS, however, always insisted that Rockefeller Foundation research be independent of other Mexican research institutes and that the effort be collaborative, not subject to unilateral control by the Mexican government.[40]

Mexican Aspirations for the Office of Special Studies

Just as many complexities shaped the motivations and behavior of the Rockefeller Foundation as it created the MAP, so, too, did many complex factors affect the Mexican government as it agreed to have the foundation open a special research office. Most of the complications on the Mexican side stemmed from the tensions and contradictions between the drive for equity as part of the revolution and the drive for productivity as part of building a modern industrial state. Understanding both the complexities and the divisions of opinion within Mexican society about equity and productivity will help clarify the significance of the high-yielding varieties of wheat and other grains.

When the Rockefeller Foundation initiated the MAP, it immediately became enmeshed in a series of changes that were then under way in Mexico. On the surface, the major change was a relaxation of the drive for agrarian reform, in which Ávila Camacho shifted priorities away from the land redistribution program of Lázaro Cárdenas. This shift, however, is perhaps better understood as part of a deeper argument among Mexicans about the future of the Mexican nation and how its natural resources would be used to achieve that future. At stake was whether Mexico would "retain" or "stagnate in" its economy of subsistence agriculture or whether it would develop the larger scale, more commercially oriented agriculture found increasingly in the United States. Historian Joseph Cotter provides compelling arguments that numerous political and scientific personnel, including those in favor of agrarian reform, wanted a more dynamic, science-based agriculture.[41]

Associated with this question was the question of whether Mexico would develop an industrial capacity, which required, among other changes, that some labor leave agriculture and migrate to the cities. If industrialization was to occur, agriculture thus had a major role to play: it must change to technologies that were more labor-efficient so that some people could choose or would be forced to leave the rural areas and become part of the industrial workforce. Agriculture would have to provide at least part of the capital for Mexican industrialization.[42] The research programs and policies needed to obtain more productive small-scale agriculture were not necessarily the same as those needed to create an industrialized agriculture that would be compatible with an urban-based manufacturing and commercial economy.

From the original survey report of the Rockefeller Foundation, *Agricultural Conditions and Problems in Mexico*, one gains the impression that the foundation believed its work in agricultural science was to alleviate the poverty of Mexico's masses. Its thinking on the subject seems remote from the debate that was then under way about restructuring the Mexican economy toward industrialization.[43] This debate is best seen by tracing the Mexican attitudes toward building an industrialized urban economy along with its necessary companion, a modern agricultural sector based on irrigation, mechanization, and the use of fertilizers and improved seeds.

Mexico's traditional economy, both before the Conquest and after, until the end of the nineteenth century, was based primarily on the rainfed production of maize and beans in the highlands around Mexico City, an area known as El Bajío. This was the location of the highly advanced Aztec civilization, and after Spanish conquest El Bajío continued as the major population and economic center of Mexico. Scattered, sparse populations were located in parts of the lowlands that had ample water, for example, the Yucatán, and in the vast semiarid and desert areas of the north of Mexico. Economic production in the northern part of the country was based either on extensive grazing or on small, easily irrigated agriculture in the river valleys. For the most part, however, few people and little economic activity occurred outside El Bajío. Mining for gold and silver plus traditional craftsmanship in items like jewelry were about the only nonagricultural forms of economic production. As a whole, the economy was not particularly productive, and the hacendados (major landowners) had no particular interest in seeing it developed. This political economic elite had an idyllic lifestyle, based on the economic self-sufficiency of the haciendas and on the subjugation of the large Indian majority.

Change began to come, however, in the latter part of the nineteenth century. Following war with the United States in the 1840s, Mexico's tiny urban class began to push for the commercial and political freedoms that would enable it to build a modern industrial and commercial economy. In addition, some landowners began to see the potential for wealth in Mexico's river valleys that ran through the arid north. With large-scale dams and irrigation projects, they saw a potential to generate wealth that would vastly surpass the meager output of extensive grazing or small-scale irrigation works. In the Porfirio Díaz dictatorship, the Mexican government began to respond to this ferment for change, primarily in the initiation of irrigation development for wealthy investors.[44]

Mexico's economy was left in a shambles by the revolution, but postrevolutionary recovery still contained the tensions inherent in the Díaz period between (1) the hacendados, who were satisfied and eager to defend their feudal-like privileges, (2) the still impoverished and oppressed landless majority of the rural areas, (3) the emerging commercial and industrial entrepreneurs of the cities and the commercially minded farmers who wanted modern agriculture, and (4) a new element with legitimacy and power, the urban industrial working class and its unions. Despite the initial efforts at land reform after 1917, by 1934 the power of the hacendados was still largely intact.[45] Perhaps the most important consequence of the Cárdenas reforms was the breaking of the power of the hacendados in Mexico forever.

Cárdenas ruled with the support of the rural peasants and working class plus the loyalty of the army.[46] Nevertheless, his reforms for the landless poor and the small urban trade union movement removed the hacendados, who were probably the biggest hindrance to the emerging middle class. Cárdenas thereby unleased the latent potential of the entrepreneurial middle class, which may not have shared his sympathies for the rural and urban working classes. Despite the major attention of the Cárdenas government going to the *ejidos* and urban trade unions, Cárdenas was by no means oblivious to the growing commercial interests of Mexico's small middle class. He made a number of decisions as president that ultimately proved extremely beneficial to this portion of the Mexican population.

First, he continued the development of economic infrastructure in Mexico that was important to the emergence of a more complex, productive, and commercial economy. Transport, communication, and irrigation works were most important, taking over 87 percent of the federal public investment. Perhaps Cárdenas's heart lay with land reform, but he did not ignore the needs of the middle class.[47]

As an example, consider his decision to build a major dam and irrigation project in the Yaqui River Valley of Sonora. As a way to maintain the support and perhaps to garner the loyalty of the former president, General Plutarco Elías Calles, Cárdenas and Calles visited Calles's native city, Ciudad Obregón in Sonora, in July 1934. Calles had been less than enthusiastic about Cárdenas's radical reforms, despite the fact that Calles has supported Cárdenas's rise to the presidency. Cárdenas may also have been considering the support he might obtain from Don Rodolfo Elías Calles, the son of Plutarco Calles and then governor of Sonora. Shortly thereafter, the Cárdenas government approved plans to build a large dam and irrigation works on the Bapispe River in the Yaqui River Valley. In moving on this project, the Cárdenas government also wanted the cooperation and assistance of the Americans, with whom his government later had serious differences.[48]

Cárdenas seems to have succeeded, at least temporarily, in his aim to keep General Calles as a supporter. When the two appeared together again later that year, Calles gave a ringing endorsement of the radical reforms under way by Cárdenas: "voracious and egotistical capitalists, conspiring with the clergy and the reaction, are intent upon presenting a problem for the Government." United States Ambassador Josephus Daniels believed this was the strongest socialist statement ever from Calles, signaling his rapprochement with Cárdenas.[49]

Although this dam and irrigation works in the Yaqui Valley may have been conceived largely as "pork-barrel" federal water development designed primarily to win political support, it was symbolic of the larger economic reform movements long under way in Mexico. The Yaqui Valley was sparsely inhabited, primarily by the indigenous Yaqui people whose culture of farming, hunting, and gathering was adapted to the torrid desert climate of northwestern Mexico. Cárdenas responded to Yaqui requests for secure landholdings and initiated other reforms to improve the living conditions of the Indians. However, federal control of the new irrigation works ultimately forced the Yaqui into commercial rather than traditional patterns of agriculture.[50] Other newly irrigated land went to middle-class farmer-entrepreneurs who wanted to emulate the commercial successes of their counterparts in the United States.

Although it could not have been foreseen at the time, the opening of the Yaqui Valley to large-scale agriculture may have prompted the government of Ávila Camacho, successor to Cárdenas, to want wheat rust as a top-priority item for the Rockefeller Foundation. Wheat was a major crop planted in the newly watered Sonoran desert, and it was subject to devastating outbreaks of wheat rust. Outbreaks in 1939–41 were particularly destructive.[51] Ávila Camacho's desire to see increased land in wheat and higher wheat production was, in fact, threatened by these severe rust epidemics. In a sense, the success of the entire project of the Yaqui River Valley development scheme depended on solving that plant disease problem.

A second decision also demonstrated that Cárdenas by no means ignored the aspirations of the developing middle classes: his selection of General Manuel Ávila Camacho to succeed him. Albert Michaels, who has studied the 1940 election in depth, argues that two chief candidates for president were active in 1939.[52] General Francisco Mugica, who had distinguished himself in the revolution, was a radical leftist with an aggressive personality. Mugica was also a leader in the oil expropriation effort and would probably have been uninterested in reaching a compromise with the American government over this issue. He was the logical ideological successor to Cárdenas and might have continued the Cárdenas reforms with relatively little change.

Cárdenas, however, gave his endorsement to General Manuel Ávila Camacho, a loyal, tactful, quiet man known as a good mediator. Twenty years after the election, Cárdenas stated in an interview that his final decision was influenced by the need for national unity in Mexico and the need for reconciliation with the United States. Cárdenas was clearly aligned against the fascist powers in Europe, and this international stance inevitably brought Mexico into closer collaboration with the United States.

Ávila Camacho's rise to the presidency symbolized that the moderates of the revolution were in control of Mexico.[53] Ávila Camacho accordingly rewrote the 1933 platform, with its emphasis on land reform and its language based in Marxist terminology and anticlericalism. In its place went concerns of nationalism and the liberal concern for individual rights, which were essential for the flourishing of entrepreneurial capitalism. Ávila Camacho emphasized that if he were elected, Mexico would be governed by technicians, not ideological reformers. His rise to power ended Mexico's antagonism to capitalism and brought the government into friendly relations with the Roosevelt administration and its Good Neighbor Policy.

Only one other candidate, General Juan Andreu Almazan, posed a challenge in the 1940 election. This was the candidate of the seriously disaffected middle class and Catholics who were upset with the Socialist Education Law of 1939. Michaels argues that the government probably committed fraud in the elections, and Almazan continued a claim for the presidency even after the election was over. He dropped his efforts, however, when the Roosevelt administration announced it was recognizing the election of Ávila Camacho and sending Henry A. Wallace to the inauguration. In any case, Ávila Camacho may have been as satisfactory to the interests of Almazan's supporters as Almazan himself.

Control of the presidency by people committed to the economic modernization of Mexico continued for decades. What is thus clear in retrospect is that the Rockefeller Foundation MAP began at the outset of a new political economic climate in Mexico. From the Mexican point of view, the foundation scientists may simply have been part of the contingent of "technicians" with which Ávila Camacho expected to govern Mexico and transform it into an industrial and commercial state. For their part, the foundation scientists gave every indication that they believed in the power of new agricultural technology to improve the well-being of the Mexican people. Foundation literature from the time paints a picture of helping subsistence farmers produce more food, which may have happened in some cases, but the Mexican government probably saw the foundation as helping transform Mexican agriculture away from subsistence farming toward the type of commercial farming that was compatible with a modern industrial state.

Symbolic of the needs of the new Mexico was the situation with wheat at the time the Rockefeller Foundation began its work. The argument developed earlier suggested that a major consideration leading Mexico to place wheat rust at the top of the research priorities for the foundation was the desire to expand the wheat areas in the newly irrigated areas of the northwest. Part of the motivation for this priority was to serve the new agricultural elite, which was commercial in character. Part was also to justify the expense of the irrigation projects. What is missing from this argument, however, is the link to the developing urban economy of Mexico. To understand this link, we have to raise the question "Why was wheat important to Mexico?"

The traditional diet of Mexico was based on maize and beans, crops that did well in all parts of the country and that together provided an inexpensive, nutritionally balanced diet. Flavored with vegetables, chiles, and other spices, it was also a tasty fare. Moreover, it was a diet that could be raised easily by subsistence growers, stored at home, and prepared for eating without elaborate equipment. In other words, maize and beans made up a diet that was admirably suited to a subsistence farming lifestyle.

Wheat was grown in Mexico after the conquest, and by the 1930s it ranked third in value of Mexico's crops, after maize and cotton. In contrast to the traditional diet, wheat was more difficult to prepare than maize. Nevertheless, consumer demand in the 1930s was starting to shift away from maize to wheat. Assured supplies of wheat at reasonable prices thus became an important issue.[54]

Without a viable, commercial agriculture in the countryside, however, Mexican consumers could not be assured of regular supplies at reasonable prices from domestic production. Cárdenas garnered much popular support from his agrarian and labor reform and his seizure of foreign oil properties. Unfortunately, agricultural production fell during his administration, quite probably as a result of the reforms. Prices of maize, beans, and wheat, the Mexican staples, rose, and the country imported part of its basic foodstuffs.[55] Heavy imports put a strain on limited foreign exchange reserves, which further dismayed the entrepreneurial elite of the urban middle class: scarce dollars (earned largely from oil, mining, and tourists) that went for wheat could not be spent on modern industrial equipment from the United States or Europe. Thus the growing preference for wheat created difficulties in the larger economy, which in turn led the emerging middle class to think that more domestically grown wheat would be a good thing. In all likelihood, it is this sort of reasoning that may have led to President Ávila Camacho's support for a policy of more wheat in Mexico, which in turn led to the Rockefeller Foundation's mandate to control wheat rust.

A second factor about wheat was that, in contrast to maize, wheat in the 1940s had a well-developed international trade. Europe in particular was a large importer of wheat from the Americas, and the commerce in this grain had well-established channels. Entrepreneurially inclined Mexican farmers undoubtedly realized that if they could grow large quantities of the grain perhaps they, too, could enter the export markets to Europe. This would give them a choice of markets, either domestic or the international, and they could sell to the place where prices were highest. Being paid in pounds sterling would have conferred advantages on both the Mexican farmer and the Mexican government. The farmer might choose to purchase British luxury goods, and the government of Mexico would see more earnings of foreign exchange. Thus promoting the production of wheat had the promise of both satisfying domestic demand and improving the international economics of Mexico.

Given this set of circumstances from the Mexican point of view, it is little wonder that President Ávila Camacho welcomed the Rockefeller Foundation and insisted that promotion of wheat production by controlling rust be a high priority for the foundation's work. The fact that Stakman was surprised about the inclusion of wheat rust at the top of the research priorities may simply have reflected that he and his colleagues from the foundation did not fully comprehend the dynamics then under way as Mexico sought to establish a new economy based on industry and commercial agriculture. In any case, Harrar and his colleagues truly fit the role of the technicians who would help fulfil the vision of Mexico becoming an industrialized commercial nation.

Even if the foundation scientists had fully comprehended the nature of Ávila Camacho's vision, there is little reason to think they would have been uncomfortable with it. After all, they were trained in the land-grant universities of the United States, they were accustomed to the notion that science promotes efficiency in commercial agriculture, and they were probably most comfortable in the social dynamics involved between scientists and commercial farmers. Deborah Fitzgerald has argued that the foundation scientists were perhaps constrained by their experiences in the United States, which led them in turn to favor dealing with the emerging entrepreneurial Mexican farmer.[56] She undoubtedly provides a useful insight in this argument, but the problem was actually much deeper and more fundamental. The Mexicans in control of Mexico after 1940 wanted to replace their traditional agrarian economy with a new economy based on active capitalism both in the countryside and in urban industries. It was understood, at least intuitively, that the two sectors were related and that neither could change without the other.

The case may have been put most explicitly by Roberto Osoyo in 1968 at a symposium, "Strategy for the Conquest of Hunger," held at Rockefeller University in New York. At the time, Osoyo was director general of agriculture in the Mexican Ministry of Agriculture, and during the 1950s and 1960s he had been a major administrator in agricultural modernization in the state of Sonora. After noting the rapid population growth rate of Mexico, then 3.2 percent per year, he went on to argue that Mexico had made great progress in removing labor from agriculture. In 1943,

65 percent of the labor force was in agriculture, but by 1968 this figure had dropped to 50 percent. Osoyo then went on to state: "We now nurture the hope that by 1970 not more than 45 percent of our active population will have to be engaged in agriculture."[57]

A similar thought emerged from staff members of the Ford Foundation in the late 1960s. Eduardo L. Venezian and William K. Gamble argued that Mexico had developed a dualism in its agriculture. The large private farms of the Pacific northwest (Sonora and Sinoloa) and the north used advanced technology and were irrigated, commercial, and export oriented. They constituted only one-fifth of all Mexican farms, but they were highly productive. The other part of the dualistic structure was the *ejidos*, which depended on rainfall, used low technology, were low in capitalization, and oriented toward subsistence maize and bean production. Venezian and Gamble felt that the small farms must "disappear if Mexican agriculture is to be fully modernized and rural misery is to be eliminated."[58]

Perhaps in 1940 neither President Ávila Camacho, his secretary of agriculture, Marte Gomez, nor the Rockefeller Foundation officers could have been so articulate about the need to reduce labor in agriculture or the need to eliminate the *ejidos* in order to eliminate rural poverty. After all, Osoyo, Venezian, and Gamble were speaking with the advantage of twenty-five years of hindsight and experience with the results of the modernization of Mexican agriculture. But the prescience of government and foundation officials is not really the issue. What is at stake is understanding the context in which the scientific research of the MAP began and the consequences of its results. It may be comforting to think that Rockefeller Foundation thinking really was oriented toward the amelioration of poverty of the rural folks of Mexico, and that the needed improvements in their lives would flow naturally from finding new varieties of maize and a way to control wheat rust. What was never spoken in 1943 at the start of the MAP, however, was that its primary Mexican proponents wanted to completely reshape the Mexican economy. In order to create the modern industrial state, labor would have to be enticed off the farm or, presumably if need be, be forced off by conditions that were so deplorable that leaving would seem a good thing to do. In addition, the fostering of a liberal political economy was essential to forging the new industrial state, which meant a focus on individualism rather than the communal values of traditional Indian villages. What was also unspoken was the need to extract or entice capital out of agriculture to finance industrial development.

In understanding the origins of the MAP, therefore, we have to realize that we can summarize the context of its founding in either or both of two possible ways:

- The MAP was motivated by humanitarian concerns for impoverished Mexican peasants. Agricultural research was to be the tool through which these farmers could obtain better yields and a better life. The Rockefeller Foundation had the humanitarian needs of the poor Mexican people uppermost in its mind and expected the fruits of its research programs to lead naturally to the empowerment of the Mexican people to make choices that would better their lives.
- The MAP was an agreement between the Rockefeller Foundation and the Mexican government. The context in which it began was complex and included (1) a tenuous and recent relaxation of severe tension between the U.S. government (which

regulated the foundation) and the Mexican government; (2) a highly dangerous world situation in which it is reasonable to believe that the foundation wanted to foster the development of liberal democratic capitalism rather than see either socialism or fascism make further inroads; and (3) a complex and dynamic struggle within Mexico over which vision for Mexico's future would prevail, with alternatives being a liberal democratic capitalism and industrial economy, a reversion to the quasi-feudal oppression of the hacendados, or the continued socialist radicalism of the Cárdenas era. The MAP was an alliance between a U.S. foundation that promoted liberal democratic capitalism and a Mexican government that was struggling to establish a liberal democratic capitalist political economy. In the long run, probably both the foundation and the Mexican government had no particular wish to improve the lives of peasant farmers *in their capacities as peasant farmers*. Perhaps neither could have articulated it at the time, but both sides probably knew that the forces of agricultural modernization would have far-reaching effects that, in their eyes, were for the better.

The second description is more complete and better accounts for a multiplicity of concerns.

Consequences of the Mexican Agricultural Program for Mexico

The MAP is now a program of the past. The Office of Special Studies within the Mexican Ministry of Agriculture closed its doors in 1960 and was transformed into CIMMYT in 1966, which has been a major force in exporting the methods of high-yielding wheat and maize production and in furthering the development of high-yielding technology. Mexico's government was very proud that it could collaborate as a full partner in the further development of wheat production worldwide, and CIMMYT's varieties quickly spread to many other countries, a story that will be developed later in this book. Mexico's economy was also restructured in the process, and it is useful at this point to consider some of the effects of the MAP.

Superficially, the major effect of the MAP was to find new varieties of wheat that raised Mexico's total wheat production and production per hectare. Mexico changed from a wheat importer to a wheat exporter in 1958, an event that was highly significant for the Mexican economy.[59] The growing preference for wheat became a foreign exchange earner, not a foreign exchange drain. Maize and beans also continued as staples, but the emerging urban economy of Mexico was well served by the transformation to wheat self-sufficiency.

Simultaneous with the development of a wheat export capacity in Mexico was the emergence of a new rural elite of commercial farmers. Based primarily in the irrigated valleys of Sonora and Sinoloa, these entrepreneurial farmers used new lands, developed with great subsidies from the Mexican government, to create a style of agriculture that was similar to the agriculture of the United States.

Venezian and Gamble argue that the farmers of Mexico, which meant the poor farmers of Mexico, were the suppliers of capital for the industrialization of Mexico. Extraction of forced savings through inflation was the primary method used to make the transfer of capital out of agriculture into the new industrial activities. For the poor farmers who were thus manipulated by the currency management of their country, it was somewhat ironic that they be the suppliers of capital for the new industry.

Their ability to benefit from industrialization was limited in that they had to leave farming to enter the cities in order to prosper. It is doubly ironic that these capital transfers were used to finance a new breed of commercially minded export farmer who outcompeted the traditional farmer at every step. Mexico thus achieved its industrial status, in both manufacturing and agriculture, on the backs of traditional farmers.

What happened to traditional farmers? Essentially, they disappeared into the cities. One consequence is that Mexico City is now perhaps the largest city in the world, and the poor immigrants of the rural areas may or may not have found much material improvement in their living standards.[60]

Not only were rural people taken advantage of in the transformation, women compared to men probably lost access and control of resources in the transformation to modern agriculture and modern industry. Neither the Rockefeller Foundation scientists nor the Mexican government was sensitive to the roles of women in traditional agriculture. When the elimination of subsistence farming became the policy of Mexico, women were not provided with new and comparable economic roles in the modern sector. Perhaps the usual fate was for a country girl to get a low-paying job in one of the new factories. This work provided neither the security nor the dignity of traditional agriculture, precarious as that system of livelihood was.[61]

Was the transformation worth it? For the middle classes and the new entrepreneurial elite, yes, of course. For the poor who supplied the capital? Probably not, except that they might see their children with better incomes and more options for education because of their new lives in the urban areas.

Consequences for the Rockefeller Foundation and for the U.S. Government

Several features of the Rockefeller Foundation's program in Mexico need to be noted, because these attributes were important in the evolution of the foundation's strategy in agriculture. First, although Rockefeller philanthropy was not a complete stranger to agriculture, the dimensions of the effort in Mexico far surpassed all previous grants.[62] Furthermore, the decision to move into agricultural science was in some ways unanticipated. In 1938 Warren Weaver had identified no agricultural fields of work in a strategic planning exercise for the natural science program of the Rockefeller Foundation. In fact, Weaver had explicitly ruled out animal and plant breeding and instead suggested a concentrated effort in basic genetic research.[63] Yet by 1945 the Rockefeller Foundation was spending nearly $100,000 per year on the Mexican Agricultural Program when the entire natural science expenditure per year averaged only $1.7 million.[64]

Second, the Mexican program was an *operational* program, not just the usual Rockefeller Foundation effort that relied on the *disbursement* of grants to others responsible for actual scientific operations.[65] Perhaps the lack of sufficiently trained Mexican nationals led the foundation to move directly into operating a research station, but this pattern became characteristic of other Rockefeller Foundation work in agriculture.

The MAP, therefore, was on several grounds a fundamentally new type of program for the foundation. The program's successes were consequently well received in the foundation's headquarters in New York. As is frequently the case, a successful venture in a pilot program lends itself to expansion. Although clear interest in expansion of agricultural assistance emerged from foundation trustees such as John D. Rockefeller III even before the MAP began work,[66] by 1950 the results of the program encouraged the foundation to initiate a similar assistance program in Colombia.[67]

Expansion of the MAP, however, takes us ahead of our story for the moment. For now, suffice it to say that program's successes, in the context of concerns about strategic security of the United States, prompted both the Rockefeller Foundation and the U.S. government to launch extensive assistance programs in agricultural development. The U.S. government's involvement began in earnest with President Truman's inauguration speech in 1949. "Point Four" of that address called upon the government to lend technical assistance in agriculture and other fields to the poorer nations of the world, which led to the "Point Four" program that eventually evolved into the U.S. Agency for International Development.

The Rockefeller Foundation's movement into a vastly expanded agricultural assistance program came with the decision in 1952 to begin the Indian Agricultural Program. These efforts, combined with the government's experiences, in turn led eventually to the conceptualization and funding of a large network of international agricultural research stations funded by many governments and agencies around the world. Strong evidence, to be developed in chapter 7, suggests that the American government saw the foundation's MAP as a highly successful model for the development of new agricultural technology, and in various ways the MAP's collaborative model of research involving both U.S. and foreign nationals was copied as the method for effectively transferring technology from the developed to the less developed world.

The MAP's inspiration of similar efforts elsewhere, however, was realized in a particular context: a fear of famine, overpopulation, and the threatened rise of communist governments in areas considered by some to be a strategic threat to the United States and its post-1945 security arrangements to contain the socialist revolution of the Soviet Union. The next chapter develops this framework of the international politics of plant breeding.

6

Hunger, Overpopulation, and National Security

A New Strategic Theory for Plant Breeding, 1945–1956

The Mexican Agricultural Program (MAP) was the catalyst that brought plant-breeding science into the arena of international relations. During the first few years of the MAP's operations, however, no programmatic framework existed to promote plant breeding on a global basis. Although a private philanthropy like the Rockefeller Foundation might support plant-breeding research, it was not clear that governments would be interested in the field as a way of achieving their international ambitions. The trustees of the Rockefeller Foundation, with their sense that success with MAP might lead to further ventures, were possibly the only group with even the rudiments of an idea about the international importance of plant breeding.

By 1970, however, plant breeding was firmly entrenched in global international relations. Extensive national research organizations in many countries, a collection of prestigious international research stations, and an international coordinating network of supporters created a complex institutional nexus within which plant breeding and allied sciences were well supported. Research conducted in this network of national and international experiment stations led to the high-yielding varieties of wheat and rice that significantly altered regional, national, and international economies.

Several factors governed the ability of plant breeding to become a "normal" part of international dealings. First, the science had to have something to offer. As described in chapters 3 through 5, by 1945 plant breeding had demonstrated that it could produce results of interest. Second, national governments wanting to extend or receive international aid in plant breeding had to have a national capacity to conduct the science. Chapters 4 and 5 provide an account of how the United States, Britain, India, and Mexico each gained this capacity.

A full comprehension of why and how wheat breeding entered the international arena requires attention to three additional points. First, what was the general intellectual and political climate that promoted the science's entry into international relations? Second, what specifically did individual countries do to participate in aid programs including wheat-breeding research? Finally, what was the research program that led to the high-yielding wheat varieties, and how was this program created? In this chapter we turn to the first of these three questions. The latter two are developed in chapters 7 through 10.

Hunger, Overpopulation, and National Security

For plant breeders the decade after the end of World War II was critical. Between 1945 and 1955 a series of events, studies, and conceptual syntheses created a climate in the United States that saw plant breeding and all of the modern agricultural sciences as a critically important adjunct in the battle between capitalist freedom and the tyranny of communism. Because the United States played a preeminent role in promoting the development of high-yielding wheat varieties, it is especially important to understand events in that country.

The American intellectual and political climate from 1945 to 1955 was critically shaped by the development of the cold war, a part of which included a theory that purported to link causally overpopulation, resource exhaustion, hunger, political instability, communist insurrection, and danger to vital American interests. As a convenient shorthand, the theory can be called the population–national security theory (PNST) (Figure 6.1).

In PNST, hunger was a symptom of overpopulation and resource exhaustion, and it in turn became a cause of further resource exhaustion and political instability. Plant breeding was seen as a remedy for hunger because the science could increase and stabilize yields. For the moment, we will defer a critical examination of whether or

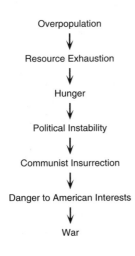

Overpopulation

↓

Resource Exhaustion

↓

Hunger

↓

Political Instability

↓

Communist Insurrection

↓

Danger to American Interests

↓

War

Figure 6.1   Population–national security theory, the postulated analytical framework developed in the United States to justify promotion of agricultural development in the third world.

not PNST provided useful insights and accurate predictions. What is important is to understand how the theory was constructed and what effects it had on guiding the formation of agricultural research institutions and programs.

Different components of PNST had historical threads reaching back many years, often to the nineteenth century. In many ways, PNST was not a remarkably new way to view the world, but it was a new way of linking several previously discrete centers of thought: demographic science, a series of famine threats that occurred during and immediately after the Second World War, natural resource conservation, and the new exercise of global power that the United States assumed after the end of the war.

What emerges from this inquiry is a sense that a relatively small group of people, linked intellectually and socially, constructed PNST. By no means were they actively collaborating on a day-to-day basis. Instead they were dispersed broadly in time and space in a number of universities, foundations, the business world, and government agencies. Individuals who were prominent in creating the theory sometimes moved between different organizations, which suggests that a social network linked them.

Some of the architects of PNST were of humble origins, but others were prominent members of the highest stratum of American economic, political, and social life. To designate PNST as the product of an American elite may suggest a conspiracy theory of conscious collusion among members of the upper class, which will not promote understanding of the origins and effects of the theory. However, an argument will be made that PNST reflected many of the values held by the American political economic elite and that it was intended to promote the global political economic arrangements necessary for the continuation of that class.

As a result of the critical synthesis in PNST, the United States became the world's major stronghold and supporter of plant-breeding science. Significantly, however, other countries also substantially increased their abilities to conduct research in this science even though they were indifferent or even hostile to the American notions that plant breeding was a bulwark against communist subversion. The postwar United Kingdom and independent India had considerably more sympathy for socialist ideology than the Americans, but they, too, came to see plant breeding as essential to protecting their interests, both domestically and internationally. Convergence toward support for this science in spite of diverging ideological justifications for it thus becomes one of the puzzles to follow in understanding why the science became so important in foreign relations after 1945.

In tracing the origins of PNST and how it affected plant breeding, we turn first to the development of modern demographic science, its legacy from Malthus, and the ideology of neo-Malthusianism. We then examine the fear of famine that occurred during and after the war. Finally, we trace the ways in which links were forged between the concepts of overpopulation, resource exhaustion, hunger, and threats to national security, and how plant breeding was seen to have a role in protecting vital national interests.

Population, Malthus, and Neo-Malthusianism

England is generally credited (or blamed!) for initiating the industrial revolution. Similarly, it was in England that a fully articulated vision first appeared for that economic system we now call capitalism. Adam Smith prophesied that it was through free-market economies that nations would find their wealth.

England also spawned a derivative of the early capitalist economy: a fully articulated theory of population. Thomas Robert Malthus (1766–1834) was educated for the clergy but spent his career as England's first professor of political economy. His theory[1] had more influence on Anglo-American images of population in relation to resources than that of any rival.

Malthus was particularly concerned in his 1798 essay with what he said was the nearly inevitable tendency of the human population to grow at a geometric (i.e., exponential) rate. In contrast, human abilities to increase food supplies grew at no more than an arithmetic (linear) rate. As a result, Malthus foresaw the fate of people as always tending to increase their numbers just to the point at which they were miserable with poverty and hunger. Any increases in food production would quickly result in more babies, not a permanent improvement in living standards. For Malthus, the wiser (i.e., wealthier) classes could avoid misery through moral restraint and avoidance of marriage, but the vast bulk of humanity (i.e., people with little property) were most likely to end up perpetually hungry and miserable.

Malthus intended his profound pessimism to deflate what he saw as misguided utopian promises for a better life through revolution, redistribution of property, and a leveling of social and political relations. In his time he drew bitter criticism from reformers and revolutionaries who saw him as, at best, an apologist for an unjust social hierarchy. His theories provoked responses and challenges for most of the first half of the nineteenth century.

A number of factors, however, conspired to move Malthusian pessimism off the intellectual agenda in the last half of the 1800s. Most important was the vast increase in agricultural supplies from North America. New land in cereals plus better land- and sea-transport technologies plus emigration from Europe made it possible to feed a rapidly growing population. Most intellectuals forgot Malthus's argument about relative rates of population growth in relation to growth of agricultural supplies.

Malthusian ideas, however, were not gone for long. When they returned, they were in different packages developed to analyze different sets of problems. Moreover, at least two tendencies of neo-Malthusian[2] thought developed in the United States and the United Kingdom. One (political neo-Malthusianism) linked high population densities to the complex workings of national and international economies. The other (ecological neo-Malthusianism) linked high population densities to the exhaustion of resources and the collapse of productive ecosystems. Both ultimately linked population inquiries to questions of national security. Significantly, most neo-Malthusians shed the pessimism of Malthus, who believed very little could be done to alleviate human misery. Neo-Malthusians tended to argue that a scientific understanding of population could lead to planning that would avoid the catastrophe of overpopulation.[3]

John Maynard Keynes (1883–1946) was a pioneer in reviving the link between population issues and political economy.[4] He was a young man of thirty-six in 1919, the year he resigned as the representative of the British Treasury at the Paris Peace Conference. He maintained that his resignation was prompted by a complete dissatisfaction with "the whole policy of the Conference towards the economic problems of Europe."[5] He proceeded to write *The Economic Consequences of the Peace*, a reasoned polemic explaining his objections.

Keynes introduced Malthusian theory to explain the behavior of people and of nation-states. At its core, his disillusionment with the Paris Peace Conference stemmed from his sense that the conferees either could not or chose not to understand the precarious nature of the European civilization that was mangled by the war. Importantly for Keynes, the inherent instability of European culture derived from a population explosion that took the populations of Germany, Austria-Hungary, and Russia to 268 million by 1914.

Keynes believed that this mass of people in central and eastern Europe, plus the western Europeans, were tied together by intimate economic relations involving the import and export of agricultural and industrial goods and investment capital. The whole of the European economy survived only because of the forbearance of the working classes from seizing a larger part of the produce from the capitalist class. In addition, cheap cereal grains from the New World were critical to the viability of this fragile political economic system teeming with people. Peace could be achieved, Keynes argued, only by rekindling prosperity (i.e., reestablishing the economic interactions of western, central, and eastern Europe and the relations between Europe and the New World). Revitalizing German industrial skill and organization was, therefore, key to a peaceful Europe.[6] A few years later, in 1923, Keynes wondered whether the material progress of the nineteenth century was a temporary aberration, to be replaced with a harsh Malthusian reality.[7]

Whether Keynes was right or wrong, and whether his remedies would have averted the further tragedy of World War II, is not at issue here. What is important is that Keynes based his analysis of prewar Europe and of postwar hopes for a peaceful future on the instabilities and problems induced by rapid population growth. Furthermore, he coupled his thoughts on population to the question of how to develop a stable supply of grain for densely populated areas. Keynes's was far from the fundamental pessimism that troubled Malthus over 100 years earlier. Nonetheless, Keynes introduced the concepts of population size, population growth, food supply, and the instabilities of complex economies as the analytical framework for explaining what nations had to or could do.

Keynes was by no means a solitary scholar in the field of population studies. Throughout the 1920s and 1930s, a group of Americans, British, Indians, and others churned out a stream of articles and books on a wide variety of issues in population size and growth rates. Their interests included adequacy of food supplies,[8] projected losses of population in Europe and North America,[9] overpopulation in India,[10] the technical issues involved in making population size estimates,[11] and the eugenic implications of changing fertility rates.[12] Despite a broad range of viewpoints and opinions about population, this outpouring of studies had only one tangible effect on the public policy agenda: a number of European countries adopted measures to promote births because of their fears of the consequences of a lowered population size.[13] Other than this one topic, however, study of population was primarily a theoretical exercise confined to academics and a variety of eugenic reformers.

One other striking feature of the population studies in the 1920s and 1930s was that at least two strands of neo-Malthusian thought developed. Where Keynes and others emphasized population issues in terms of political economy, the second strand of neo-Malthusianism drew on images of naturally limited ecological systems. In contrast to Keynes, this other thread of "ecological" neo-Malthusian thought emphasized limits in the natural world and tended more to a catastrophic vision of population exceeding food supply and a collapse of civilization or war. An American, Edward Murray East (1879–1938), was an early spokesman for this viewpoint. He also was pivotal in bringing this strand of ecological neo-Malthusian analysis into the community of applied plant breeders.

East began his career in 1900 as an assistant chemist at the Illinois Agricultural Experiment Station, where he worked on the chemistry of maize (corn) seeds under Cyril G. Hopkins. During five years at Illinois, East became progressively more interested in plant breeding, and in 1905 he became a plant breeder for the Connecticut Agricultural Experiment Station in New Haven. There East, along with George H. Shull of Cold Spring Harbor and Donald F. Jones of the Connecticut Agricultural Experiment Station, found a way to produce hybrid maize seed, a major early invention in the development of high-yielding cereal production.[14] East went on to become professor of plant morphology at Harvard in 1909.[15]

Mankind at the Crossroads (1923) was East's effort to see the question of food supply broadly in terms of biology, economics, and politics. In this book East moved from being simply an astute student of Mendel and applied breeding to being a ecological philosopher attempting to explain political economy. He believed that the science of genetics allowed an understanding of the human condition, which was at a crisis point. Malthus's day of reckoning was surely to arrive, East believed, because little hope attended the possibilities for indefinite further increases in food production.[16] Overpopulation was likely to cause exhaustion of resources, hunger, and misery.[17]

Population theories and research in the early twentieth century, such as East's, were closely linked to studies in human genetics and eugenics.[18] As Kenneth Ludmerer and Daniel Kevles have argued, however, a deep-seated racism and social prejudice permeated eugenic studies.[19] In a similar vein, Malthus's social prejudice against the poor left its legacy in population studies. Ultimately, demographers pushed population studies into a more mathematical science without these overt prejudices.

East was involved as an adviser in this effort, as a founding trustee of the Population Reference Bureau (PRB), one of several institutions created in the 1920s and 1930s. Others included the Scripps Foundation for Research in Population Problems (1922), the Milbank Memorial Fund (1928), the Population Association (1932), and the Office of Population Research at Princeton University (1936). Although each of these institutes was founded on the presumption that population was a "problem," each was a locus for reform of population studies.

Among the first of the new scientific demographers was Warren S. Thompson, director of the Scripps Foundation at Miami University in Oxford, Ohio. Thompson received his doctorate in sociology from Columbia University in 1915 and published his dissertation as *Population: A Study in Malthusianism* (1915).[20] One of Thompson's major technical interests was differential fertility rates among different groups of people and the long-term implications of changing fertility rates for future population sizes. E. W. Scripps, the newspaper magnate, endowed the research unit in 1922.[21] Until his retirement in 1953, Thompson, his associate Pascal K. Whelpton (1893–1964), and a number of staff assistants put out a steady stream of reports that established the Scripps Foundation as a major center for demographic research.[22]

Thompson, however, had interests beyond the technical dimensions of differential and changing fertility rates. Like Keynes, Thompson argued that population growth trends created a public policy issue that needed to be discussed as a strategic political problem. In 1929 he published *Danger Spots in World Population*, which argued that increasing populations with limited resources were a key variable in causing war. He particularly believed that the western Pacific, the Indian Ocean area, and central Europe–Italy were the places of near explosion.[23]

He continued this line of thought in his 1946 book, *Population and Peace in the Pacific,*[24] which was an extensive revision of the 1929 book. He argued that postwar possibilities for peace in eastern Asia depended on the United States' recognizing the pressure put on the natural resource base by the large populations of China and Japan. Thompson was thus a crucial pioneer in linking overpopulation, resource exhaustion, and threats to peace, three of the major components of the population–national security theory. Perhaps his 1929 prophesies, which arguably came to pass in the Second World War, gave his 1946 study on eastern Asia high credibility.

Frank Notestein (1902–1983) was a second pioneer whose work contributed to the crystallization of PNST. Notestein received his doctorate in social statistics from Cornell University in 1927.[25] He, too, was interested in differential fertility rates, and also was an early student of the use of contraception to promote birth control.[26] He joined the Milbank Memorial Fund in 1928 and moved to take charge of Princeton's Office of Population Research in 1936. During the war years, Notestein took on the task of estimating demographic trends for Europe and the Soviet Union under the sponsorship of the League of Nations. After hostilities ceased, he took a leave from Princeton to head the United Nations Division of Population. Notestein's interests thus included both theoretical and applied aspects of the new demographic science.

Through the work of Thompson, Notestein, and others, demography emerged during the 1930s and 1940s as a respected field of academic study in the United States. More important, by 1945 American demography was becoming integrated into strategic thinking. Much of the support and coordination of demographic work occurred through the Rockefeller Foundation because of that organization's prime role in funding population studies before, during, and after the war. Both Thompson and Notestein, for example, drew significant amounts of their support from the foundation.[27] As described in the final section of this chapter, the Rockefeller Foundation was at the center of the network that produced PNST.

Famine: Searing the Consciousness

When food is plentiful and its supply seems secure, then food and agriculture are interesting subjects of conversation only to farmers, a few intellectuals, and policy makers. When the food supply is insecure, suddenly just about everyone finds food and agriculture an important topic. When food is unavailable, no other topic of conversation exists. Thus is the fickleness of people's interest in food. Famine can sear the human consciousness, but generally for no longer than the generation that has experienced it.

During the twentieth century, famine became a less common event than it had been during most of human history. The industrial revolution created conditions for food abundance, but for most of the period after 1900 abundance was confined to the industrialized countries and a few others. Other areas, like India and China, continued to have famines caused primarily by social conditions until after 1950, but their ferocity tended to diminish compared with those of earlier periods.

The major exception to food abundance in the industrialized countries was wartime. Combat disrupted production, trade, and storage. Despite the infrequency of famine and its clear links to the "abnormal" condition of war, however, when famine or food shortages occurred they retained their ancient power to capture people's attention. Thus a few episodes of famine or its threat were crucial shapers of a climate supportive of increased food production. Plant breeders and their otherwise prosaic work on wheat thereby derived increased attention and support from both the public and government agencies.

Four episodes were critical in bringing attention to plant breedes. First, the United Kingdom suffered vulnerability from lack of domestic food production during the Second World War, just as it had during the First World War. Second, India suffered the Bengal famine in 1943, an event that influenced policy for years, both before and after independence from Britain. Third, Mexico's close relationship with the United States created conditions of severe shortages of food and the threat of famine in 1943. Finally, a series of severe food shortages in the immediate aftermath of the Second World War affected many countries. This chapter describes the latter two cases and shows how they affected strategic policy in the United States during the 1940s.

Mexican Food Shortage, 1943

In 1943, a shortage of food in Mexico involved the highest decision makers in both the United States (President Franklin Roosevelt) and Mexico (President Manuel Ávila Camacho). Mexican maize production declined at the very time that the Mexican government launched a cooperative effort with the Rockefeller Foundation to improve agricultural productivity (see chapter 5). The lessons from this shortage emphasized the importance of agricultural science as a component of national security planning.

Reports of maize shortages in the Yucatán came to the U.S. State Department in June 1943, but a formal request for a loan of 15,000 to 20,000 tons of maize did not arrive from the Mexican embassy until 22 July. A request for an additional 50,000 tons came from George Messersmith, the American ambassador to Mexico, on 4 September.[28] Messersmith emphasized that the request was the result of his meeting with Ávila Camacho, Minister of Agriculture Marte Gomez, and the minister of foreign affairs. Involvement of these top leaders suggests that the request was extremely important to the Mexicans. Messersmith agreed with their assessments and noted that cooperation between Mexico and the United States was now at an unprecedented level compared with several years earlier. Messersmith believed the crisis was real and urged the State Department to take the requests seriously.

It is important to note here that the food crisis was in part the result of an effort by the Mexicans to mesh their agricultural production with the wartime needs of the American economy. As early as 1940, then Secretary of Agriculture Henry Wallace had indicated that the United States would be interested in purchasing more "complementary" crops from Mexico, for example, rubber.[29] Complementary crops were those that would not compete with those of U.S. producers. USDA sent two scientists from the Bureau of Plant Industry to Mexico during the summer of 1941 to survey the potential for production of complementary crops,[30] but little came of this venture until 1942.[31] In July of that year, an Inter-American Conference on Agriculture was held in Mexico City, and subsequently the American embassy in Mexico worked with the Mexican Department of Agriculture to create a national agricultural production plan. The Mexicans would reduce cotton production and increase production of crops like oil seeds for export to the United States. In return, the United States would increase exports of cotton to Mexico.[32] Although the Mexicans may not have reduced their cotton plantings, the evidence is quite clear that they considerably increased plantings of oil seeds like sesame, peanuts, and linseed. The supplies of these three industrial crops increased dramatically, at the expense of maize, which was a basic food grain.[33]

Apparently, Ambassador Messersmith received little satisfaction from his dispatch of 4 September, so he sent another long letter on 23 September directly to Secretary of State Cordell Hull and a separate letter directly to President Roosevelt. His plea was passionate: "The situation with respect to corn here is serious, and when I say serious, I mean very serious."[34] For his efforts, Roosevelt on 13 October authorized the release of 60,000 bushels (about 1,800 tons) of maize to Mexico, which was increased by State Department and USDA officials to 5,000 tons on 15 October. Messersmith emphasized on 16 October that 5,000 tons was not enough.[35]

At a cabinet meeting on 15 October, Messersmith's urgency was discussed in some detail. Roosevelt personally was concerned and authorized the shipment of additional maize. By 9 November Messersmith knew that a total of over 25,000 tons could be sent to Mexico, which was close to a new estimate of the minimum needed to avoid a serious famine.[36] Food supplies continued to be tight in Mexico during 1944, but a recurrence of the serious shortages of 1943 were prevented.[37]

Mexico's food shortage in 1943 was serious enough to attract the attention of decision makers at the highest levels. From U.S. State Department records, it is clear that American strategic planners in the midst of war saw food supplies as crucial to the stability of Mexico. From the Mexican point of view, the shortages were destructive of the country's goals to create a modern industrial state and to ensure the stability of government authority. In a sense, one of the messages was that increase of industrial and export crops, however valuable that might be for economic development, could not be sustained in the absence of a secure supply of basic food crops. The entire episode solidified a feeling that the prospects of increased yields from plant-breeding research must be a good thing.

Postwar Food Shortage

Feeding the densely populated British Isles with imported food was one of the major problems solved by the Allied governments during World War II. Despite the effectiveness of the War Agricultural Executive Committees, North American food supplies, from the United States and Canada, were crucial. The Canadians emphasized wheat, while the Americans led in supplies of eggs, milk, meat, and fats and oils.

After 1944, American agricultural planners began to prepare for the end of the war and the likely drop in demand for American export crops. Thus as early as 1944, wheat supplies in the United States were being converted into animal feed stocks for producing eggs and meat. Pork was unrationed in September 1944, even though some planners believed that grains should be stockpiled in case of later need.[38] As a result, the carryover stock of 1 July 1945 was only 280 million bushels of wheat, compared with 630 million in 1942. During 1945, pessimistic predictions about the hunger looming in Europe had no influence in the United States.[39]

Despite the optimism in American agricultural production planning, concern about food supplies grew during 1945 in Britain. By May 1945, the *Economist* was reporting deplorable situations in Europe, noting that although both the United States and Canada had food supplies adequate for intakes of over 3000 kilocalories per person per day, Britain had enough for only about 2900 and liberated countries in Europe for no more than 2000.[40] Particularly difficult problems faced defeated Germany because the bulk of German food production was in the Soviet-occupied zone but a majority of people were in the British-, French-, and American-occupied zones. Labor was needed to go east, and transport was needed to bring food supplies west. In addition, German railroads were not designed for these movements because they had not been built to match the military zones of occupation.[41]

Bread shortages and queues also appeared in London in June, the first since 1939.[42] Nevertheless, the full magnitude of the looming shortage still was not fully understood, even in Britain. Churchill's coalition government resigned on 23 May 1945 in favor of a caretaker government pending elections in July.[43] During July the Ministry of Agriculture noted problems of transport and harvest labor but believed that world wheat supplies were adequate to remove mandates on British grain production.[44]

The newly elected Labour government took office in late July, expecting to face a severe balance-of-payments deficit for several years. Nevertheless, the Labour government anticipated that it could restore and increase British exports over a period of several years, assuming the continuation of the lend-lease arrangements from the Americans and a continued war against Japan. Both the Japanese war and lend-lease, however, unexpectedly ended within a month of the Labour government's arrival in power. The new Labour government had no choice but to negotiate a loan with the Americans to continue the imports needed for British reconstruction and current living expenses, including food.[45]

Not only was the new government immediately faced with a financial crisis but the optimism about food supplies expressed by the caretaker government in July 1945 was an illusion. By December the Ministry of Food had prepared "World Wheat

Supplies," a discussion paper for the cabinet, which forecast a need in the first six months of 1946 for 19 million tons of wheat by the importing countries compared with estimated exportable supplies of about 12 million tons, and thus a shortfall of 7 million tons. Spurred by the stark realization of the consequences of such a shortfall, the Labour government went into a period of dramatic activity that lasted throughout 1946.

Despite the holiday, the cabinet was in session on 1 January 1946 to review the issues in "World Wheat Supplies." At that time it agreed that Prime Minister Clement Atlee would send a telegram to President Harry Truman seeking U.S. cooperation to alleviate the hunger it saw looming. In anticipation of more information from the British occupied zone in Germany, the cabinet also agreed to send immediate shipments of wheat to Germany and that the claims for scarce supplies for Germany had to rank equally with others.[46]

Two days later Field Marshal Bernard Montgomery reported to the cabinet that the existing ration in Germany was only 1500 kilocalories per person per day, which was a bare minimum to prevent starvation and disease. Montgomery went on to say that if wheat imports to the British zone ceased, the ration would drop to 900 kilocalories per person per day, about one-third the existing ration in Britain. The army was making arrangements for handling an epidemic; Montgomery believed that if it occurred it would probably spread to Britain.[47]

The situation in India was also dire and of enormous political concern to the Labour government. Near the end of January a telegram from the viceroy indicated that the wheat shortage was substantially worse than previously estimated.[48] On 30 January a discussion paper noted that the estimated deficit was at least 2 million tons. Political consequences were unsavory: "The political situation in India will in any event be critical this year; the recurrence at the same time of famine conditions would inevitably provoke widespread disorders all over India and would probably remove the last hope of obtaining an orderly solution of the Indian problem."[49]

Domestic concerns were as compelling as the international situations with Germany and India. No elected government can long survive a reputation that their policies were incompetent to assure secure food production, equitable food distribution, and a stable currency. Atlee's government had to balance a number of competing claims.

Consumers' hopes for increased quality and variety at reasonable prices were not necessarily congruent with farmers' wishes for freedom to return to the more lucrative production of milk, eggs, and meat rather than the wheat and potatoes demanded by the War Agricultural Executive Committees under the direction of the cabinet. In addition, farmers wanted stability of supplies of feedstuffs for their livestock enterprises. Decreased production of grain and potatoes in order to favor livestock production, of course, meant a need for higher imports of these commodities, which had to be paid for with foreign exchange that was in extremely short supply. The cabinet was forced into long and complicated efforts to balance these competing claims, and the minister of food, minister of agriculture and forestry, and chancellor of the exchequer were frequently at odds as each argued for controlling the problems faced by their respective ministries.[50]

By 31 January the cabinet felt it was dealing with a situation as serious as the battle of the Atlantic, which was the effort to secure steady imports to Britain from North America against the destruction from German submarines. However, it reached a preliminary set of decisions in late January and early February that shaped its management of the food situation through 1947.[51]

Britain would remain with its existing policy to promote farm production rather than institute an emergency premium to encourage the planting of spring wheat in 1946. Livestock growers, however, were to lose expected supplies of feedstuffs because flour extraction rates from wheat were to be increased. This resulted in a coarser bread for consumers and less remains for livestock feed. India was to receive more imported wheat, and food imports were to be decreased to save foreign exchange. In addition, domestic stocks of wheat were to be reduced to a level that meant the country would be totally dependent upon imported wheat during July and August, a risky proposition that put the British food supply at the mercy of foreign shipments of grain.[52]

Success of the British plan depended on appropriate agreements with the United States and Canada, the two largest wheat surplus countries in the world. Amid sentiments that the North Americans were not taking the British viewpoint seriously, the cabinet spent considerable effort during the remainder of February and March preparing a strong negotiating position to enlist the support of Washington and Ottawa.[53]

Ben Smith, minister of food, returned from Washington in late March with a sense that he had obtained a commitment from the American government to provide more assistance toward food security in Europe.[54] In a lengthy report to the House of Commons at the end of March, Smith outlined the steps the cabinet had taken to conserve the tight supplies of wheat. These included continued increases in the rate of extraction of flour from grain, continued control of agricultural land use through the War Agricultural Executive Committees, a reduction of livestock rations from one-fourth the prewar levels to one-sixth, and a decrease in cereal allocations to alcoholic beverage manufacture.[55]

Despite the initial cautious optimism at the end of March about forthcoming help from North America, during April the British cabinet began to think that neither the Americans nor the Canadians were really acting with the seriousness that the food crisis demanded. The British knew that the Americans wanted them to reduce their reserve stocks of wheat. In addition, they felt that Americans may have favored allocations of American wheat to the United Nations Relief and Rehabilitation Agency, which was providing relief assistance in central and eastern Europe outside of Germany.[56] For its part, the cabinet suspected that American farmers and speculators were holding back wheat supplies in anticipation of steep price increases. The cabinet also believed that, in terms of rational planning, the Americans had no real concept of how to control land use and food supply.[57]

Matters began to shift in mid-April with a change in position in the American government that the British had previously considered impossible. The breakthrough may have been a British proposal to ration bread, provided the United States also agreed to rationing.[58] Although full-scale rationing came to neither country, the Americans felt even the proposal of it was so formidable that their attention was fully captured. In a great flurry of activity, including a radio broadcast by President Truman on Friday, 19 April, and a two-hour cabinet meeting on Saturday, 20 April, the American government finally seemed to fully grasp the seriousness of the situation that had been perceived by the British for about four months. Despite the appearance of considered action, however, the president did not know what he was going to announce until less than one hour before the broadcast.[59] Despite the last-minute decisions, Mr. Truman gave an impassioned plea:

> America is faced with a solemn obligation. Long ago we promised to do our full part. Now we cannot ignore the cry of hungry children. Surely we will not turn our backs on the millions of human beings begging for just a crust of bread. The warm heart of America will respond to the greatest threat of mass starvation in the history of mankind. We would not be Americans if we did not wish to share our comparative plenty with suffering people. I am sure I speak for every American when I say the United States is determined to do everything in its power to relieve the famine of half the world.[60]

In the broadcast, Truman announced the appointment of former president Herbert Hoover to head a Famine Emergency Committee. The group, which traveled 35,000 miles and visited twenty-two countries, estimated the most severe crisis time to be from May to September 1946, when the 1945 harvest was about gone but before the bulk of the 1946 harvest was available. The group's study reached a conclusion that the gap between needs and likely supplies was 1.5 to 3.6 million tons, substantially lower than the 7 million tons originally projected by the British or the 11 million tons projected subsequently by the Combined Food Board.[61]

This amount of deficit could be met by altering the existing patterns of wheat use in the United States. Accordingly, the U.S. government ordered a series of steps, including higher extraction rates in milling flour (which produced a coarser flour for "white" bread), diversion of wheat away from livestock feed, brewers, and other food manufacturers, and near-total elimination of wheat from alcohol manufacture. Rationing was considered by the Famine Emergency Committee, but it recommended that eliciting voluntary compliance was likely to be more effective than the reestablishment of cumbersome rationing machinery.[62] An extensive advertising campaign by the Advertising Council took the message to the American public of the need to send wheat to Europe and Asia.[63]

The campaign for wheat conservation and export, sparked by the British and agreed to by the Americans, helped Europe and Asia limp through 1946 on supplies of wheat that were meager but sufficient to prevent widespread starvation and the outbreak of massive disease epidemics. Continuing bad weather and drought in Europe kept grain supplies limited in 1947 as well, but by that time the victorious Allied powers were more accustomed to providing the strategic supplies than they had been in 1945 and 1946. Nevertheless, historian Amy L. Bentley believes the United States essentially failed to fulfill its potential for alleviating hunger during this time, despite a willingness of American women to manage with less plentiful food supplies at home.[64] In contrast, the proposal by Secretary of State George Marshall in 1947 of a massive relief and rehabilitation program for Europe presaged a major change in American foreign policy toward the use of foreign assistance as a policy tool.[65]

Nevertheless, the food crisis of 1946 was so complex that the direct involvement of the highest levels of government in both the United States and Britain was needed to find a solution. Their sensitivity to food supplies was undoubtedly altered, and schemes to increase agricultural yields must surely have appeared to be worthwhile, even in America, with its long history of food surpluses.

Demography, Resources, Agriculture, and National Security

Shortages of food were clearly but briefly on the public policy agenda immediately after the end of World War II. Shortly thereafter, European production recovered and American concerns about food and agriculture reverted to the overwhelming importance of surplus production in relation to economic demand.[66] In terms of policies to address the situation, the U.S. government did not see overpopulation and food shortages as one of the problems it had to solve. A small group of experts in the postwar years kept up a steady barrage of analyses and projections, claiming that the American government needed to take additional steps. This group eventually succeeded in placing its claims on the policy agenda, and the fortunes of plant-breeding science were tightly linked to these issues. Understanding how population and food shortages were handled, therefore, is the fundamental clue to understanding the origins of high-yielding varieties of wheat.

In the victorious countries, the major analytical framework for building the public policy agenda, with its myriad issues demanding attention, was the national security framework. It demanded that suggested solutions to all problems, including population and famine issues, would be judged by their estimated contribution to ensuring national security. In the aftermath of the most extensive and destructive war in human history, nothing else mattered.

Individual countries, of course, had different concepts of what provided national security. For the United States, most decision makers sought security through neutralization, and sometimes obliteration, of German National Socialism and Japanese military circles, and confinement of the Soviet Union.[67] American leaders were also quite delighted with the prospects of being a large, highly industrialized country that was relatively unscathed by the war and thus poised for a remarkable expansion of its economic markets, the power of its currency, and the extension of its political and military influence.

High-level leaders were not intimately familiar with the technical details of population growth, food production and distribution problems, and questions of resource exhaustion. These subjects were the province of scientists, intellectuals, philanthropists, and reformers. Even though President Harry Truman and Prime Minister Clement Atlee had been drawn personally and intimately into the 1946 world food crisis, they were typical of most political leaders: they could respond to the information brought to them by experts, but they generally did not take initiative on these subjects by themselves.

Thus experts who were deeply and passionately concerned with population, agriculture, and conservation had a major problem: How could they capture the attention of leadership at the highest level and thus assure that their issue received proper attention? Success for the experts was achieving policy-agenda status for their issue *before* it became a disaster that even nonexperts could recognize.[68] Because the driving political concern in the postwar world was national security, the solution to the experts' problem lay in formulating issues in terms of national security. If a community of experts could persuade political leaders that they (the experts) had intellectual command of an issue with national security implications, then leaders would pay attention to their arguments.

Learning how to put questions of population growth, resource exhaustion, food production, and famine in terms of national security issues became the problem that had to be solved if these issues were to gain access to the postwar policy agenda. A number of experts in universities and philanthropic foundations learned how to do this in the decade after 1945, and they were thereby successful in capturing the attention of top political leaders in the United States. The formulation that worked was what was earlier called the population–national security theory (PNST).

Emergence of PNST came by incremental efforts to understand the significance of population growth, destruction of natural resources, world hunger, poverty, and the political turmoil that continued after the end of the war. Over a period of time, these increments linked all of these issues, pointed to their consequences in terms of danger to the United States, and offered reasons why research in plant breeding could help protect vital American interests. Because PNST posits overpopulation as the prime cause of problems, the developments within demographic science are the appropriate place to begin tracing the emergence of this theory.

Although much of early-twentieth-century demography was concerned with differential fertility rates among domestic groups within the United States and Europe, a number of demographers began moving away from this orientation. As noted earlier, Warren Thompson of Miami University was one of the first with his efforts to understand the strategic importance and environmental consequences of population growth.[69]

Dudley Kirk of the Office of Population Research at Princeton University made an important statement in 1943 that population growth in Asian countries, combined with the countries' increasing mastery of industrial technology, would result in shifts of political economic power away from Europe and America to Asia. Kirk continued that these developments would inevitably challenge the white supremacist notions underlying imperialism:

> We are not going to see again a world in which huge areas inhabited by non-European peoples may be casually regarded as the political playthings of Western European and American powers. The day is rapidly passing when a handful of Europeans, equipped with superior weapons and a complacent and somehow contagious faith in white supremacy, can expect indefinitely to dominate the half of the world that is occupied by the colored peoples.[70]

Kirk's brief analysis was important because he made two important linkages: population growth rates in the less industrialized world would cause political instability, and the United States could choose to respond, either by force of arms or by enlightened self-interest, to encourage the transfer of modern technology to the less industrialized peoples. Kirk, however, was not specific about what sorts of technology ought to be transferred or how it should be done.

Kirk's analysis was a criticism of capitalist countries that engaged in imperialism, and he clearly was criticizing the European imperialist powers (Britain, France, the Netherlands, and Belgium) and the United States. This critical dimension of his analysis never became part of PNST, but the notion that enlightened transfer of modern technology was the appropriate response to population growth became integral and included plant breeding as a prominent technology. Peter J. Donaldson, a population professional, later credited Kirk with outlining the rationale for American efforts to help control population growth in the third world, a movement that did not achieve government policy status until President John Kennedy's administration in the 1960s.[71]

Despite the solid academic reputations of scholars like Thompson and Kirk, neither was institutionally located in a place where he could advance the thinking on population beyond theories and recommendations into programmatic actions. Only governments and foundations had an ability to create action programs. Furthermore, Thompson at Miami and Kirk at Princeton were not connected physically or intellectually with schools of agriculture. Possibly for this reason, neither made much effort to link their thinking on population with considerations of agriculture and food supply.

Until the Kennedy administration, the U.S. government avoided any movement into population control because of its reluctance to endorse birth control methods and practices.[72] The Rockefeller Foundation, however, had a long interest in population studies and had been a major supporter of demographers like Thompson at Miami and Notestein at Princeton. In addition, foundation trustee John D. Rockefeller III became personally committed to the efforts to study and control population growth. He was active both as a trustee of the foundation and as an individual donor from the Rockefeller fortune that was not controlled by the foundation. These factors, combined with the foundation's work in the Mexican Agricultural Program, made it reasonable for the foundation to play an important, but circuitous, role in fostering the construction of PNST.

Studies on the human population were of interest to the foundation before World War II, but its early grants centered on eugenics, the genetics of mental deficiency, and studies of population redistributions.[73] In 1946 population concerns at the foundation took a radically different course. Raymond Fosdick, president of the foundation, became concerned about criticisms, probably by trustee John D. Rockefeller III, of public health programs that the Rockefeller Foundation had supported for more than forty years.[74]

Critics called public health protection measures unethical if they resulted in a massive increase in the population with no prospects for feeding the people. Fosdick agreed with the idea that public health science should pay attention to population size, but he was troubled by the corollary, which implied that the Rockefeller Foundation should not sponsor public health projects. He felt at a loss to rebut the arguments:

> I confess . . . [the criticisms sound] faintly unethical to me, and I am not convinced that it is the course to follow; but I don't know exactly how to answer the argument. I have always had a feeling that a country like India in a sense represented a vicious circle. You have an enormous population, with the result that food supplies are inadequate. Consequently you have the always-present problem of undernourishment and starvation. Out of this comes the impossibility of providing adequate educational sys-

tems or the basis of an industrial life, and because you have no organized industry and no education, you have an overcrowded population.[75]

Fosdick shared his thoughts with George K. Strode, director of the foundation's International Health Division (IHD), which guided the work in public health. Strode agreed that specialists in public health science should be concerned about the population increases caused by their work and that simply ending public health work was inappropriate. He specifically felt it was difficult to argue that India was worse off because of increased efforts in public health, despite its comparatively rapidly growing population. However, Strode had no suggestions for how the IHD could begin to attack the question of population growth.[76] Trustee John D. Rockefeller III raised the issue again a year later in December 1947.[77] Strode promised to bring the question of population and public health to the scientific directors of the IHD.[78]

In June 1948 Strode reported that Marshall C. Balfour, a physician and longtime staffer directing IHD programs abroad, had agreed to take on the task of drafting a plan for population research as it might interest the IHD. In addition, Strode announced that the foundation was sending a delegation headed by Frank Notestein and including Balfour to survey the population situation in Japan, China, Formosa, and possibly Java and the Philippines. This trip, combined with Balfour's report, was to provide the blueprint for future IHD work in population and public health.[79] Subsequently, Strode asked Marston Bates, an ecologist, to work with Balfour on the population question. Bates readily agreed, and the team gave both the medical and ecological perspectives.[80]

It was the combined results of the Balfour-Bates report, and its follow-up, and the Notestein trip report that drove further foundation thinking on population for several years. Strode, Balfour, and Bates remained for some years the focus of Rockefeller Foundation thinking on the question of overpopulation. The report from Balfour and Bates, in November 1949,[81] was partly an effort to articulate a conceptual framework for the field of "human ecology." The latter part of their proposal outlined a field study of human fertility and demographics in relation to economic and cultural factors, possibly to be conducted in Ceylon (now Sri Lanka). Field operations would be under the direction of the IHD and ultimately might have as their objective the control and manipulation of fertility rates and population size.[82]

Other staff and some trustees sharply criticized the report. Warren Weaver, head of the Division of Natural Sciences, which included the foundation's agricultural work in Mexico, objected to the possibility that the scientific directors of the IHD might send recommendations to the board of trustees without thorough discussion among all divisions of the foundation with clear interests in population and resources.[83] Trustee Henry A. Moe was highly critical because the report did not answer what was to be done, by whom, and at what cost.[84] President Chester I. Barnard defended the report as adding great clarity to the issues at hand. He told Moe that such negative reactions indicated the trustees did not know what was happening in the field of population. Barnard was especially defensive because the Balfour-Bates report was the result of a request by a special trustee committee on policy and program, and John D. Rockefeller III was the prime protagonist in that request.[85]

Less contention greeted the report of the Notestein trip to Japan, Taiwan, Korea, China, Indonesia, and the Philippines between September and December 1948.[86] Although the study clearly rejected a simplistic Malthusian interpretation of the Far East, the demographic team fit nicely into the neo-Malthusian framework: populations were high because death rates had dropped but fertility rates had not. Although Japan showed clear evidence of the potential for a demographic transition, the team came to the firm conclusion that the population was too large for food self-sufficiency and that the Japanese must therefore industrialize far beyond what they had done prior to the war if they were to attain a prosperity sufficient to bring about a completion of the demographic transition. Japan also needed to reduce fertility rates. To fail to connect demographic reality with political economic planning was, in the survey team's eyes, a threat to the goals of the occupation.[87]

The Rockefeller Foundation, as a result of its internal staff work to create a program on human ecology and its sponsorship of the Notestein investigation in eastern Asia, was in the forefront of sensitivity to a literature that saw finite limits to agricultural subsistence and high population growth rates that would create conditions of poverty. This argument was the heart of ecological neo-Malthusianism, but the demographic team report went one step farther: populations reaching the limits of food subsistence were politically dangerous. Their fertility rates needed to drop, and they must have the potential for industrialization in order to pass through a demographic transition to a new state of lowered death rates, lowered birth rates, and zero to low overall rates of population growth.

Subsequent discussions by the officers of the foundation never led to a consensus on the matter of a human ecology program, especially one that could lead to a foundation-operated program on population control.[88] Nevertheless, the debate on population firmly entrenched the subject in the thinking of foundation staff members.[89] Grants to other organizations continued,[90] but it was not possible to study population growth by itself to develop effective foundation programs.

Other books, written independently of the foundation, were important additions to the argument that population could grow so large as to threaten ecological resources. Particularly important were William Vogt's *Road to Survival* and Fairfield Osborn's *Our Plundered Planet*, both of which argued that environmental and political collapse of American civilization, and of cultures elsewhere, was likely if changes in behavior were not made. These books were the heart of the connection between resource exhaustion and hunger in the PNST scheme. Even though they were prepared outside the Rockefeller Foundation, their authors were in various ways connected socially to each other and to foundation efforts. In addition, Vogt's book was read by the foundation's president, Chester Barnard, as a challenge to the Foundation's efforts in agricultural research. The book thereby stimulated further efforts by foundation staff to articulate a coherent theory to justify programs such as the ones being conducted in Mexico.

Fairfield Osborn (1887–1969) had the simpler argument because it was primarily one for conservation of resources, especially soils. For Osborn the major components of the planet upon which civilization depended were water, soil, plants, and animals. Soil was critical because it was the complex material in which plants grew and upon which all animals depended. Osborn's attitude toward the soil reflected his sense that its health was crucial to the health of people. He was not a vitalist in the sense of attributing qualities to the soil that could not be explained by science, but Osborn firmly believed that soil was so complex biologically and chemically that only living processes could maintain it in good fertility. Artificial creation and restoration of soil by chemical technology alone were unlikely ever to be practical.[91]

Osborn was well placed personally to have an influence on a wide range of people. He was born into a well-to-do New York family, and Osborns and Rockefellers met socially. Fairfield was the son of Henry Fairfield Osborn, professor of paleontology at Columbia University and president of the Museum of Natural History. Groton and Princeton were his schools, and he entered the investment business in mining, manufacturing, and oil after service in the First World War. From an early age, however, he had a fascination with animals, which never left him. Osborn left the business world in 1935 to work with the New York Zoological Society, of which he became president in 1940. From this position, Osborn popularized biology and conservation for twenty years, hosting as many as 3 million visitors per year at the zoo. Regularly, Osborn tracked down the wealthy to support his operation. In 1947 he became founder and president of the Conservation Foundation.[92]

Osborn's book was temperate in tone, issued no stunning denunciations of any institutions, and emphasized soil conservation rather than the population explosion. Still, he argued that the number of people here on earth was surely to be a problem in years to come, especially if soil were not conserved. His vision was perhaps best characterized as "mildly" apocalyptic, or serious but not hysterical in any sense. In many ways he behaved in his book with the gentility that marked his place in the upper stratum of New York society, its businesses, and its clubs.

Probably of more influence was Vogt's *Road to Survival*, which also came out in 1948. In contrast to Osborn's *Our Plundered Planet*, however, Vogt's book was more complex in argument, considerably more polemical in tone, ready to tackle capitalism as a foolish and foolhardy institution (but just as critical of socialist governments), and probably of greater influence.

William Vogt (1902–68) grew up in and around New York City but not in the same moneyed circles as Osborn. Nevertheless, he, too, eventually became associated with interests supported by the likes of the Rockefellers and the Osborns. His undergraduate studies were in languages, but his interest in ornithology took him by 1930 into the world of professional science and ecology. He edited and wrote on bird conservation and served as field naturalist and lecturer for the National Association of Audubon Societies in the late 1930s.

Vogt went on to become a consulting ornithologist for a Peruvian guano company and then associate director of the division of science and education of the Office of the Coordinator of Inter-American Affairs, headed by Nelson Rockefeller, during World War II.[93] After the war, Vogt prepared his book, which became a bestseller, was translated into nine languages, and propelled him into a new career. He became the national director of the Planned Parenthood Federation of America in 1951. Vogt served ten years there before becoming secretary of the Conservation Foundation, headed by Osborn, from which he retired in 1967.[94]

Vogt's interest in ecology provided the major analytical framework of the book. He used the equation $C = B:E$, where C was carrying capacity, B was biotic potential, and E was environmental resistance. His equation, together with his argument that all human energy, especially for food, was derived from the sun's work on green plants, led him to conclude that the ability of the land to support people was limited. Vogt argued that when population became too high, then E increased and caused a fall in C. It was the fall in C, either through of human deaths or lowered levels of civilization, that formed Vogt's apocalyptic vision of the future. Only a collective appreciation of the limited resources and a conscious effort to curb population growth rates could save humankind from barbarism.[95]

Both Osborn and Vogt, but especially Vogt, were at the heart of the neo-Malthusian tradition. Population was a driving influence on the conditions of human culture, and too much population spelled collapse in both political economic and environmental systems. In keeping with the tradition of Keynes before them, these two American neo-Malthusians were essentially optimistic, in contrast to Malthus's nearly irreconcilable pessimism. Both Osborn and Vogt saw a role for science and reason to preserve humankind from the apocalypse, so long as reason could lead to a diminution of the population growth rate.

Vogt's book was read by the officers of the Rockefeller Foundation. Chester I. Barnard, the new president, was so concerned by Vogt's message that he asked foundation officers why the organization was sponsoring a program in Mexico to raise agricultural yields.[96] Barnard felt that Vogt's depiction of humans already exceeding the carrying capacity of the earth made it of dubious value to raise more food, presumably because only more population would result.

In order to understand the significance of Barnard's question, however, it is necessary to see it as a pivotal event in the evolution of the agricultural program of the Rockefeller Foundation. Foundation officers had made a major commitment to the development of experimental biology in the 1930s, and, from this effort, in 1941 explored a possible new venture in agricultural science in Mexico. No concerns were voiced about population when the Mexican program was started, but by 1950 the foundation's agricultural program was based on a neo-Malthusian vision of the future. More productive agriculture was one of the keystones that was to help humankind avoid collapse. The intellectual framework of neo-Malthusianism became connected to the programs of the foundation through a complex pathway.

Part of the links almost surely occurred through a network of personal associations and friendships. As noted earlier, Edward Murray East, one of the inventors of hybrid maize seed, believed too much population was a cause of difficulties in the human condition. East also was an influential teacher. One of his students, Herbert Kendall Hayes, joined him at the Connecticut Agricultural Experiment Station in 1908 and later became a professor of plant genetics and chief of the division of agronomy and plant genetics at the College of Agriculture, University of Minnesota.[97]

Hayes's interests in wheat and rust resistance brought him into close collaboration with two other colleagues, Elvin C. Stakman, a plant pathologist at the University of Minnesota, and Norman E. Borlaug, a graduate student of Stakman's who studied plant breeding with Hayes. In 1941 Stakman chaired the exploratory expedition to Mexico for the Rockefeller Foundation, and his committee's report recommended how the foundation should establish an agricultural experiment station in Mexico to work on wheat and maize. Borlaug was hired in 1945 by the foundation's

Mexican Agricultural Program and within a short time was in charge of wheat breeding in Mexico. Thus some of the key actors in foundation agricultural programs had at least a link of professional associations and friendships, dating back to East, that probably prepared to sympathize with the apocalyptic vision of the population explosion.

Overpopulation was not, however, much of an issue when the foundation established its initial program in Mexico. To be sure, perhaps foundation officers believed Mexico was overpopulated, but the internal planning documents prepared as part of the foundation's decision-making process did not frame the argument in these terms. Only after the publication of Vogt's book, and Barnard's question about the Mexican program, did foundation officers incorporate the issue of population into their thinking.

Warren Weaver, in collaboration with Stakman and others from the MAP, produced a report entitled "The World Food Problem." The subsequent impact of this report resulted from its linking of population growth, susceptibility to communist agitation, and the role of agricultural science in creating the conditions for political stability (meaning thwarting communist overtures on terms favorable to the United States). First, "The World Food Problem" was based on the concept that global tensions stemmed from "the conflict between population growth and unequally divided and inadequate resources." Overpopulation, in other words, was at the root of basic human problems.[98]

In addition, Weaver persuaded the committee to argue that agricultural science had an important political role to play in the emerging struggle between the United States and the Soviet Union:

> The problem of food has become one of the world's most acute and pressing problems; and directly or indirectly it is the cause of much of the world's present tension and unrest. . . . Agitators from Communist countries are making the most of the situation. The time is now ripe, in places possibly over-ripe, for sharing some of our technical knowledge with these people. Appropriate action now may help them to attain by evolution the improvements, including those in agriculture, which otherwise may have to come by revolution.[99]

Weaver and the other agriculturalists at the Rockefeller Foundation succeeded. The trustees received the report with enthusiasm in June 1951. One trustee, Karl T. Compton, president of the Massachusetts Institute of Technology, suggested that India was indeed a place for the Rockefeller Foundation to engage in agricultural science efforts: "I suspect that India may be fertile ground for activity in this field. The overpopulation, the low living standards and the threat of communism are of course well known."[100]

Weaver and his committee had succeeded in articulating what may have been one of the most complete expositions of PNST. Chapter 7 explores another part of the origins of PNST from the foreign policy of the Truman administration, but the Rockefeller Foundation may have been more articulate in spelling out the dimensions of the theory: overpopulation set up a dynamic interaction between resource exhaustion and hunger, which in turn led to instability, followed by dangers to American interests and threats to world peace. Plant-breeding science and other allied agricultural sciences were brought to the forefront in order to block what was seen to be an inevitable series of unfortunate occurrences. By raising and stabilizing yields, plant breeding could alleviate hunger and help stop the progression envisioned in PNST.

This specific report was instrumental in leading the Rockefeller Foundation to start its Indian Agricultural Program. In addition, efforts by the foundation in plant breeding were an important model for subsequent efforts sponsored by the even larger resources of the U.S. government. American plant-breeding science thus became part of the cold war's defense of capitalist political economies.

7

Wheat Breeding and the Exercise of American Power, 1940–1970

American power at the end of World War II was paramount. The usual image of this might, however, is formed more by the array of military and industrial components of American culture than by something as seemingly mundane as wheat breeding. Nuclear-tipped missiles, airplane and tank factories, engineering prowess, and motivated soldiers are more generally assumed to be the components of military strength, not scientists patiently crossing one strain of wheat with another and searching through the progeny for a better variety.

In the direct exercise of military power, of course, the weapon systems and soldiers are the most important elements of power. Armies, however, exist only on the foundation of food supplies that are adequate for both the military personnel and their civilian support force. American strategists in both world wars were acutely aware of the role of agriculture in the projection of military might, and they considerably amplified agriculture's importance in the aftermath of World War II. Specifically, through a variety of public and private initiatives, wheat breeding and other lines of agricultural science became an integral part of postwar American strategic planning.

Put somewhat differently, after 1945, wheat breeding by American scientists became more than just an exercise in the modernization of agriculture. Old motivations for seeking new varieties did not disappear, but new motivations arose to justify expenditures. In addition, American scientists came to do their work not only in the United States for American farmers but overseas for foreign governments. Wheat breeding acquired ideological dimensions more elaborate than simply "the promotion of progress." Instead, wheat breeding and other agricultural science became part of the "battle for freedom." In the process, many countries moved to new relationships with each other and with their own natural resource base.

How did wheat breeding get caught up with strategic and national security considerations? It is necessary to follow a somewhat convoluted trail to answer this question, and the story can begin with the status of the United States after the collapse of Germany and Japan in 1945. Of all the major participants in the war, only the United States emerged with its industrial infrastructure intact and a monopoly on nuclear weapons. Its armed forces, bloodied and battle-hardened but not exhausted, were in command of many strategic locations around the world. Perhaps most important, the American president, with bipartisan support, was psychologically predisposed to exercise power and influence abroad. More than anything else, the president and indeed a major segment of the American people were convinced that they had a mission in the world. Wheat breeders were as much a part of this movement as anyone else.

At the end of the war, the United States was preeminent in agricultural science, including plant breeding. Federal laboratories in the U.S. Department of Agriculture (USDA) were active, and each of the forty-eight states had a land-grant college with an agricultural experiment station and a cooperative extension service. Coordination within this complex was high. In addition, private industry was busy pouring out new machinery, chemicals, and seeds. No other country in the world had a comparable network of facilities for agricultural science and technology.

Plant Breeders Gain an International Vision, 1937–1942

Before World War II, American agricultural scientists oriented their work toward increasing the efficiency of production on domestic farms, which were becoming fewer in number and higher in capitalization. Frederick D. Richey, chief of the Bureau of Plant Industry, USDA, and president of the American Society of Agronomy in 1937, captured this sentiment in his presidential address. He felt the most important past achievements included finding varieties of wheat that were adapted to the American Midwest, which had been settled by Euro-Americans about seventy years earlier. Plant breeders, Richey noted, had also successfully developed varieties of sugar beets that were resistant to curly-top virus, sugarcane bred with wild varieties for viral disease resistance, maize more suited to mechanical harvest, and new crops such as soybeans.

Richey's talk was premised on the need for plant breeders to defend themselves against the sentiment that their research had been responsible for the surpluses and depressed prices of the Great Depression of the 1930s. He completely rejected that criticism and pointed instead to the ongoing role that plant breeders and other agronomists could play in the continuing economic development of the United States and the stability of its agricultural industries:

> In the past much of plant research has consisted in obtaining plants that were reasonably adapted to an environment which did not change rapidly. With a greater intensity of agriculture, [and] more rapid transportation, . . . the environment changes more

rapidly. It never was static, but it has become kaleidoscopic. If man is to win, he must be as versatile in his defense as nature is in her attack. This implies adequate ammunition that continued plant research alone can supply.[1]

Richey advocated a nature controlled by human technology, but his vision was confined to the domestic chores of the plant breeders. The chief agronomist of the nation did not see a role for his science that went beyond a pattern that had been set thirty years earlier, when the American Society of Agronomy had been formed in 1907. Increasing the efficiency of production and rationalizing domestic agriculture were the prime contributions.

Visions of international opportunities and obligations were not a part of plant breeding and other aspects of agronomy before the 1940s. American agronomists, for example, rallied to the cause of higher food production during the days of the First World War, but W. M. Jardine, president of the society in 1917, still focused his presidential address on the domestic responsibilities of agronomists.[2] Only the question of overpopulation and the world's food supply occasionally crept into the thoughts of agronomists in those days.[3]

In contrast, Richard Bradfield, head of the Department of Agronomy at Cornell University and president of the American Society of Agronomy in 1942, opened an entirely new vista. He spelled out a need for American technical expertise to help rebuild the war-shattered countries and to help countries that relied on primitive agriculture:

> I am also convinced that American agronomists have a very important international service to perform. . . . When the war is over, there will be millions to feed, large communities of people to be resettled, and farms to be supplied with seed, fertilizer, machinery, and livestock. A roster of qualified personnel . . . is already being prepared. . . . In addition to these emergency problems at the close of the war, there will be a need for American agronomists to help many countries with a primitive agriculture and, in many cases, a population larger than they can support at a satisfactory level. . . . American agronomists can be of great service to the governments and educational institutions of such countries. The movement was spreading before the outbreak of the war. It will be resumed at accelerated speed after the war.[4]

Bradfield, of course, was already an active "international agronomist" by the time he gave his presidential address. As noted in Chapter 5, he was a member of the Rockefeller Foundation's study committee that resulted in the Mexican Agricultural Program. Also, he became head of the department of agronomy at Cornell University in 1937, which brought him into contact with Cornell's work in Chinese agriculture before World War II.

Agricultural scientists like Bradfield were important promoters of an expanded vision of the work of professional scientists. For the most part we must speculate that their motivations were a complex mixture of humanitarian ideals, a desire for intellectual challenge and the prestige of global consulting, and a sense of American patriotism to spread what would eventually become a gospel of "international Americanism." What is important to note, however, is that advocacy of international responsibilities for agricultural scientists was necessary but not sufficient for such duties to become common.

From Vision to Policies, 1943–1949

Political and economic commitments by government and groups like the Rockefeller Foundation were also needed, and for these people the motivations were different from those of the scientists. Humanitarian ideals played a role in the commitment to international technical aid, but other factors were also present. Probably the most important was a growing belief that the security and prosperity of the United States was dependent upon favorable relationships with other countries. Traditional American isolationism completely crumbled during the course of World War II. Plant breeding and the other agricultural sciences, along with all other facets of American life, were strongly affected by the changed political climate.

Several earlier starts were made to transform ideas about international work in agriculture into organizations with missions, budgets, personnel, and programs. Two threads were most important before the outbreak of World War II. First were the efforts to form an international organization that would promote the economic well-being of agricultural producers. The International Institute of Agriculture dated from a conference in 1905 in Italy and prompted member governments to exchange statistical information on agriculture and to promote the "common interests of farmers and . . . the improvement of their condition." By 1934, seventy-four countries had joined, but the institute did not offer many solutions to the drastic financial crises of agriculture during the Great Depression of the 1930s.[5]

A second line of activity also promoted the international significance of agriculture before World War II: attempts to form international commodity agreements, either among exporting countries or between exporting and importing countries. Wheat was the most valuable and important commodity in world trade, and rapid expansion of production in the late 1920s prompted several efforts in the 1930s to rationalize the world markets. Conflict between the interests of domestic producers in importing countries (who wanted tariff protection) and the producers of exporting countries (who wanted free access to all markets), however, sank several international wheat agreements.[6]

The outbreak of war temporarily halted both types of efforts, but they were soon replaced by new arrangements that solidified agriculture and agricultural science as an important part of international relations, especially in the United States. Formation of the Food and Agriculture Organization (FAO) of the United Nations began with an international conference in Hot Springs, Virginia, in 1943. FAO's origins stemmed from British and Australian work at the League of Nations and from President Franklin Roosevelt's Four Freedoms, one of which was freedom from want. FAO was intended to link the provision of adequate nutrition with the economic well-being of agricultural producers. A strong provision of FAO's initiating resolution was that improvement of nutrition and agriculture in every country was a responsibility of all countries.[7]

FAO began its formal existence at a conference in Canada in 1945, where its constitution was signed by member governments. American participation in FAO, in contrast to the earlier U.S. refusal to participate in the League of Nations, symbolized the emerging internationalism of the U.S. government's stance toward agriculture. Further governmental commitment to active internationalism came in the form of the Marshall Plan in 1947. Both the FAO and the Marshall Plan, however, were but mere preludes to the resolution of purpose announced by President Harry S. Truman in 1949, the Point Four Program. Point Four was the critical break with past practices, and Truman's language was so resolute and sweeping that it merits a detailed presentation here.

Truman, the surprise victor in the 1948 elections, gave his inaugural speech on 20 January 1949 and made four points. Point Four was billed as a "bold new program for making the benefits of our scientific advances and industrial progress available for the improvement and growth of underdeveloped areas."[8] In understanding the Point Four proposal, however, it is important to remember its context and companion, Point Three. It is also important to understand more of the origins of the Point Four idea because the ultimate shape of the proposed program reflected its origins. First, however, we examine the arena created by Points Three and Four.

Point Three was the call for a collective defense arrangement in the North Atlantic area, a call that eventually resulted in the formation of the North Atlantic Treaty Organization (NATO). In essence, Truman's entire speech was a call to arms to resist what he felt was an alien, unacceptable philosophy that posed a threat to postwar recovery and peace. He used his address to

> proclaim to the world the essential principles of faith by which we live, and to declare our aims to all peoples. . . . From this faith we will not be moved. . . . In the pursuit of these aims, the United States and other like-minded nations find themselves directly opposed by a regime with contrary aims and a totally different concept of life. . . . That regime adheres to a false philosophy which . . . is communism.[9]

Where Point Three was the overt military component of his program, Point Four was the effort to spread American influence in the less industrialized countries, not by force of arms but by the transfer of technology and the political economic philosophy of capitalism:

> More than half the people of the world are living in conditions approaching misery. Their food is inadequate. . . . Their poverty is a handicap and a threat both to them and to more prosperous areas. . . . The United States is pre-eminent among the nations in the development of industrial and scientific techniques. . . . Our imponderable resources in technical knowledge are constantly growing and are inexhaustible. I believe that we should make available to peace-loving peoples the benefits of our store of technical knowledge in order to help them realize their aspirations for a better life. And, in cooperation with other nations, we should foster capital investment in areas needing development. . . . The old imperialism—exploitation for foreign profit—has no place in our plans. What we envisage is a program of development based on the concepts of democratic fair-dealing. . . . Greater production is the key to prosperity and peace. And the key to greater production is a wider and more vigorous application of modern scientific and technical knowledge. . . . To that end we will devote our strength, our resources, and our firmness of resolve. With God's help, the future of mankind will be assured in a world of justice, harmony and peace.[10]

Point Four, therefore, was ostensibly a humanitarian venture of enlightened self-interest set in the midst of a call for building military might to repel perceived aggression from the Soviet and communist insurgencies. Put another way, Point Four was the technocratic front of the cold war, which lasted from 1949 to 1991. Point Four became the American program of foreign aid, which, as developed in the following, had a tremendous influence on the creation and spread of high-yielding agriculture. The sincerity of the humanitarian ideals need not be doubted. Nevertheless, it is essential to understand that the origins of international assistance in agriculture were rooted in the initiation of the cold war.[11] Military treaties like NATO and technical assistance like Point Four were merely opposite sides of the same coin: determination that the values embedded in American individualism and capitalism would prevail.

Point Four indeed was a bold new venture. At the time Truman proposed it, neither he nor anyone else really had much of an idea about how to go about providing technical assistance to less industrialized countries. Should the aid be in agriculture since because less industrialized countries relied heavily on their agricultural economy? If so, what sorts of agricultural expertise were most useful? Alternatively, should aid assist in developing industry? Should American experts be sent abroad to work, or should foreign nationals be brought to this country to learn from American expertise? What does it mean to transfer a technology? Is it simply a matter of transmitting bodies of scientific theory and machines? Or does it also entail transfer of concepts of property, law, political theory, and attitudes toward nature?

At Point Four's debut in 1949, no answers to these and other questions existed. Nor were any programs, budgets, agencies, or personnel in place to turn President Truman's ideas into action. During the decade following the proposal, however, Americans learned how to deliver elaborate programs of technical assistance to what was later called the third world. Their "school" was a complex interactive cooperation, which was not entirely deliberate or even well coordinated, between the American government and the Rockefeller and Ford Foundations. Others have written broadly of this segment of American political life,[12] but our attention is focused on the example of assistance in food, agriculture, and the green revolution in Indian wheat production. Through complex pathways described here, wheat breeders in Britain also benefited from the programs initiated from Point Four and the work of the foundations.

From Policies to Programs, 1947–1950

Chapters 5 and 6 introduced three of the most important events that constituted the learning experience for government and foundation personnel: the Mexican Agricultural Program, the postwar food crisis of 1946, and the initiation of programs by the Ford and Rockefeller Foundations in India in the early 1950s. In addition, the launching of the European Recovery Program (Marshall Plan) in 1947 was a major departure from past practices in foreign policy for the United States. Although Europe did not need technical assistance, the Marshall Plan taught the United States a good deal about providing government-to-government assistance.[13] We will now take a more systematic look at the changes in America that were rooted in Truman's launching of the Point Four initiative. These were the experiences in which the United States learned about India and how to promote change in its agriculture. Chapter 8 will look at these changes from the Indian point of view and will provide an account of that government's desire to acquire technical assistance in agriculture from America.

Before India's independence in 1947, American involvement with that country was limited and generally channeled through the British. Americans may have disapproved of Britain's perception of "owning" India, but American foreign policy toward India was premised on the legitimacy of Britain's power. Thus most Americans, both in and out of government, tended to have few or no ideas about India, and involvement between citizens of the two countries was minimal. This lack of interaction started to change with the events of the Second World War, largely because of India's strategic geographic location as a base for repelling Japanese advances in Southeast Asia.[14]

More direct contact between the U.S. and Indian governments began to emerge after the war. Although Britain's Labour government was more involved than the United States with the situation in India during the food crisis in 1946, former president Herbert Hoover's Famine Emergency Committee visited New Delhi in April 1946 as part of its worldwide tour and noted that 230 million people were at risk if adequate supplies of wheat could not be found for India.[15]

Possibly because of attention from the Hoover committee, India received shipments of American wheat that were important in its management of a serious shortage.[16] An estimated total of 890,000 tons of food grains were imported by India in 1946–47. Imports of food grain had been typical for British India since about 1925, which reflected the steady downward trend in food grain production per capita during the last twenty years of British rule.[17]

Although American wheat was part of the imports in 1946, American involvement with the Indian food situation remained low-key in the months leading up to partition and independence in August 1947. For their part, Indian officials felt it necessary to make urgent pleas for American grains[18] and to defend their estimates of shortages against unnamed critics who believed India (and other food-importing nations) overestimated their needs.[19] For their part, members of the diplomatic mission of the United States in India sent a steady stream of dispatches to Washington that indicated India's need for food imports was genuine.[20]

Not only did India ask for grain in 1946 and 1947 but the government of India also initiated requests for technical assistance in agriculture. In these inquiries before Point Four, it was apparent that the U.S. government had no way to provide the help requested. Approaches to the U.S. Department of Agriculture by the State Department prompted the reply that career civil servants with the USDA would have to resign their positions with the department in order to take an assignment in India. Not surprisingly, no volunteers were found.[21]

In September 1946 the Indian Ministry of Agriculture indicated it knew of the Cornell University–Nanking University exchange in plant breeding (see Chapter 5) and asked whether the U.S. government could arrange a similar program for India. American diplomats in New Delhi noted that a bill to allow secondment of American civil servants to a foreign country had failed in Congress but wondered if the Department of State could "interest one of the philanthropic foundations, such as the Rockefeller Foundation, in financing the expenses of a state agricultural college professor . . . for a period of one year?" Washington replied that it would be better to wait for the bill to pass Congress rather than approach the private foundations.[22] In other words, "No, and we don't think there is any way to help you." Records of the interchange gave no clue as to why the State Department opposed approaching the Rockefeller Foundation on this matter.

In a similar set of requests for a soil conservation expert made in late 1947, the governments of the two countries negotiated for over a year, but arrangement proved difficult despite India's willingness to pay at least part of the needed salary money. When a suitable candidate finally emerged in early 1949, India was no longer interested.[23] Reasons for the failure were not clear, but these episodes strongly indicated that without a formal mechanism to provide technical assistance, it was not likely to be forthcoming.

Point Four, a formal initiative that stemmed from the highest political source in America, the president, finally made technical assistance possible. Within three months of Truman's 1949 speech, Clifford C. Taylor (agricultural attaché in the American embassy in New Delhi), Wolf I. Ladejinsky (USDA), and S. T. Raja (under secretary, Ministry of Agriculture of India) discussed a proposal by the central government to establish an agricultural extension service. Taylor initiated the suggestion that Point Four might be a source of help from the United States and that a request from Raja's minister to the American ambassador was the way to proceed. The cold war dimensions of the situation were also present: Raja, as reported by the embassy, believed India would go the way of China (i.e., go communist) within five years unless more rapid progress were made on food production.[24]

Over a year was to pass before the formal initiation of the Point Four Program, when Capus M. Waynick was appointed the first director of the Technical Cooperation Administration (TCA) in May 1950.[25] Nevertheless in May and June 1949, Taylor and other officers of the American embassy in New Delhi held a series of conversations with civil service and political officers of the Indian Ministry of Agriculture. A number of questions emerged that apparently did not yet have answers:

- What would be the likely conditions of work that American experts would find in India?[26]
- Would American experts be welcome on a political basis, or would the India take the suggestion of Britain's Lord Boyd-Orr, first director of the FAO, who advised India not to accept foreign technicians? When this issue was raised, the American ambassador to India noted that the technical specialists would "promote our own interests in India [and] would tend to offset the efforts of communists . . . and would not be harmful to British interests."[27]
- How should an effective technical extension service for India be designed?[28]
- For what purposes would expenditures be permitted, and under what conditions could funds be obtained?[29]

Answers to questions such as these were vital to the operation of any technical aid program, but answers could not be developed in the absence of the creation, staffing, and funding of the TCA. Although the Truman administration was able to take some initial, minor steps to bring Point Four to life before Congress agreed to a systematic program, it was not until September 1950, twenty months after his inauguration speech, that the president had complete authorization and funding to proceed.

Public Law 81–535, the Act for International Development, approved the philosophy of technical foreign assistance and was signed in June 1950. An appropriations bill passed in September provided $34.5 million for the first year's operation of Point Four. Truman's order launching the program also created the International Development Advisory Board, as provided for in Public Law 81–535, which was intended to propose a broad operating philosophy for Point Four.[30] It was though this board that the fundamental ideology of Point Four was solidified, which in turn created the arena for many other public and private activities, including plant breeding.

Refinements of the Point Four Program Ideology, 1950–1952

The International Development Advisory Board, chaired by Nelson Aldrich Rockefeller (1908–79), outlined a broad, visionary use of government and private grants plus private, for-profit capital investment as the foundation for ensuring American preeminence (domination?) in the years to come. Nelson Rockefeller's role in this, however, did not begin in 1950. He had already played an important role as a catalyst for an internationalist foreign policy in the previous ten years. Before examining the principles for operating Point Four articulated by the advisory board, it is necessary to trace the earlier work of Nelson Rockefeller.

Nelson Rockefeller made his debut in Washington politics in 1940 as a brash young man of thirty-two who convinced President Franklin Roosevelt that he (Rockefeller) was the right person to head a new government agency to coordinate U.S. government programs related toward Latin America. Rockefeller became interested in Latin America partly through one of his first ventures into the business world as a director of Creole Petroleum, the Venezuelan subsidiary of Standard Oil of New Jersey (the latter company was the source of his family's fortune). Concern about penetration of Axis power into Latin America was the other source of his motivation. Although the United States was not yet at war, the Roosevelt administration had clearly tipped its favor toward Britain and France, and prevention of German influence in Latin America was an emerging goal of American policy that the president felt was not yet adequately met.[31]

Rockefeller served as the coordinator of Inter-American Affairs (CIAA) until 1945, when he became assistant secretary of state for Latin America. He worked in the State Department until August of that year, when he resigned under pressure for the controversies he had started by defending the admission of Argentina to the United Nations.[32]

Rockefeller's work as CIAA was important primarily because he articulated the position that the United States, meaning both public and private sectors, had to be concerned about the welfare of Latin American people if American business was to prosper and American military security was to be maintained. More significantly, however, he served as a catalyst for what eventually became the Point Four program.

Some of the actions in which he was involved were simply personal. Soon after his arrival in Washington, for example, he was engaged as a regular tennis partner with Vice President Henry A. Wallace. Perhaps this is how Wallace became the crucial link between the American and Mexican governments and the Rockefeller Foundation in establishing the Mexican Agricultural Program (see chapter 5).[33]

After Rockefeller left government service in 1945, he continued promoting his dual policy—that both government and private business had to contribute to the well-being of foreign peoples in order to protect the overall interests of the United States. In 1946 he led a group of other New Yorkers in the formation of the American International Association for Economic and Social Development (AIA). AIA was a dual-purpose organization. One part of it was to invest private capital in Latin America in profit-making ventures. Those profits, or at least some of them, were then to be funneled into a foundation that would promote programs of technical assistance and social betterment in Latin America. Within a year, U.S. laws forced the divorce of the dual-purpose corporation into profit and not-for-profit segments.[34] Rockefeller's ideas of private capital and nonprofit activities interacting for the supposed benefit of common people, however, was a strong theme in AIA and in all of his projects.

Seeds of Point Four may well have been laid in Rockefeller's work as CIAA, the work of the Rockefeller Foundation in Mexico, and the ideas embodied in AIA. In fact, the formal idea for Point Four, as it was expressed in Truman's inaugural address, came from Ben Hardy, who had worked for the press office of CIAA and moved to the State Department after the end of the war. Hardy had continued contacts with Rockefeller and was much interested in the idealism he saw in AIA. As Truman's speech was being drafted, Hardy put in a proposal for technical assistance to underdeveloped countries. This, he felt, would aid prosperity and thwart communism.

Hardy's proposal did not excite the State Department hierarchy, and it was jettisoned before the draft went to the White House. Truman, however, apparently was dissatisfied with the draft of the speech and asked his aides to put something more interesting into it. Clark Clifford, an aide to Truman, obtained Hardy's material, and that section became Point Four.[35] It is overstating the case to credit Nelson Rockefeller with originating Point Four, but his various roles in and out of Washington catalyzed the formation of a new type of foreign policy, aimed at creating national security through a partnership between American government and capitalism.

Joe Alex Morris, Rockefeller's biographer, noted that Nelson Rockefeller was delighted with the appearance of Point Four in Truman's speech. Probably Rockefeller wanted to head up the government agency that would administer Point Four,[36] and his name was floated in the press as the possible appointee.[37] However, Henry Garland Bennett, president of the Oklahoma State College of Agricultural and Mechanical Arts, won the appointment to head the Technical Cooperation Administration.[38] Rockefeller was relegated to the job of chairing the International Development Advisory Board, which would advise, not implement, the Point Four effort.[39]

Rockefeller's report in March 1951[40] came in the midst of the Korean War, and its major theme was that American defense depended as much on the social and economic progress of the underdeveloped world as it did on military alliances and a powerful armed forces. Although the board made it clear that development efforts would be justified for humanitarian reasons alone, the overall tone of the document made it clear that the original linkage of military policy and foreign aid policy (Points

Three and Four in Truman's inauguration speech) was even more necessary in 1951 than it had been in 1949.

A second key philosophical point in the report was that neither private capital nor government program was sufficient to the challenge. Both were required, and the job of the Point Four administrator was, in the report's words, "to hitch all the horses into a single team."[41] Administratively, the advisory board advised the president that foreign aid was so important that a new agency should be created, the U.S. Overseas Economic Administration (USOEA). USOEA would report directly to the president and would implement foreign aid programs under the general policies constructed by the State and Defense Departments. Anything short of such a high-level, highly centralized position would be inadequate to the challenges, argued the board.[42]

One problem was identified as the "first major objective — to cooperate with these [underdeveloped] countries in a vigorous food-production drive which would break the back of famine and hunger."[43] Although the report urged a broad array of methods to accomplish this objective, it was clear that increasing the intensity of production, not just increasing the area of cultivation, was key: "For the immediate future a small improvement, such as the introduction of an improved variety of a single crop, could do more to increase food output than opening up of a new land area."[44] Here the board was urging a fundamental transformation of the methods of agriculture in the underdeveloped areas, not just an expanded application of existing technology.

Point Four was a proposal for a new type of partnership between private capital, private philanthropy, and government policy, all aimed at creating security, prosperity, and American influence abroad. Its promoters believed the good times would come both in the United States and abroad. Teamed with NATO and Point Three, the American government in 1949 thus took an original step toward creating a new type of foreign policy. Despite the genuine enthusiasm for Point Four by President Truman, it had a difficult time at its start-up. Over two years passed from Truman's first announcement of Point Four to the report of the Rockefeller advisory board in March 1951. Point Four's accomplishments were certainly meager at that time, and little but apparent bureaucratic resistance came in the eight months following the advisory board's report.

President Truman never accepted the board's recommendation to create a new superagency, the U.S. Overseas Economic Administration, to consolidate all foreign economic policy as an integral part of strategic planning. However, he incorporated the recommendation to increase funding in 1951–52 for Point Four projects to about $500 million per year.[45]

Administratively, Truman proposed merging the TCA, housed in the State Department and home of the Point Four program, into the larger Economic Cooperation Administration (ECA), an independent agency. Rockefeller endorsed the move as consistent with the advisory board's recommendation to consolidate foreign economic assistance,[46] but the move must have appeared as more of a bureaucratic shuffle than a grand reorganization of economic foreign policy. Further bickering centered on whether military and economic aid should be controlled by the same agency and whether Point Four's emphasis on agriculture in the TCA would be subverted by the ECA's emphasis on aid for industrial development.[47]

Further demoralization could not have been avoided after Rockefeller's resignation as chairman of the advisory board in November[48] and administrator Bennett's death in a plane crash in Iran in December.[49] Nevertheless, in Truman's state of the union address in 1952, a full three years after Point Four's debut, he still had a fiery faith in it and used the opportunity to eulogize his friend Bennett:

> This last year, we made available millions of bushels of wheat to relieve famine in India. But far more important . . . is the work Americans are doing in India to help the Indian farmers themselves raise more grain. With the help of our technicians, Indian farmers, using simple, inexpensive means, have been able since 1948 to double the crops in one area in India. . . . This is Point Four—our Point Four program at work. . . . We have recently lost a great public servant. . . . Dr. Henry Bennett and his associates died in line of duty on a Point Four mission. It's up to us to carry on the great work for which they gave their lives.[50]

Foundations and Agricultural Assistance, 1950–1956

Despite Truman's perseverance, many advocates of foreign economic assistance felt the government was not rising to the challenge. It was at this point that private foundations, particularly Rockefeller and Ford, came in with reinforcements. In a pattern that characterized the task of completing a foreign policy based partly on economic assistance, the apparent logjam on expanding foreign agricultural assistance to a large scale was broken by a tactical shift of employment: Paul G. Hoffman, administrator of ECA (the Marshall Plan) from 1948 to 1950, moved to become president of the Ford Foundation from 1951 to 1953.

Hoffman, who headed the Studebaker Corporation before moving to ECA, was a convinced cold warrior who brought an articulate vision of the necessity for all American institutions to collaborate against the perceived threat of aggression from communism. Shortly after assuming the helm at the Ford Foundation, he won an appropriation for $5 million from the trustees for use on foreign projects similar to those envisioned in Point Four. In early August 1951 he and four other Ford staff members set off on a trip to Europe, India, and Pakistan in order to start the process of Ford involvement in those areas. In a press conference upon departure, Hoffman emphasized the importance of India as "of great importance to the United States from the standpoint of maintaining peace in the world."[51]

Hoffman had requested his invitation to India through Prime Minister Nehru's sister.[52] At the time of his visit in August 1951, a minimal amount of movement toward a program of technical assistance from the United States to India had already occurred. Point Four had three USDA advisers in India,[53] one of whom was Horace C. Holmes, who was already in direct contact with and receiving support from Nehru.[54] Thus Hoffman and the Ford Foundation entered an India that had already begun technical cooperation in a small way.

Hoffman, however, wanted to move quickly. After five days in New Delhi, Nehru had issued an invitation to open a foundation office in order to assist India in ways that were mutually agreeable. Upon his return to the United States, Hoffman hired a young extension specialist in USDA, Douglas Ensminger, to head the New Delhi office. By December, a mere four months after Hoffman's visit, the foundation and

India had signed an agreement, and the trustees of the foundation had approved a grant of $2.225 million. Of this sum, $1.2 million was to help on a broad-based rural community development program, $94,000 went to the Allahabad Agricultural Institute for an extension service project, and $85,000 was meant to support a Gandhi Memorial Community Center.[55]

Ford Foundation action on agricultural development in India originated in a short time, but the Rockefeller Foundation was simultaneously having an internal debate about its possible involvement. Staff conversations since 1947 had danced around what might be done and whether the Rockefeller Foundation had sufficient funds to do anything worthwhile.[56] By 1951, as noted in Chapter 6, sentiment among Rockefeller Foundation staff was clearly oriented toward a cold war interpretation of their work. Until October 1951, however, the Rockefeller Foundation had not yet made a concerted commitment to agricultural science other than in Mexico and a small program in Colombia.[57]

Once a commitment to agriculture was made, the Rockefeller Foundation went through a deliberate process of considering its potential involvement in India. At about the time the Ford Foundation was making its first commitments to India, the Rockefeller Foundation committed itself to sending a study team to India—J. George Harrar (first director of the Mexican Agricultural Program and a plant pathologist), Paul C. Mangelsdorf (member of the study committee that recommended the MAP and a plant geneticist), and Warren Weaver (head of the Division of Natural Sciences and Agriculture and a mathematician).[58]

Harrar, Mangelsdorf, and Weaver's "Notes on Indian Agriculture"[59] solidified the case that the Rockefeller Foundation would commit funds to Indian agriculture. Their argument was a careful weaving together of the themes of overpopulation, hunger, political instability, and the threat of communism.[60] Harrar, Mangelsdorf, and Weaver believed that an infusion of modern, Western knowledge was capable of overcoming the massive problems faced by overpopulated India. Specifically, they recommended that the Rockefeller Foundation could usefully support (1) improvement of wheat and rice varieties, (2) reform of agricultural education to make Indian agricultural universities more like the land-grant universities of the United States, and (3) some participation in village improvement projects involving extension education. They envisioned both a participatory program involving Rockefeller scientists working with Indians and a grant program to enable Indian scientists to travel and do research.[61]

"Notes on Indian Agriculture" launched the Rockefeller Foundation's India Agricultural Program (IAP), but a series of meetings between Indian officials and foundation officers were needed to create a specific agenda.[62] Reaching agreement with India took until early 1956, nearly four years after Harrar, Mangelsdorf, and Weaver completed their original fact-finding mission. In the intervening time, a few small grants were made to various Indian institutions to support indigenous activity, but a major program involving a foundation-sponsored research laboratory took longer to materialize.

Part of the problem stemmed from differences over the specific objectives of the most important research. Harrar and Weaver wanted basic research, especially on rice and wheat. Moreover, they came to see the research needed on rice as a problem that applied to all of Asia, not just India. The Indians, led by Badri Nath Uppal and B. P. Pal, however, wanted a short-term project on maize, which would lead to field trials and the establishment of a hybrid maize seed industry. They may not have opposed thinking about rice in other Asian countries, but their immediate thoughts were on India's situation.[63]

Another factor in the negotiations between the Rockefeller Foundation and India was the entry of other major actors. The Ford Foundation's $1.2-million grant of 1951 was joined in early 1952 by an even larger agreement between the U.S. TCA and India. The United States agreed to support Ford's community program and also extensive programs to improve rural infrastructure. The U.S. contribution was to be $50 million, with India more than matching that with 410 million rupees (about $86 million).[64] Ford's community development project emerged from work of the Grow More Food campaign (see Chapter 8). Community development had the full endorsement of Prime Minister Nehru and reflected some of the egalitarian ideals of independent India's new government. Neither the Grow More Food campaign nor the community development effort was based on new scientific knowledge as a means to agricultural assistance.

A pilot community development project began in the Etawah District of Uttar Pradesh in 1948.[65] The Etawah project was well respected by the Rockefeller Foundation officers, but they were highly skeptical of the ability of the U.S. TCA and the Ford Foundation to expand the work in that one district over 600-fold and to increase the amount spent in each of India's districts 10-fold. Weaver felt that the project could crumble for one or more reasons: lack of trained personnel, technical problems that would soon demand answers, or insufficient genetic variety in the crops planted. In other words, Weaver was firmly convinced that new scientific research was essential to genuine change in Indian rural life. He was also critical of the caliber of personnel representing the TCA.[66]

It is possible that the Rockefeller Foundation's insistence that the most important task at hand was basic research led to the four-year delay in establishing an operational program in India. India, the U.S. TCA, and the Ford Foundation were more interested in using existing knowledge for community development. At the start of the Mexican Agricultural Program, the Rockefeller Foundation scientists believed that extension developments would not be helpful because no really useful knowledge existed to extend.[67] In some ways George Harrar, Warren Weaver, and the other foundation scientists had much the same attitude as they designed a program for India: they did not believe that the appropriate knowledge existed, so a scientific agriculture for India had to be created almost from the beginning.

Despite the lengthy negotiations needed to launch Rockefeller Foundation research in India, agreement was reached in late 1955, and the formal contract was signed in April 1956. India asked for help in developing three agricultural institutes, each with a cooperating researcher. In addition, the Indians sought assistance to improve hybrid maize and wanted the foundation to coordinate its research with the larger program financed by the TCA.[68]

The foundation granted assistance on two fronts. It agreed to help India develop the Indian Agricultural Research Institute in New Delhi into a modern facility and postgraduate educational institution granting master of science and doctoral degrees.

In addition, the foundation agreed to enter into a "cereals" improvement program, a terminology insisted upon by Harrar, probably to mask the foundation's sense of disappointment that maize was the focus of Indian interest when the Americans really wanted to work on rice and wheat.[69] The board of trustees approved a grant of $1.38 million in April, and the Rockefeller Foundation was firmly established with an operating agricultural program in India.[70]

America and India Fully Embraced through Agricultural Assistance

The agreement between the Rockefeller Foundation and India was the capstone of the American capacity to conduct foreign technical assistance in agriculture. To be sure, the agreements between India, the American government, and the foundations were not the only programs in which the Point Four idea had been realized. An extensive survey by the *New York Times* in January 1953 indicated that the allotment to Point Four in the two years 1951–53 totaled $276.6 million, $165.5 million of which was for Asia, including $98.2 million for India. Thus over one-third of the expenditures of the TCA for Point Four were in India, which was the largest single recipient and over double the next-largest recipient, Iran, at $43.7 million. By 1953 over 2,000 technical experts were working in thirty-five different countries.[71]

Expenditures of the U.S. government dwarfed those of the Ford and Rockefeller Foundations. Nevertheless, the programs of the foundations were to have critical, catalytic effects that were far more important than the size of their budgets suggested. The Rockefeller Foundation's promotion of scientific research and scientific education was a key to developing entirely new methods of agriculture, particularly in plant breeding, and these new methods were the basis of the green revolution. The Ford Foundation's strategic analysis of Indian agriculture and its instigation of new organizational structures for agricultural assistance were critical for launching the green revolution. United States government funds, in contrast, tended to be spent on the bulk infrastructure of high-yielding agriculture, for example, construction of tube wells and provision of training for many extension agents.

Although the Ford and Rockefeller Foundations were American organizations operating with American funds and American staff, they perceived themselves as different from the U.S. government. They prided themselves on their independence of action and worked hard to convince their Indian hosts that they were distant from, not adjuncts to, the American embassy in New Delhi. Officials of the Indian government accepted the differences and probably found them useful for their own purposes. The foundations could act more quickly than the staff at the embassy, whose every action was in theory subordinate to the larger foreign policy objectives of the secretary of state and the president.

Despite the separateness of the foundations and the embassy, however, there were no fundamental differences between the Americans who worked for the government and those who worked for the foundations. Everyone accepted the basic premises of the cold war announced by President Truman in 1949. They all saw their work aimed fundamentally at thwarting a perceived threat of communist subversion and keeping India from going the way of China. Overpopulation as a cause of hunger and political instability was not a disputed theory, nor did any disagreements center on the use of science and technology to raise productivity of Indian agriculture. Increased production was universally seen as the key to solving India's most severe problems. Agreement along these lines between the Americans working in India was of far more significance than any supposed differences between government and private organizations.

Officials in India agreed with many points made by the Americans. At the very highest levels of government in independent India, Prime Minister Jawaharlal Nehru had an enthusiasm for modern science and its necessity to build a modern, industrial state. Nehru made it perfectly clear that he supported and took a personal interest in the technical assistance spawned by the Point Four program. Nehru distinguished between the science needed to run modern industry and the science needed to run high-yielding agriculture; he was more of an enthusiast for the former and a reluctant, ambivalent supporter of the latter. Nehru's distress at the imperatives of agricultural science was deeply rooted in his perceptions of the Indian countryside and a Ghandian philosophy of frugal self-sufficiency as the path to dignity for the Indian peasant. These dimensions of the problem are explored in Chapter 8.

Concluding Remarks

Before leaving the story of events in the United States, it is important to note that the changes induced by Truman's Point Four program were paralleled by the development of American agricultural science for domestic purposes. Even though American agriculture in 1945 had some of the highest levels of labor productivity and highest yields per hectare in the world, its "development" was by no means finished. Disruptions to farm life by mechanization and other inventions before World War II had by no means run their course. Farmers in the postwar years were fully prepared to continue their quest for profits by adopting newer, bigger, and more labor-efficient machines. In addition, a series of new chemical inventions unleashed many new and cheaper fertilizers and pesticides after the war.

Farmers knew that extra nitrogen fertilizer would stimulate their crops to higher levels of production, but they were faced with an immediate dilemma. Grain crops, especially wheat, when given a heavy dressing of nitrogen, will produce a tall, luxuriant plant with a heavy head of grain at the top of the stem. Unless the stem is sturdy, the result of heavy fertilization is that the grain head lodges, or falls into the dirt. A farmer's potentially larger yields and larger profits thus end up in the mud.

Plant breeders rose to the challenge to find varieties that would be able to make use of the now cheaper fertilizer, a story presented in chapter 10. For the moment, suffice it to say that the combination of new fertilizers, new machinery, new pesticides, and other factors created a full and demanding agenda for wheat breeders in the United States in the two decades after the end of World War II. The most technologically astute and agressive of the farm businessmen absorbed, indeed avidly sought, the new developments of agricultural science in order to further their enterprises.

A second set of factors affecting domestic plant breeding after World War II was the renewed atmosphere of fear that prices would collapse as the large war-generated export markets evaporated. Thus most agricultural policy discussions in the latter days of the war and the immediate postwar period addressed the subject of price supports: what crops, what level, and accompanied by what restrictions?[72] Advocates of research also revived an effort that began in the 1920s under the banner of "chemurgy." If chemical research could find new industrial uses for agricultural goods, then the surplus problem would diminish, perhaps vanish.[73] Passage of the Research and Marketing Act of 1946 (Public Law 79–733) was a direct outcome of the chemurgy campaign and sought to promote market utilization research.[74] Secretary of Agriculture Charles F. Brannan addressed the Fifteenth Annual Conference of Chemurgy with a strong pitch that agricultural research helped both domestic growers and the Point Four Program.[75]

As a result of technical changes, particularly the widespread use of fertilizers, and government price support policies, American agricultural production continued a steady rise in the years after 1945. Wheat was one of the major leaders of increased production levels, and by the 1950s this grain and many others were in chronic surplus. Price support programs kept farmers insulated from the downward pressures on prices that would have resulted; consequently, the U.S. government became the owner of massive amounts of grain and other commodities. Arrival of the new administration of President Dwight D. Eisenhower in 1953 brought with it promises of doing something to lower the financial burden of the price support payments and expenses of storing such large amounts of goods. Thus was born Public Law 480, the Agricultural Trade Development and Assistance Act of 1954, a device for unloading surplus grains in markets that could not otherwise pay for them. Shipment of P.L. 480 wheat stocks to India had far-reaching consequences for India and Indian agriculture. More important, in 1959 P.L. 480 assistance grew into the Food for Peace program. From the 1950s through the 1970s, American crop surpluses were consciously used in strategic foreign policy decision making.[76]

Thus American plant breeders became internationalized after 1945. Their science was taken into the cold war efforts by incorporation into foreign aid programs. Domestically, the fruits of plant breeding, crop surpluses, became a bargaining chip for U.S. strategic negotiations. Liberty Hyde Baily had urged the farmer to conquer his farm, but his science of plant breeding went on to play a role in grander adventures: the exercise of American power in all parts of the globe.

8

Wheat Breeding and the Consolidation of Indian Autonomy, 1940–1970

Between 1940 and 1970, India vastly increased its wheat-breeding efforts, which in turn became part of the country's capacity for high-yielding agriculture. These changes in agricultural science and production practices were by no means incidental to Indian politics. On the contrary, the embrace of science and high-yielding practices was the result of intense debate and experimentation with a number of policies. On the surface, the arguments were about how best to increase food production, but the debates had a far deeper meaning. They touched the very heart and soul of what independent India was to be, and on their outcome rested India's ability to be an autonomous nation.

The debate sharpened its focus at independence in 1947, when a single question became paramount: Should India aspire to be an industrialized, urban society? Or should India create a more prosperous but agrarian society based on hundreds of thousands of largely self-sufficient rural villages? Finding an acceptable answer to this question posed far more difficult problems for Indian politics than the central question that had existed for nearly a century: how to get the British to leave.

During the three decades from 1940 to 1970, India at various times (1) expanded the amount of land devoted to food production; (2) managed food shortages through a system of price control and state procurement of grain; (3) launched community development programs in an attempt to raise the well-being of villages in an egalitarian way; (4) obtained grants and low-cost sales of surplus wheat from the United States and elsewhere in order to cover its food production deficit; and (5) embraced the promise of science to increase yields, reluctantly at first, then enthusiastically. In the end, it was the science, particularly the key contributions of plant breeding, that tipped the balance toward higher levels of wheat production. Repercussions of the ideological clashes involved in making this choice continue within India even today.

This chapter covers the events in three phases. First, at independence India was already in a food crisis caused by the collapse of food production in the last years of the British raj and the effects of partition. Second, India's first decade of independence (1947–57) was a transition to a new order in which building social equality was the guiding ideology, even though the achievement of egalitarian prosperity was frustrated. Third, India's second decade of independence (1957–67) was marked by a continuing food crisis, which was finally resolved by an embrace of the fruits of plant breeding and an abandonment of the quest for egalitarian development.

Last Legacy of the British Raj: Decline of Food Production

Production of food in India declined after 1920, despite the fact that India was not a major industrial manufacturer and had limited means with which to import food. Between 1891 and 1946, the production of rice, India's biggest crop, steadily dropped, especially after 1921. For the entire period, the average annual rate of change was –0.09 percent per year. Wheat, gram (a pulse), bajra (a millet), and maize showed average annual increases; wheat was the highest at +0.84 percent per year. Aggregate statistics indicate that food grain availability per capita declined: population increased at about +0.67 percent per year from 1891 to 1946, while aggregate food grain production increased at only +0.11 percent. Imports helped make up the deficits, but even they were not sufficient to keep consumption levels steady after 1921. Food grain availability per capita, including imported grains, increased between 1891 and 1916, but from 1921 to 1946, the net rate of change was –0.15 percent per year.[1]

In the years before independence, rice, jowar (sorghum), gram, barley, and ragi (a millet) all had negative growth rates. Wheat, bajra, and maize showed some increases, but the aggregate, average annual change in food grain production in the last twenty-five years of British rule was only +0.03%, far less than the average annual population increase of 1.12%.[2] India changed from being a net food exporter to a net food importer after 1919.[3] Against this rather dismal picture for food, however, the Indian economy as a whole grew by about 1 percent per year, about the same amount as the population increase, but no per capita increase in total economic wealth occurred.[4]

R. N. Chopra, who rose to chair the Food Corporation of India in the 1970s, argues that after 1921 population increased more rapidly, the rate of increase of the area devoted to cultivation shrank, and commercial crops began to replace food crops, all leading to food shortages in India.[5] Based on figures collected by George Blyn, the aggregate supplies of rice and wheat per person in British India in 1896 were about 298 pounds per year. In 1941 the figure was about 228 pounds per year, or 70 fewer pounds.[6] In terms of food energy, 228 pounds per year can deliver about 1000 kilocalories of energy per day.[7]

The number of kilocalories needed per day to avoid starvation varies with age, size, physiological status (such as pregnancy and nursing), and activity. For adults, symptoms of food deprivation will start to appear if diets drop to less than 1900 kilo-calories per day.[8] Thus the two major staples of the Indian diet could not by them-selves provide adequate nourishment to the Indian population. Of course, other grains, potatoes, fruits, nuts, fish, meat, and dairy products were also available. In addition, some food supplies may have been unrecorded in the Government of India's statistical reports. Nevertheless, the decline of statistically recorded supplies suggests that at least some of India's people suffered a steady deterioration of their diet in the last three decades of British rule.

The nadir of this declining trend, the Great Bengal Famine, came in 1943. Responsibility for the famine was laid squarely on the Governments of India and Bengal by the Famine Inquiry Commission's final report in 1945.[9] From the period of December 1942 into 1944, at least 1.5 million and possibly up to 3 million or more people perished in the province of British India known as Bengal. This region included what is now the state of West Bengal in India and the new nation of Bangladesh. Starvation and the outbreak of disease induced by malnutrition were the proximate causes of death.

Different interpretations of the causes of the famine can be found. Was the famine the result of lack of physical supplies of food grains, especially rice? Alternatively, was the famine produced by a series of policy decisions in which the Governments of Bengal and India mismanaged the procurement and delivery of supplies, which in fact were adequate for the population? Yet another interpretation, from economist Amartya Sen, argued that the famine was not caused by a lack of physical supplies of grain but by a drop in the purchasing power of the wages of laborers dependent upon purchased grain.[10]

According to Henry Knight, a member of the Indian Civil Service who served as adviser to the governor of Bombay from 1939 to 1945, the basic food problem in India was feeding those who did not raise their own food: city dwellers, shopkeepers, artisans, clerks, pensioners, and landless laborers. Only about 30 percent of the food grain produced in India came onto the market, and these supplies plus about 1 million tons of imported food grains (about 1.4 percent of total supplies) were the life-blood of the dependent population.[11]

In 1954 Knight argued that in 1943 the Government of India had inadequate knowledge and records of agricultural production plus an assumption that India could feed itself, save for the few imports. This situation, combined with the division of powers begun in 1919, and especially after 1935, between the London-dominated Government of India and Indian-elected provincial governments, led to disaster.[12] The British had hoped that granting of some autonomy to provincial governments would stem the tide of nationalist aspirations. In the case of food management, tension between provincial and central governments may have rendered India bereft of competent administration.

Other factors of possible importance included the occupation of Burma by Japan in April 1942, which eliminated Burma's regular exports of rice to Bengal. In addition, a cyclone in October followed by outbreaks of fungal diseases destroyed per-haps one-third of the grain that was supposed to be harvested in November and December. Confusion caused by the imposition and removal of controlled market-ing during 1943 sent mixed signals to farmers, dealers, and consumers. In addition, the government of Bengal implemented the plan advocated in the Bengal Chamber of Commerce Food Stuffs Scheme, which placed Calcutta's industries and workforce as priority customers so as not to divert attention from wartime production goals; Calcutta was thus a haven of food security, in contrast to the scarcity areas in the countryside.[13]

Regardless of the causes, it was surely no coincidence that Jawaharlal Nehru, the future first prime minister of India, used the famine in Bengal to initiate his story, *The Discovery of India*, written in Ahmadnagar Fort Prison between April and September 1944, just as the famine was coming to its awful conclusion. Nehru's pain upon learning of the events in Bengal went straight to the heart of why Indians wanted the British gone from India:

> Famine came, ghastly, staggering, horrible beyond words. In Malabar, in Bijapur, in Orissa, and, above all, in the rich and fertile province of Bengal, men and women and little children died in their thousands daily for lack of food. They dropped down dead before the palaces of Calcutta, their corpses lay in the mudhuts of Bengal's innumerable villages and covered the roads and fields of its rural areas. Men were dying all over the world and killing each other in battle; usually a quick death, often a brave death, death for a cause. . . . But here death had no purpose, no logic, no necessity; it was the result of man's incompetence and callousness, man-made, a slow creeping thing of horror with nothing to redeem it, life merging and fading into death, with death looking out of the shrunken eyes and withered frame while life still lingered for a while.[14]

Not surprisingly, food policy in independent India had its roots in the conditions that created the Great Bengal Famine. With the outbreak of war in Europe in 1939, the Government of India moved to monitor and, by December 1941, control prices and movement of grain, particularly wheat. Establishment of a separate Department of Food in the Government of India in December 1942 completed the change of the Indian food economy from completely laissez-faire to highly centralized control with the government empowered to procure food grains at fiat prices.[15] With these mechanisms, London hoped to manage existing supplies of food in ways that kept India a valuable contributor to British war efforts.

London also tried to increase the supplies of food in India. "Grow More Food" (GMF) emerged as a slogan and formal policy after a conference of central, provincial, and princely state government officials on 6 April 1942. This group had gathered in the aftermath of the invasion of Burma by Japan, and the Government of India announced the formal GMF campaign in the summer of 1942. Major activities under GMF were encouragement to replace cash crops with food crops; to use more irrigation, better seeds, and manures; and to expand the arable land base. No targets for enhancement were set, however, so it was never possible until after 1947 to judge the effectiveness of GMF.[16]

Crop production in 1943–44 was much improved, which permitted food deficit areas of India to be supplied with little inconvenience. In 1945–46, however, severe drought plagued the province of Bombay and areas of the southwestern coast of India, which precipitated the postwar food crisis on the subcontinent (see Chapter 6). Memories of the Bengal famine of three years earlier, plus the new administrative machinery in the Department of Food to procure and ration supplies, enabled

India to move domestic and imported food grains to where they were needed.[17] This comparative success in the management of scarce food supplies, combined with the independence movement, created a strong national momentum for rural development.[18] At independence in 1947, therefore, India's central government was presumed to have a strong role in the management of food and agriculture, despite the fact that constitutionally the affairs of agriculture had been devolved to the provinces since 1919.[19]

Independence and the Shattering of British India's Economy

Indian nationalists were delighted to see the British leave in 1947, but partition of British India into the Republic of India and the Islamic Republic of Pakistan was a disaster for the Indian agricultural economy. India lost major areas of irrigated wheat land in the west, vast rice-producing areas in the east, and important agricultural research and education facilities.

Punjab was the center of wheat production in British India prior to independence. This province had been annexed by the British in 1849 and remained under British rule for ninety-eight years until 1947, when it was partitioned between Pakistan and India. In some ways, Punjab's experiences under imperial rule were different from other parts of India. Of particular importance were extensive irrigation works developed in the late nineteenth century.[20]

Punjab, meaning "The Land of Five Rivers," had, since at least 4,000 years ago, been a civilization based on irrigation from the Indus, Jhelum, Chenab, Ravi, and Sutlej Rivers. By the nineteenth century, most agriculture in this area was confined to areas near the rivers, with easy access to the water. Punjab was too far west to benefit from the monsoonal rains of the Indian subcontinent, and its population was sparse because of a lack of developed water supplies.[21] Archaeological remains, however, suggested that ancient civilizations had developed, and then lost, more extensive canals that carried water farther into the *doabs*, or "lands between the rivers."[22]

Imran Ali, a Pakistani-Punjabi historian, argues that with the accession of Punjab into the British Empire, Britain developed a three-pronged management plan that led it to promote new agricultural settlements or colonies in the *doabs*. Consolidation of political and military control was the first agenda, which was accomplished by forming partnerships with existing elites in the area and by enhancing wealth production through nine irrigation colonies after 1885. British habits of individual landownership replaced community ownership patterns, further tying landed elites to British interests.

Enhanced revenue extraction (taxation) from the Punjabi agrarian economy followed. Revenue extraction in India was a primary source of support for the military forces that Britain used to control its empire in India and elsewhere. In addition, Punjabi men constituted about half of the British Indian army, and the British came to call the Punjabi people a martial race. Britain also used the Punjab to supply the army, and the newly irrigated land in the colonies was given as grants to retired, loyal members of the army.[23]

Britain's pattern of imperial management in Punjab thus created a major problem at partition in 1947. Before independence, the food grain production of western Punjab, especially wheat, passed easily to the more densely populated areas of British India, and the whole Indian economy was adjusted to this division of labor: western Punjab grew wheat, the rest of India concentrated on other work.[24] Unfortunately for the new Republic of India, about 50 percent of the canal irrigation works in Punjab went to Pakistan. India, however, obtained over 80 percent of the total population of the subcontinent. Moreover, the irrigation works in eastern Punjab, now Indian Punjab, were intended only to prevent complete crop failure, not to substantially increase productivity. Thus "partition had magnified the normal food deficit in India created by the separation of Burma and the War."[25]

Not only was the most productive land of Punjab given to Pakistan, but the western part of Punjab also contained the bulk of the agricultural research and education facilities of this part of British India. Of particular importance was the Agricultural College and Research Institute at Lyallpur. India's section of eastern Punjab had virtually no facilities for teaching and research.[26] Although Punjab was not the only source of agricultural science in British India, the fact that Punjab's facilities went to Pakistan left India with little in the way of facilities for research on wheat, the most important food grain of northwest India. Only the Indian Agricultural Research Institute at Delhi provided expertise on this crop.

Transition to a New Order, 1947–1951

Decline of food production in the last quarter century of British rule, mismanagement of supplies creating the Great Bengal Famine, and the loss of major resources at partition all served to create an extraordinarily difficult start for independent India. Difficult as these events were, however, they were combined with an institution that strongly shaped the patterns of agricultural development after 1947.

Independent India inherited an assumption about the moral purpose of farmers, which had its roots in ancient Hindu culture and had been modified by the rule of Muslim invaders and then the British: elites believed the moral purpose of farmers and rural workers was to produce for the greater good of the king or the state but without necessarily expecting immediately to benefit and prosper. This institutional stance toward those who worked in agriculture had counterparts in British and American life, but the subjugation of India by foreign imperial powers for over 400 years had left a legacy of poverty, hunger, and oppression among the rural masses that had less of a counterpart in the Western countries.

Division of Hindu society into the four major castes (Brahmans, Ksatriyas, Vaisyas, and Sudras) embodied this institution. Brahmans were the repository of the spiritual ideas of society and the leaders by dint of their spiritual, not physical, powers. Ksatriyas, who formed the military components of society, were expected to defend the land and people and keep the internal peace. Vaisyas were knowledgeable about matters of economics and commerce; agriculture and trading were their domain for the organization of wealth production. Sudras, owning little or no land, were the rural labor of India, the hands for producing the wealth.[27]

Islamic texts also justified the transfer of wealth from those who directly produced it to those who ruled. Mogul domination of India from the sixteenth to the eighteenth century imposed a strong state in which subservience to the emperor in Delhi was an obligation held by all subjects.[28]

British rule, both by the East India Company and, after 1857, by the British government, continued the traditional belief that subservience to state authority was the duty of all Indian subjects. Because most Indians were rural villagers engaged in an agrarian economy, this meant that the duty of farmers was to send tribute, called revenue, to the central authorities. This tribute paid for the military might needed to keep internal peace in India and to support the imperial ambitions of Britain elsewhere.[29] Most important, the state was not to serve the citizen; instead, the subject was to serve the state.

Many of Britain's mandarins, the Indian Civil Service, continued to serve after independence. Although India's government was no longer the creature of London, its administrators did not immediately change a long habit of assumed authority over the rural masses of India.

Perhaps the tone of governance from Delhi is best captured in a statement from 1951 of N. C. Mehta, who had been secretary to the Imperial, then the Indian, Council of Agricultural Research:

> Briefly the State will have to take charge of the individual farmer and raise the standard of his economic health by persuading him if possible, and compelling him where necessary, to adopt the new methods of agriculture, for the country can rise as a whole only if our agricultural economy with its millions of farms and *lakhs* of villages is to revive with a new sense of energy and well-being.[30]

At the same time, a perhaps somewhat gentler way of outlining the role of the farmer in India was given by Sardar Bahadur Sir Datar Singh, who had served as joint vice chairman of the Imperial Council of Agricultural Research:

> For centuries, the Indian peasant has used the old plough and sickle; for centuries, he has suffered from inhuman toil and drudgery. The time to repaint the picture has come. The primitive peasant economy dating from five thousand years old Mohenjo Daro period must be replaced by scientific development and scientific technique, so that the peasant gets leisure for educational and cultural pursuits.[31]

What Mehta was in effect saying was that the peasant farmer would help the state as a whole rise, hopefully by persuasion but if necessary by compulsion. Singh was just as keen to replace peasant agriculture with new, scientific methods. While Mehta mentioned no immediate reward for the cultivator, Singh believed that the peasant farmer would gain new leisure.

As developed later, in the years immediately following independence, central government planners and many policies favored the extraction of capital and resources from the rural areas in favor of industrial development in the cities of India.[32] Although all countries that have developed have used this strategy of capital extraction, in the Indian context the policies created a situation that discouraged domestic production of food for about twenty years, perhaps more, following independence. It was this instrumental use of the Indian farmer and farm laborer that provided the most profoundly important context for the unfolding of India's agricultural development and the green revolution.

At independence, Prime Minister Jawaharlal Nehru intended to lead India as a socialist democracy committed to a secular, egalitarian state. His chief rival for leadership of Congress, and thus of India, was Vallabhbhai Patel (1875–1950), who represented the more conservative elements of Congress. Nehru and Patel had their disagreements but managed to work together until Patel's death in 1950 left Nehru as the unrivaled leader of Congress.[33] Their tensions, however, reflected the larger tensions within India about the public and private control of wealth. Governing the wealth generated in agriculture in turn affected the productivity of individual farmers, the use of the wealth generated, and the total, national agricultural production.

From independence in 1947 until 1951, India gyrated through a complex series of decisions designed to solve its most pressing problems. Partition from Pakistan was the foundation from which the short-term problems stemmed, because India at independence had to import even more food than was considered normal before 1940, much of it from what was now Pakistan. This situation was quickly aggravated by an outbreak of hostilities with Pakistan in 1947 over the control of Kashmir. In addition, Nehru's government devalued the rupee in 1949. Partition, hostilities, and currency devaluation thus formed the context within which Nehru tried to promote his vision of India's future compared to the more conservative tendencies within the Congress Party.

Immediately he had to solve three specific problems: (1) the guarantee of adequate food supplies and distribution to prevent famine, (2) land reform to eliminate the most odious forms of inequality from colonial India, and (3) promotion of better practices to increase prosperity in the villages.

Programs to deal with food distribution came from the Foodgrains Policy Committee, appointed before independence in July 1943, during the Bengal famine. The committee's report in September 1943 proposed a complex mix of government procurement and distribution of food grains. The committee rejected both the laissez-faire of earlier years and establishment of a government monopoly in food grains trade. Included in the plan were rations of one pound per day per person in all towns of greater than 100,000 population and in selected rural areas likely to be deficit in food grains, even if imports were required. The government would procure grain at fixed prices if traders did not bring supplies to market.

Price control was a key issue of the report, and the committee held that controls would be successful only if accompanied by control over supplies. The committee recommended a food grain reserve to be held by the central government expressly for the purpose of defeating grain speculators who would hold grain off the markets in anticipation of rising prices. Importantly, however, the committee also held that while farmers' costs of production had to be considered in setting prices, it would be unwise to permit an unlimited rise in prices.[34] This last provision was the fundamental weak spot in Indian agricultural and food policy for many years, because the prices were routinely set low in order to keep grains cheap for people who did not raise enough of their own food grains. In the efforts to protect landless consumers, India created a lack of incentives for landowners to produce a surplus.

The Department of Food was implementing the Foodgrains Policy by February 1945. Despite the poor harvest of 1945–46 and the ensuing postwar food crisis of India, the department arranged for the importation of over 2 million tons of food grain in 1946, much of it from the United States (see chapter 6). Hunger still reigned in India, but the gruesome famine conditions of 1943 did not return.[35]

When partition came in 1947, India had no way to feed its population without importing wheat from what was now Pakistan or from some other source. Cotton and jute needed by Indian mills also were now grown in Pakistan and thus in short supply domestically.[36] Imports from any source, however, required foreign exchange to pay for them, a situation that quickly became a severe problem. On 18 September 1949 India followed Britain's lead when the latter devalued its pound sterling from $4.03 to $2.80 per pound.[37] India, despite its independence, was economically still part of the sterling trade area and felt it also must devalue. Pakistan, however, did not do so, so India's imports from Pakistan immediately became more costly.[38] Within one year India changed trade balances from deficit to surplus with countries in both the dollar and sterling trade areas.[39]

Devaluation can be a sensible, indeed essential, move. Nevertheless, the necessity to devalue erodes the purchasing power of a currency for imports. India's need to devalue may have made importing food grains more difficult. In any case, being forced into currency devaluations while relying on imported food sent a signal to central government authorities that a more productive agriculture was needed. Reliance on an "enemy" (Pakistan) for imports of food exacerbated the foreign exchange problem. Although perhaps not predictable in 1949, Indo-Pakistani conflicts initiated in the war over Kashmir in 1947 began three decades of tension, which flared into open warfare again in 1965 and in 1972.[40] Thus the continued reliance on Pakistan for food imports created a situation of ongoing embarrassment for India.

The food situation in India's first four years of independence, therefore, can be characterized as follows. A pattern of increasing reliance on imports of food grains was standard behavior in the sense that trade channels in use since the 1920s expected to make the imports in order to satisfy consumer demand. At the same time, India was acutely conscious of its vulnerability on the food issue. Failure to assure adequate supplies at acceptable prices would have damaged the legitimacy of the new national government as surely as the Bengal famine had tarnished the British. Imports of food could satisfy demand, but they were a drain on foreign exchange, which was needed for other projects. In addition, Indian leaders surely did not like to be dependent for supplies on Pakistan.

Despite the new government's programs of procurement and rationing, the necessity to do so was distinctly undesirable to many political leaders. As a result, the government's policies tended to shift erratically in cycles of crisis followed by compromise followed by a new crisis. In September 1947 the government appointed a new Foodgrains Policy Committee, which recommended that the central government gradually reduce its commitments for rationing and distribution of food. The committee's general sentiment was that India, with better practices, could raise enough food for its people. For example, the committee concluded that the Grow More Food campaign, begun just before the Bengal famine in 1943, was inadequate as implemented but could be made useful. The committee recommended a planned reduction of imports over a period of five years and an increase of 10 million tons production per year (about 15 to 20 percent of total Indian production at that time). In addition, the committee urged the government to build a reserve of up to 1 million tons of grain, mostly wheat and rice.[41]

Although the committee recommended a gradual withdrawal of the central government from the control of grain purchases and distribution, the moral objections of Gandhi to control of the food trade led the new government to end suddenly the control of food grain transactions in December 1947. Gandhi was assassinated in January 1948, and thus was removed from Indian politics. Gandhi's absence and a continued deficit in domestic production were probably the key considerations to reimposition of central government control of food grains in September 1948.[42] Under the policy, the central government would prohibit imports and exports of grains between provinces except on a government-to-government basis. Provincial governments would procure grains at prices set by the central government, all dealers would be licensed, and rationing would be extended to cover more people.[43]

Independent India's new government thus regulated the grain trade, a policy that was successful in famine prevention but less than successful in ending the everyday poverty of many Indians. India's poverty at independence, however, must be understood on three levels: low productivity in agriculture, a suppressed industrial sector, and a skewed distribution of the wealth produced, largely from agriculture. India's agricultural wealth was controlled by large landowners who extracted their riches from rents, sharecropping, and the maintenance of low wages for landless labor. Genuine alleviation of poverty thus required at a minimum that the surplus value siphoned off by large landowners be redistributed to the tenants, sharecroppers, and landless workers.[44] Reduction of poverty also required development of more industry.

Nehru and his supporters within Congress wanted a genuine redistribution of wealth in the rural areas, which meant some type of land reform.[45] For this reason, the Constituent Assembly, which wrote India's constitution between 1947 and 1950, spent hours debating how much land reform, how much compensation, and how to pay the current owners. J. C. Kumarappa, a longtime associate of Gandhi, chaired the Agrarian Reforms Committee in 1948, which prepared a detailed set of recommendations to the Congress Party.

Although the Agrarian Reforms Committee was successful at one level of creating a land reform policy, it was stymied at another level by the complexity of Congress politics. Its successes came in the recommendations to end the role of intermediaries between the tiller of the soil and the new state of India. Abolition of intermediaries was not just an effort at economic reform; it was also the abolition of the privileges of those Indians who were the main collaborators with the British. Thus abolition of intermediaries was as much a policy of purification of Indian politics as it was of poverty alleviation. In the process of purifying India, Congress created a new political power base of landowners, who would be both the decision makers on agricultural practices and the organizers of power and votes in the rural areas.[46]

Abolition of intermediaries dismembered the land tenure system instituted by the British at the end of the eighteenth century. For example, in Bengal and other parts of north central India, Britain had instituted a system of rule that gave zamindars, the revenue collectors of the Mogul Empire, the ownership of the lands on which they collected taxes from cultivators. This arrangement encouraged the zamindars to be loyal to the British raj and enabled them to live at a comfortable level on the portion of the tax income they were allowed to keep. As titled land owners, they were free to extract as much revenue from their tenants as they could so long as they paid their dues to higher authorities.[47]

Two considerations drove the land reforms of the Congress Party.[48] First was simply the question of social justice. Zamindars, for example, did little or no work, were not interested in the welfare of their tenants, and often did not even live in the rural areas. Instead they lived in great or comparative luxury in the cities and towns of India while the rural villagers eked out a meager and often insecure living on the land. Considerations of equity alone suggested that the cultivators of the soil receive more of the wealth they produced, which meant that zamindars had to be removed. Enfranchisement of the rural villagers also meant, of course, that the Congress Party might benefit at the polls for abolishing the institution of zamindars.

Second, an exploited cultivator was unable and unwilling to take up improved methods of farming. Insecurity of tenure meant that long-term land improvements such as leveling and draining would not be done. Exorbitant rents meant that cultivators would have no way of financing the purchase of fertilizers, irrigation equipment, and other inputs that could increase yields. Thus, at least in theory, placing cultivators in an ownership position, or at least in a secure tenancy position, was a prerequisite for increasing food production. Political scientists Lloyd and Susanne Rudolph argue that Nehru counted on land reform as a way to activate a tremendous potential for increased agricultural production from India's millions of cultivators.[49]

Implementation of land reform after independence was complex because of differences between provinces in how land was held and rented. For the most part, the central government had only indirect control over the pace and method of land reform in the new states of India formed from the former provinces of British India. Only in the princely states, which had not formally been incorporated into British India, did the central government have direct powers to promulgate land reform measures, but even there the central government in New Delhi moved cautiously. Nevertheless, by 1951, abolition of zamindars and other types of landlords was well on its way to completion over all parts of India.[50]

In the first five-year plan begun in 1951, India announced its dedication to combining the justice of land reform with the promise that a more equitable society made to the prospects for development:

> The future of land ownership and cultivation constitutes perhaps the most fundamental issue in national development. To a large extent, the pattern of economic and social organisation will depend upon the manner in which the land problem is resolved. Sooner or later, the principles and objectives of policy for land cannot but influence policy in other sectors as well.[51]

Food procurement and distribution and the abolition of intermediaries were programs with immediate benefits. Neither, however, had much effect on increasing India's agricultural production, despite the hopes that land reform would stimulate production. In a third set of reforms, the new Indian government aimed to bring better practices based on science to agriculture.

The oldest of the programs to increase production through better uses of existing resources was the Grow More Food campaign, originated as a war emergency in 1942 and continued after independence. Implicit in the effort was a critical assumption that India had the natural resource base to be self-sufficient in food, provided those resources were mobilized properly. Some elements of the program sought higher production levels through expanded amounts of land in agriculture, and other components promoted better scientific practices in agriculture.

Despite a desire to achieve higher production, however, the Grow More Food program from 1943 to 1951 was only partially successful.[52] Between 1947 and 1951, the effort led to an increase in domestic production of food grains of about 3 million tons per year, but imports held steady or increased in that same period and reached 4.8 million tons in 1951.[53] Failure of the Grow More Food campaign to increase production substantially led the government in 1949 to hold extensive conversations with the United States on the possibilities of low-priced or barter trades of Indian minerals for wheat.[54]

Grow More Food was the oldest program designed to achieve rapid increases in production, but it was not the only set of technical reforms begun in India. As early as May 1943, a reconstruction committee on agriculture, forestry, and fisheries began to outline a postwar all-India policy for technical improvement of agriculture.[55]

Between February and July 1944, the Advisory Board to the Imperial Council of Agricultural Research (ICAR) undertook to plan India's participation in the newly organized Food and Agriculture Organization (FAO) of the United Nations. The board noted that over 30 percent of Indians were underfed (over 100 million people) and that better scientific practices could considerably expand food production. The board went on to advocate a Department of Agriculture in the central government, an expansion of research at the Indian Agricultural Research Institute, commodity research stations in different regions, and initiation of an extension service to reach each village.

Research was to expand substantially in the advisory board's recommendations, but it is difficult to interpret exactly how much. At least four different mechanisms were used in British India to organize and fund research: ICAR institutes, ICAR ad hoc schemes, central commodity institutes, and central research institutes (Table 8.1).[56] Uniform and comparable budget estimates for these different methods are not readily available. Nevertheless, the advisory board recommended increasing annual research expenditures to 18.8 million rupees.[57] Such a sum was presumably for all agricultural research, under a centralized Department of Agriculture based in Delhi. This would have represented more than a six–fold increase over the estimated 1.8 million rupees spent in 1940–41.[58]

Despite the vision of creating a much grander agricultural research establishment for India, the advisory board's recommendations were only partially adopted in the last years of British rule. Total research expenditures for ICAR institutes rose to over 3 million rupees per year in the two years between the end of the war and independence. With an estimated 1.5 million rupees per year for ad hoc schemes, the total for this period was about 4.5 million rupees per year.[59] This was about a 50 percent increase over the prewar level of expenditure (uncorrected for inflation).

Congruent with the rise in expenditures was a rise in the number of agricultural scientists employed in ICAR institutes, from 104 in prewar years to 178 by 1948,[60] an increase of about 70 percent. Nearly half of the increase in scientific personnel

Table 8.1   Support of Indian Agricultural Research at the Time of Independence in 1947

Method	Basic features	Origins and examples	Estimated expenditures at end of World War II
ICAR institutes	ICAR effectively governed the work of these institutes	Indian Agricultural Research Institute (IARI) was organized in 1905 and was the major site of ICAR work on foodgrains	All ICAR institutes spent 2.2 million rupees on research in 1945–46. IARI probably took between 30 and 40% of these funds because about 35% of ICAR scientists were at IARI
ICAR ad hoc schemes	ICAR could decide which schemes to finance	Work was performed at ICAR institutes, agricultural universities in the provinces, and general universities	In 1929–38, an average of six per year were organized in agricultural science, and an average of 400,000 rupees was spent per year. In 1939–48 the number of schemes dropped to just over one per year, and expenditures were about 1.5 million rupees per year
Central commodity committees	Research institutes organized around a specific crop; located in the main growing areas of that crop. These institutes were managed by semi-autonomous governing boards	Indian Central Cotton Committee, 1921, Bombay; Indian Lac Cess Committee, 1931, Ranchi; Indian Central Jute Committee, 1936, Calcutta; Indian Central Sugarcane Committee, 1944, Delhi	In 1929–38, the Indian Lac Research Institute in Ranchi had 12 scientists, out of a total of 104 ICAR scientists. This placed it as the third-largest research station, after the Imperial Agricultural Research Institute and the Indian Veterinary Research Institute
Central research institutes	Organized around specific crops, under the direct control of the government of India, Ministry of Food and Agriculture	Central Rice Research Institute, 1946, Cuttack; Jute Agricultural Research Institute, 1948, Barrackpore; Central Potato Research Institute, 1949, Simla	These research stations were insignificant in British India and grew to prominance only after independence

occurred at the Indian Agricultural Research Institute in Delhi, which bore the primary responsibility for research on food crops, especially wheat.[61]

Although expenditures rose, the organization of agricultural research was left in the fragmented condition it had been in since the devolution of agriculture from the central government to the provinces in 1919. By the end of World War II, the ICAR was still largely a body of scientists and nonscientists who had recommendation powers to the central and provincial governments but no ability to control them. It was served by only two technical staff, one for agriculture and one for animal husbandry. This situation of weak central planning of Indian agricultural research persisted through the transition to independence; a substantial increase in the ICAR's authority to plan and govern research did not come until 1965.[62]

Postwar reconstruction of Indian agriculture also had effects on provincial research and agricultural education. Government authority had been given to the provinces starting in 1919 and even more so with the Government of India Act of 1935. Major duties of local agricultural agencies had, since the nineteenth century, included gathering statistics, famine relief, and agricultural improvement. States before independence, however, had seldom done much about promoting new practices in food production.[63]

As an example, consider the budgets for farm demonstration work in the Punjab. In 1906–7, a total of 258,000 rupees were spent, and this figure rose only about tenfold in forty-one years, to 2.66 million rupees, a rate of increase of about 6 percent per year. In contrast, only two years after independence these types of expenditures were up fourfold, to 11.1 million rupees in 1949–50, an annual rate of increase of about 100 percent.[64]

Some plant-breeding work occurred in the provinces after 1905. Researchers in the Punjab, for example, selected promising wheat varieties using a pure-line method before 1925 and hybridization followed by selection of recombinants after 1925. By 1947 the Department of Agriculture for Punjab had released seven new varieties, all with good grain quality but very susceptible to rust.[65] Hybrid maize breeding began in Punjab after 1945.[66] Nevertheless, most wheat breeding before independence was done by the Imperial Agricultural Research Institute, a central government agency. Quite possibly a major hindrance to development of research programs in the provinces was the lack of agricultural universities.

Five agricultural colleges were created by 1908. They granted diplomas until the early 1920s, after which degree courses were available. Postgraduate agricultural education was available at the Imperial Agricultural Research Institute after 1923 and elsewhere after 1930. These offerings were slim, however, and seventeen institutions enrolled only about 1,500 students in agricultural degree courses in 1947. Establishing local educational facilities became an aim of the central government when in 1949 the University Education Commission recommended the creation of rural universities in each of the states.[67] This idea languished for a number of years, perhaps in deference to Gandhian ideas about simple village self-sufficiency. As noted later, it was not until the mid-1950s that the creation of these universities became possible.

Despite these increases in research activity, India at independence did not have an unequivocally positive image of science, a position that contrasted India with both the United States and the United Kingdom. Gandhi's ideas about science showed the complexity of the issues involved.

Although Gandhi made it clear that he disapproved of price controls and a government-manipulated system of food distribution, he was in favor of self-reliance, both for the Indian nation as a whole and for individuals. Gandhi also believed that India could make substantial gains in food production by using improved methods and growing fewer commercial export crops,[68] but his thoughts were more complex than his stand against government control of food distribution. He favored better production methods, especially improved seeds and the use of manures. He had, however, strong feelings that India was and probably should continue to be a rural civilization based in the villages. Moreover, it should be a society of equality among all individuals.[69]

Perhaps as a legacy of Gandhi's influence, during the first and second-five year plans (1951–61), India promoted various efforts to increase its agricultural output with some but limited successes. At no time during this period, however, did the government make a concerted effort to promote higher production by increased use of science. Instead, the assumption was that land reform, additional land, better use of underemployed labor, and extension advice about existing practices would be sufficient to bring India to self-sufficiency in food grains. Not until 1967 did India discard its ambivalence toward agricultural science and fully embrace the promotion of high-yielding agriculture through science. When it did, agricultural advancement was no longer tied to any goals for equality.

Planning, Industrialization, and Agriculture, 1947–1957

Despite Gandhi's preference for an agrarian India, most Indian leaders at independence wanted two changes: creation of an industrial sector and improvement in the lives of its people, most of whom lived meagerly in rural villages. Essentially, India wanted to change its society from largely agricultural and rural to substantially more industrial and at least partially urban. In the process of change, the Congress Party also stated its advocacy of equality among its citizens, a daring goal given the vast historical inequities among Indians based on religion, language, caste, sex, ethnicity, and landownership.

What India reluctantly discovered in the course of the 1950s was that forming an industrial sector placed potentially stifling demands on the rural economy. Essentially, if industrialization was to occur, the wealth of the countryside had to finance most of industrial development. No other source of capital of sufficient size was available, not even aid from such countries as the United States. Urban leaders worked to mobilize rural capital for the industrialization project, perhaps regardless of the rural sentiments. It turned out, in fact, that unbalanced promotion of industrialization interfered with increased yields and prosperity in the rural areas.

Two policy threads must be followed to understand the efforts to achieve an industrial economy and a more prosperous agricultural sector. First, at independence, planners under the leadership of Jawaharlal Nehru advanced policies intended to draw surpluses from the agricultural sector of the economy into the industrial. Several devices were used, including mandatory procurement of food grains at low prices and management of savings and foreign exchange reserves to promote industry.

Second, after independence, India found itself drawn inexorably into a relationship with the United States on issues of food imports, agricultural practices, and foreign policy. The partnership that emerged from this seemingly unlikely set of concerns was sometimes happy, sometimes tempestuous, with occasional outbursts of acrimony and mutual recrimination. Nevertheless, the governments of both countries found that they needed each other to promote their own passionate interests.

Although the major threads of the story are clear, reconstructing the events of the 1950s is hindered now by the same problem that plagued both the Indian and the

American government at the time: How accurate were the statistics on food grain production in India? The partition of British India into Pakistan and India clearly removed important granaries from India, and the Indian government was convinced at the time that increased supplies of food grains were essential, to be obtained either from domestic production or from imports. What is less clear is why and how much additional supply of food grains was necessary.

During years with bad weather and poor harvests, were more food grains needed to prevent hunger, malnutrition, and famine? That is, were supplies of wheat and other grains physically inadequate to meet physiological needs? Alternatively, were additional supplies needed to keep prices low for the urban population (a small minority of Indians), consistent with the public policy of low wages in order to promote industrialization? If the latter were correct, then increasing food grain supplies was not so much a matter of physical supplies as a question of political and economic management of the marketing system for food grains. One estimate by the secretary of the Indian Ministry of Agriculture was that one-third of the crop was not even reported,[70] an amount large enough to create severe difficulties in managing the food supply, especially in poor harvests.

Whatever the true situation with grain supplies, by 1951 India found itself in an increasing entanglement with the Americans over grain imports. American involvement with India's food situation began in 1946 during the postwar food crisis (see Chapter 6). One year later, in 1947, discussions were still occurring between the American and Indian governments over the question of whether the Indians had overestimated their need for emergency grains.

In their own defense, Vishnu Sahay, secretary of the Food Department of India, explained why they had requested 4 million tons of grain in 1946, received only 2.5 million tons, yet avoided famine. Sahay explained that the suggested overestimation of need by 1.5 million tons came from the Indian desire to reduce semistarvation, to have at least two months' grain supply to cushion against difficulties in moving grain internally in India, and to allow for a poor 1946 harvest (the estimate of 4 million tons was made in February, before the main Indian wheat harvest was in for that year).[71] Local shortages of wheat continued to be reported by the Indian government in 1947,[72] but the severe scare of 1946 was not repeated. Nevertheless, the fact that Sahay felt he must defend his government's estimates of need for imported food grains indicated a tension over the issue between the Americans and the Indians.

In 1949 an extended series of serious discussions again commenced between the American and Indian governments. In July, just two months before Britain and India both devalued their currencies against the dollar, India requested to buy at reduced prices an extra 1 million tons of wheat from the United States. This was in addition to 1 million tons the government was already committed to buy from Canada, Australia, and the United States. Response from State Department staff was not encouraging, possibly because of the estimate at that time that the American crop was not likely to be in heavy surplus.[73]

One month later, on 13 August, a conversation took place between R. L. Gupta, secretary of the Ministry of Food, and C. C. Taylor, agricultural attaché of the American embassy in New Delhi.[74] This discussion illuminated some of the basic complexities of both Indian and American positions on Indian agriculture, so it is important to understand what each party believed at this time.

Taylor reported Gupta's request for an extra 1 million tons of American wheat, a repeat of what had been broached one month earlier in Washington. Gupta believed that the extra grain, used to increase the ration level, would bring out hoarded supplies of grain that were already in India, held primarily by big landowners hoping for higher prices. Gupta believed the landowners were getting too much of the national income and were profiting at the expense of urban workers and industrialists. Thus urban industry was being deprived of profits for capital investment, and urban workers were not working at top efficiency because there was not enough low-cost food. India wanted the extra grain to arrive in a steady stream for a year, to be followed by similar sales for the following two to three years.

Taylor had a series of reactions to Gupta's request. He gave some responses immediately (in August) and some about three weeks later (3 September)[75] after a follow-up conversation with Gupta. Taylor believed increasing the ration level would kill the notion that India could become self-sufficient in food grains by 1952. Taylor's reasoning was that although lower prices might not discourage farmers from producing more, mandatory procurement would. He noted that India was currently procuring about 4 million tons per year, or about 8 percent of the total domestic production. Procurement price was less than 50 percent of the open market price.[76]

Taylor also worried how political stability could be enhanced by raising the prosperity of industrialists at the expense of 250 million rural people. Significantly for ensuing events, Taylor fell into the trap of dreaming about the ability of food supplies to force compliance with political goals: "[F]ood is a defensive weapon against Communism which the Government of India can use effectively to our mutual advantage."[77]

In August Taylor did not report any real encouragement to Gupta and the Ministry of Food about their request for an extra 1 million tons of grain. By September, however, he urged a reconsideration of American willingness to send the extra grain at concessionary prices. He specifically focused on the political turmoil that could arise in India if urban workers were ill fed and the financial exchange problems India had with all imports.

Thus in the two-month period from mid-July to mid-September, American policy toward the food and agriculture situation in India underwent a fundamental transformation. Before July, requests for food imports, especially at reduced prices, were met cooly, as were requests for technical assistance, discussed earlier (see Chapter 7). By September, American officials were more interested in discussing grain supplies for India at reduced prices.

What caused the difference? The most likely answer is the fact that globally the American government saw serious challenges to its policy of containment of communism. Explosion of a nuclear bomb by the USSR ended the American monopoly over atomic weapons. In China the last resistance of Chiang Kai-shek's Nationalist government disintegrated in mid- to late-1949, and Chiang fled to Taiwan, leaving Mao Tse-tung and his Communist forces in control of the Chinese mainland. Taylor's references to the need to prevent unrest in India's cities resonate with this explanation. Also suggestive of larger geopolitical themes at work were conversations in October between the State Department and the Indian embassy over the possibility of bartering wheat for manganese, a strategic mineral for the American armed forces.[78]

Even though the sale or trade of the extra wheat did not occur, because of an inability of the two governments to agree on terms,[79] the die had been cast that cemented the agricultural situation of India to the foreign policy of the United States. For many years the Americans would be preoccupied with how American grain could keep India from going the way of China, and the Indian government would seek cheap American grain to keep its industrial and urban working classes happy. Never would foreign and agricultural policies between the two countries be separated.

The situation remained difficult in 1950 as drought struck at the heart of the Indian food supply. Early in the year, poor harvests and food shortages developed in southeast India, and local American consulates emphasized the need for imports of American food grains.[80] Severe drought in the north of India in October compounded the national government's problems, and estimates were that Bihar's grain production was off by about 2 million tons,[81] about 4 percent of India's annual production.

Nehru's government, buffeted by the vagaries of weather, made a number of adjustments in 1950 and one fateful decision that solidified the mutual intertwining of America and India begun the year before. At the administrative level, India consolidated the Ministry of Food and the Ministry of Agriculture into one unit under the leadership of Kanaiyalal M. Munshi (1887–1971). Munshi, in turn, made changes in the civil service that were welcome to the American embassy. Munshi removed the former secretary of the Ministry of Agriculture, who the Americans regarded as uninterested in getting technical assistance from the United States to India.[82]

In addition to adjustments in the ministries, India launched a major effort to reclaim 5 million acres of land in central India for food and cotton production. This land was estimated to be producing 1.7 million tons of food grains within six years,[83] an amount that would be a significant portion of the 3.8 million tons of food grains imported in 1949.[84] With new leadership in the Ministry of Agriculture and Food, plus plans for bringing extensive new land into cultivation, Nehru's government was pushing incrementally on the domestic production limits.

In addition to the internal administrative developments in 1950, Nehru also made a decision that in retrospect must be recognized as a significant turning point for India. In December, Madame Vijaya Lakshmi Pandit, the Indian ambassador to Washington and sister to Nehru, met with Secretary of State Dean Acheson to request an emergency allocation of 2 million tons of wheat. This request was significant because it was the first time Nehru had officially appealed to the United States for economic assistance. Drought had shrunk the Indian harvest projected for 1950–51, and the Indian government believed it needed an extra 2 million tons, to start delivery in mid-1951.[85]

Nehru's reasons for making an official request to the United States were undoubtedly complex. Foremost among them, however, was probably the continued desire to keep grain prices low in urban India. In making the formal approach at a high level, however, Nehru fully opened the door to American desires to contain global communism. In papers prepared for a meeting with Madame Pandit, State Department officials recognized that this official request for help provided an opportunity "to explore with the Government of India various problems of great urgency involving the prevention of further communist expansion in Asia and the status of Indo-U.S. relations."[86]

Nehru's request for 2 million tons was granted, and grain imports to India rose to 4.8 million tons in 1951, the largest amount of imports to that time. More American shipments followed in the ensuing twenty years. For its part, the Indian government from 1951 until at least 1961 based its economic planning and development upon a bedrock of steady and usually low-cost American food grain imports. At no point during the 1950s did India make a strong commitment to intensification of practices, the strategy advocated by many Indian and American agricultural scientists.

Internally, India went back and forth during the 1950s between heavy and minimal central government control of the food grain trade. Domestic agricultural production went up and down primarily as a function of the weather and some additional land brought into cultivation. As a result, the average wheat yields ranged from about 700 to about 900 kilograms per hectare until 1965, roughly the amount that could have been expected when the Mogul Empire first established itself in the sixteenth century.[87]

American grain imports took a decided upturn after the signing in 1956 of the P.L. 480 program. This agreement stemmed from the passage in 1954 of the Trade Development and Assistance Act, a law that allowed the American government to dispose of surplus American grain abroad at little or no cost to the recipient government. Although the primary motivation of the U.S. Congress in passing this bill was the support of U.S. farm prices by taking surpluses off the market, the program quickly became an integral part of the larger American policy of containing communism (see Chapter 7).

Indians paid for P.L. 480 grain in rupees, which were held by the American government in India. In the 1970s the P.L. 480 payments gave the United States control of as much as one-third of the money supply in India. It was almost as if the Americans had bought a big piece of India with grain. To make matters worse, the huge supplies of American grain that flowed into India during the 1950s and early 1960s accomplished the function intended by Nehru's government, to keep Indian grain prices down. In fact, prices were so low that Indian domestic production stagnated. Indian farmers simply could not compete against grain sold at a loss by the American government, so they stopped trying and Indian production failed to rise fast enough to meet increasing domestic demand.

Low grain prices were considered desirable by the Indian government in the decade after independence, but low prices were not the heart of planning efforts. Instead, in the first plan (1951–56), the government promoted land reform, as noted earlier. As efforts began in 1954 to assemble the second plan (1956–61), Nehru was able to dominate the planning process and push it in two directions: more consciously toward socialism, with state ownership of industry, and large-scale industrialization that could end India's dependence on imports. Draining the capital formed in the agricultural sector into the emerging industrial sector was the way Indian planners proposed to foster industrial development. For this reason, Nehru considered higher productivity in agriculture essential to the second plan's success. Unfortunately for the planners, resistance in the Congress Party to agrarian reform was high. As a result, the party forced compromises in the second plan that preserved the rights of large landowners. On top of resistance to agrarian reform, technical assistance to intensify production in agriculture was also slighted.[88] The combination of delay in both social and technical reform in agriculture ensured continued low yields and thus continued dependency on imported grains.

Intensification of Agriculture 1952–1961

Although the second plan slighted agricultural intensification, the ideal of egalitarian development guided the main effort in rural development, the community development program, begun in 1952. These projects originally represented social reform more than technical reform, but some technical reform was always present in the form of advice on use of current knowledge. Although community development was abandoned as the major program for rural enhancement by the mid-1960s, it served as one of the initial vehicles to introduce more intensive agricultural practices. The story of community development began in the projects at Nilokheri and Etawah.

Nilokheri was the resettlement camp founded for Hindu and Sikh refugees coming to India after the partition from Pakistan in 1947. Located in the Punjabi plains near Kurukshetra, north of Delhi, Nilokheri had a privileged location in ancient Hindu mythology. Kurukshetra was the place of the mighty battle recounted in the Mahabharata. In a way, the planned community of Nilokheri had a similar challenge — to forge out of a forested wilderness a place for a new life for people ripped from their homes by the disintegration of the old British Empire. Egalitarian and self-sufficient development in agriculture and industry were the guiding lights at the building of the Nilokheri new town, a project that captured the personal attention of Nehru.[89] What is important to recognize about Nilokheri is that it envisioned an influx of new knowledge and skills to create the potential for a new way of living.

At about the same time, in 1948, the Uttar Pradesh state government launched in Etawah District a planned improvement of all aspects of village life. Under the direction of D. P. Singh, this project was run by Albert Mayer, a city planner and friend of Nehru's, and Horace C. Holmes, an agricultural extension specialist. Meyer and Holmes worked with ninety-seven villages covering 61,000 acres. Their staff included four specialists in agronomy, engineering, and other fields, as well as twenty-four village-level workers. The program promoted better seeds, green manure, and other practices and reported increased yields of 15 to 30 percent. The Etawah project was probably the single most important model for all subsequent development work in India, and Nehru took a personal interest in Holmes's work. Successes at Etawah, however, were primarily technical in nature, not social.[90]

Both the Nilokheri and Etawah projects were started before the U.S. Point Four program and the Ford Foundation became involved in India, and the models developed there were highly influential in starting the community development program in 1952. S. K. Dey, in charge of Nilokheri, became the administrator of the Community Projects Administration within the Planning Commission. Holmes went from Etawah to become the Point Four leader on extension service development in India's central government.

When Paul Hoffman, president of the Ford Foundation, came to India at Nehru's request in 1951, his group toured both Nilokheri and Etawah.[91] In turn, Ford's first activities in India were to support the work of the community development program. Nehru's aspirations were that community development would bring a better life to the villages. V. T. Krishnamachari, who served at the time as vice chairman of the Planning Commission, described community development in these terms:

> The objective of the movement is, briefly, to ensure that there are concerted efforts to improve all sides of village life and to mobilize local initiative and resources for the betterment of rural conditions. The basic principles on which it lays stress are three: Firstly, all aspects of rural life are interrelated, and programmes of improvement should be comprehensive, though there might be emphasis on groups of activities. Secondly, the motive force for improvement should come from the people themselves. The movement is built upon the principle of self-help; the State only assists with supplies and services and credit. The vast unutilized energy lying dormant in the countryside should be harnessed for constructive work, every family devoting its time not only for carrying out its own programme but also for the benefit of the community. Thirdly, the co-operative principle should be applied for solving all problems of rural life.
>
> The directions in which change in outlook is needed and is to be worked for are: Firstly, increased employment and increased production by the application of scientific methods in agriculture, including horticulture, animal husbandry, fisheries, etc., and the establishment of subsidiary and cottage industries; secondly, self-help and self-reliance and the largest extension of the principle of co-operation; and thirdly, the need for devoting a portion of the unutilized time and energy in the countryside for the benefit of the community as a whole.[92]

While the ideological theory behind community development, as articulated by Krishnamachari, is interesting, it is probably more instructive to see how central government expenditures were made in support of announced principles. Table 8.2 provides the expenditures in rupees and the proportion of total expenditures for community development, agriculture, and irrigation during the first (1951–56) and second (1956–61) five-year planning periods. What is apparent is that although about 80 percent of the people lived in villages and worked in agriculture, a much smaller and declining proportion of public expenditures was devoted to improvements in

Table 8.2   Expenditures (rupees) for agricultural development in the first and second five-year plans, India

Item	First Five-Year Plan (1951–56)		Second Five-Year Plan (1956–1961)	
	Rupees spent (in crores)	Percent	Rupees spent (in crores)	Percent
Agriculture and community development	291	15	530	11
Major and medium irrigation	310	16	420	9
Total	1,960	100	4,600	100

From Sib Nath Bhattacharya *India's Five-Year Plans in Theory and Practice* (New Delhi: Metropolitan, 1987), p. 49. One *crore* = 10,000,000.

agriculture. This was the reality behind the statement that the wealth of the rural areas was being tapped for industrial and urban development.

Although these early projects at Nilokheri and Etawah and then community development brought new technical knowledge to Indian villages, they were not intended to use research as an engine for change. Rather, they were designed more to tap the wealth-creating potential of social reform, supplemented by the spread of existing knowledge. Other events in the first years of independence portended more elaborate technological change based on research. In the change of personnel in the Ministry of Agriculture in 1950 that brought K. M. Munshi to the ministership, a most important shift was the appointment of Benjamin Peary Pal (1906–89) to be the director of the Indian Agricultural Research Institute (Figure 8.1).

American embassy officials simply noted that Munshi had placed the "quietly competent" Pal in the leadership of IARI in the place of J. N. Mukherjee. Pal had become imperial economic botanist in 1937, four years after completing his doctorate in agricultural botany under the direction of Rowland Biffen (see Chapter 3) and Frank Engledow (see Chapter 9) at the University of Cambridge in England. He

Figure 8.1   Monkombu Sambasivan Swaminathan (*far left*) Benjamin Peary Pal (*second from right*), and Norman E. Borlaug (*far right*) inspecting dwarf wheat in India derived from the Mexican Agricultural Program. Courtesy Rockefeller Archive Center, North Tarrytown, New York.

became assistant director of IARI in 1946 under Mukherjee. A prolific writer and researcher on many aspects of Indian agriculture, he brought a powerful new voice of science to IARI. In addition, his training with Biffen and Engledow at Cambridge created a direct link that connected independent India's plant-breeding work with that under way in the West. In 1965 Pal went on to become the director general of the Indian Council of Agricultural Research.[93] Thus his appointment in 1950 as head of India's largest and most important research center on food grain production placed a unique voice in a position to shape the contributions of new technology to Indian agriculture.

Shortly after Pal's appointment came two other developments that, at the time, did not create immediate change but over the following ten years made an enormous difference in India. First in 1951 was the invitation from Nehru to Paul Hoffman, president of the Ford Foundation, to start a Ford Foundation program in India (see Chapter 7). The Ford Foundation representative, Douglas Ensminger (1910–89), began a long and influential career as liaison between the foundation and India. He developed a good working relationship with Nehru and other members of the Planning Commission and the ministries.[94] Ensminger's first projects in India were connected to community development and thus were aimed at increased food grain production through social reform with secondary attention paid to the spread of existing technologies but only minor interest in new technology. In time, however, this set of priorities changed radically, and the Ford Foundation became one of several organizations pushing hard for the development of new scientific knowledge.

The second development shifting toward technological change in agriculture was the formal initiation of the National Extension Service (NES) in 1953. Like the community development program, NES also was organized by blocks. Community development blocks were originally envisioned as temporary, lasting for three years, but the NES blocks were intended to be permanent.[95] In a way, formation of the NES was a smaller version of the community development block and was introduced in response to local calls for assistance before the community development program could be expanded to all of India.[96]

The NES involved village-level workers and extension officers in each community development block, as well as technical support from state departments of agriculture.[97] A lack of technically qualified personnel to work in the NES undoubtedly limited the program's effectiveness in its first years,[98] but for the first time India had a national program to bring new technical instruction to farmers. Institutionally, therefore, this was an important development indicating the profound shifts then under way.

Aside from Pal's appointment and the creation of the NES, Nehru's government for the remainder of the first five-year plan (1951–56) continued to place heaviest emphasis on social reform as the path to higher agricultural yields. Only irrigation development and the building of India's first fertilizer plants suggested that more intensive practices were really having an effect. Good weather, plus some effects from the first social and technological reforms, brought better harvests in the later part of the first plan period.

Toward the end of the first plan, however, yet two more changes were made in India's stance toward capital-intensive agriculture. In 1955–56 India accepted American advice to create an Indo-American team to study the structure and functioning of Indian agricultural universities. This committee recommended that India organize its agricultural education and research efforts more along the lines of the American land-grant universities.[99]

Also in 1956, India signed an agreement with the Rockefeller Foundation in which the foundation agreed to supply scientific advisers and funds to improve cereal production in India and to upgrade the Indian Agricultural Research Institute (IARI).[100] Whereas the Ford Foundation's first efforts were aimed more at supporting the broad-based community development program, these movements to establish agricultural universities and enhance the research program at the IARI marked important steps toward reliance on new technology rather than social reform. It was only as the second five-year plan was under development in 1955, however, that the Planning Commission made a fateful decision that unraveled the Gandhian and socialist reform agenda.

P. C. Mahalanobis, the cabinet's statistical adviser from 1955 to1958, provided an analysis for the Planning Commission in 1955 that India should deemphasize agriculture in the second five-year plan in favor of industrial development.[101] The figures in Table 8.2 show how agriculture, community development, and irrigation went from 31 percent to 20 percent of the plan's public sector outlay. Although the money allocated to these rural functions went up from the first to the second plan, their share as a proportion of total outlay shrank, an indication that the government's emphasis was toward heavy industry.

Unfortunately, as political scientist Francine Frankel has analyzed in great detail, the rapid push for industrialization was incompatible with the agricultural efforts.[102] Financing the industrialization campaign depended on a highly productive agriculture as a source of capital. Yet the drain on public funds for industry made it impossible to increase agricultural production by enhancing the capital allocated to agriculture. In addition, the plan demanded continued low-cost food grains, so India used both imports and P.L. 480 grain from the United States, which forced down the prices of domestic grains and thus provided no incentive for domestic producers. This dedication to social rather than technical reform in agriculture derived from high officials of the Indian government, including Nehru and Mahalanobis, examining the case of China's agriculture. They concluded that the Chinese had been able to achieve substantial yield increases through altering the socioeconomic structure of agriculture.

It was the Chinese experience that moved Nehru and others on the Planning Commission to speak increasingly often of more cooperative management of agricultural land, under the democratic control of each village, with every adult having a vote. Although Nehru remained a sincere democrat in his thoughts and actions, "cooperative" management turned into a code word for an attack on private property rights, an idea that instantly alarmed all larger owners of property in India.

Increasingly radical rhetoric coming from the highest levels of government probably exacerbated the anxiety of conservative elements, which combined with a drought in 1957 to produce a plunge in the production of 1957–58. With food shortages and the threat of higher prices, Nehru moved in August 1957 to substantially increase the procurement of domestic grain production at low prices.[103] Thus at the very time when agricultural producers were being buffeted psychologically and by the weather, the central government made it clear that further downward pressure on prices was to be expected. Adding further fuel to the unrest in India, the elections of 1957 brought a shift in that the largest opposition party in several states and in the center were the communists (Communist Party of India). Moreover, the state of Kerala elected a communist-led united front government to power, which immediately began to move on the land reform legislation that Congress had been reluctant to touch.[104]

Douglas Ensminger of the Ford Foundation recounts that in 1957, the very year in which radical rhetoric of Congress, bad weather, downward pressure on prices, and elections with a shift to the left all converged, he cut short a home leave and spent three months traveling the countryside of India in an effort to learn more about the extent to which producers were using improved practices. After this trip, he wrote a briefing paper for Nehru and spent about half an hour with him. It was from this meeting, according to Ensminger, that the next and nearly decisive event developed: a team of American experts organized by the Ford Foundation to prepare a study on the problems of Indian agriculture.[105]

Ensminger put together a team chaired by Sherman Johnson of USDA. Their 1959 report, *India's Food Crisis and Steps to Meet It,* was a milestone that shifted India's agrarian strategy from one based on social reform (albeit implemented with fatal internal contradictions) to one based on new technology, to be adopted by the growers and landowners most prepared to adopt the new practices. The essence of the report was that India must move away from thoughts of cooperative agriculture and from policies of draining agriculture to finance industrialization.

In their place, Johnson and his colleagues urged the identification of farmers who had access to enough land and a secure water supply. To these growers, India was urged to deliver improved seeds, fertilizers, better irrigation equipment, credit, technical advice, and a guaranteed price that would be sufficient to provide an incentive for production. With state assistance, these growers were to unleash the productive might of capitalism to increase India's agricultural yields.

Interestingly, the Johnson report framed the problem in terms of increasing food production sufficiently fast to provide food for India's rising population:

> India is facing a crisis in food production. . . . The crux of the problem is food enough for the rapidly increasing population. . . . Although there is considerable emphasis on family planning in India, no appreciable slowing down of population growth may be expected during the Third Plan period. *This means that food will have to be provided for 80 million more people by the end of the Third Plan.* . . . The present population places severe pressure on food supplies, and unfavourable crop conditions create an immediate crisis. (emphasis in original)[106]

The quotation is drawn from the first four paragraphs of the report, so the committee grounded its argument in the arena of population growth and food supplies, suggesting that all that counts is the physical production of food. How different the report might have been if it had seriously tried to grapple with the severe social inequities of India and how these affected the agricultural production practices and yields. By placing the problem in the population arena, the report was linked to the what was described earlier as the population–national security theory (PNST): population growth in the third world, according to the theory, was causing instability, and increasing food supplies, while providing birth control was the way to end that dangerous political mix that could lead to communist revolution (see chapter 6).

Ensminger recounts that when Nehru was briefed on the report at his home one evening, he didn't like the word "crisis." Sherman Johnson showed a chart with trends in agricultural production since 1947–49 and a second line with trends in food needs. Using these forecasts, India was predicted to be short of grain by 20 million tons by 1970, an amount that could not be made up from imports. It was at this point, Ensminger reports, that Nehru dropped his objections to the word "crisis."[107] That the Ministry of Agriculture and the Ministry of Community Development and Cooperation published the document suggests that at least some elements of the Indian government either approved of the report or at least wanted it placed into the policy arena for active discussion.

The Ford Foundation report of 259 pages was a general philosophical statement about what India needed to do. In order to make its recommendations concrete, the government asked the Ford Foundation to develop a specific plan for how to implement the recommendations of the report. Ensminger quickly assembled a new group, also chaired by Johnson, which prepared a ten-point program that promoted the "package" approach to increasing India's agricultural yields. On a trial basis of seven districts, India would attempt to marshal all of the inputs, to be made available to capable farmers, needed for intensive high-yielding practices. Use of improved seeds, fertilizer, irrigation, and pesticides was indispensable. Also needed were adequate credit facilities, technical advice, and a guaranteed price that would provide the grower an incentive to take the risk of trying new technology. This report was the foundation for the Intensive Agricultural District Program (IADP), the organizational framework for the green revolution.[108]

Nehru's government endorsed the IADP program, but according to Ensminger the enthusiasm for implementation was low. The person designated to be the lead for working with the states on the IADP left for a new post with the Food and Agriculture Organization of the United Nations. Ensminger expressed great frustration that his replacement was very slow to appear, a fact that gave Ensminger problems in convincing the Ford Foundation headquarters in New York that Ford should support the IADP. Only when Ford Foundation Board Chairman John J. McCloy came to India on a tiger hunt was Ensminger able to convince him that the IADP was essential to India's future.[109]

Even Ford Foundation assistance in getting the IADP launched, however, was not entirely sufficient to guarantee its future. Significant dissent within the Congress Party remained over whether the IADP violated the Gandhian or socialist reform agenda for India. (It did, of course.) Nehru himself may never have really been enthusiastic about the scheme, either because he actually did favor draining the rural areas in order to finance industrialization or because he didn't entirely understand that the plan to use the rural areas to promote urban development in fact rendered the rural areas incapable of accomplishing their mission. War with China in 1962 also undoubtedly distracted government attention from the efforts to increase agricultural yields.

Regardless of the causes, the IADP had limited successes until 1964 in terms of serving as a powerful catalyst that would ignite scientific agricultural development over all of India. It helped some farmers in a few districts to improve their yields, but India remained dependent on imported food grains. Nevertheless, with the IADP, India had made an institutional commitment to a different path for agricultural development.

From Institutional Commitment to Green Revolution, 1961–1971

Three events triggered India's final shift to green revolution technologies. First, Nehru died in 1964, and his successors were more enthusiastic about the new technology. Second, severe droughts in 1965–66 and 1966–67 drastically decreased India's production and forced it to import 10 million tons of foreign grain, with considerable damage to Indian pride and with compromise to its national autonomy. Third, new wheat seeds developed by Norman Borlaug at the Rockefeller Foundation in Mexico (see chapter 5) were capable of doubling and tripling wheat yields in India. Increases of this magnitude are rare in agricultural science, and it was impossible to ignore the payoff from using them. It is to these final episodes of the green revolution's origins in India that we now turn.

Nehru's Passing, a Drought, and New Seeds

Jawaharlal Nehru, born in Allahabad in 1889 and long a prominent leader in India's quest for freedom, suffered a stroke on 8 January 1964 and died four months later on 27 May. With Nehru's passing, India lost its most prominent voice for the mixture of Gandhian and socialist ideology. To be sure, various Gandhian groups continued to work in the countryside on issues like *bhoodan*, or land gift from the rich to the poor, a voluntary land reform to improve the soul of the landowner. In addition, India's socialist and communist political parties had a place in Indian politics, but except in a few states they were marginal to Congress. Nevertheless, Nehru was unique and the stature he gave to India was irreplaceable.

Despite Nehru's personal prestige, his influence eroded during his last few years in office. Strong attacks from both the Left and the Right had kept Nehru on the defensive. He was successful in keeping India focused on the project of developing heavy industry in the public sector. At the same time, he continued to push for rural development through land reform, agricultural cooperatives, and community development.[110] He avoided allocating public resources to promote more capital-intensive methods of farming and also stood steadfast against the use of incentive prices to encourage more grain production. Instead, he relied on the American P.L. 480 shipments and the threat of procurement of grain at low prices to keep domestic grain prices low.

After Nehru's stroke the question of a successor became the dominant issue in the many factions of the Congress Party. While it is possible that Nehru personally wanted his daughter Indira Gandhi to succeed him, he did not push this agenda. In her place, Lal Bahadur Shastri (1904–66) emerged from the tangle of Congress politics, perhaps in the eyes of some to be merely a placeholder prime minister until someone with greater charisma and forcefulness would arise to capture the top position. Or perhaps, in the eyes of others, Shastri would occupy the leadership position until Indira Gandhi was ready to assume it.[111]

To the surprise of everyone, Shastri, perhaps driven by pragmatism and common sense, made a series of appointments, reorganizations, and decisions that dismantled Nehru's development schemes and substituted a plan to push the intensification of farming. In a major reorganization, Shastri downgraded the power and influence of the Planning Commission, which Nehru had used for making decisions that he then asked the cabinet to ratify. Shastri made the Planning Commission advisory to the cabinet and devolved on his ministers the responsibility and power of making decisions in their portfolios.[112]

To the Ministry of Food and Agriculture, Shastri appointed C. Subramaniam (b. 1910) as minister, a critical appointment because of a chronic food crisis that had stymied the Planning Commission's work with stagnant yields since 1960–61. During this time, Nehru's government had accepted the analysis by the commission that the problem stemmed from hoarding of grain supplies by farmers and merchants hoping for a rise in prices. Shastri and Subramaniam, even though they did not immediately challenge this postion, wanted government actions to quickly create a betterment of the rural masses, an outcome they believed required an increase in food grain production. That increase, they further believed, depended on more government support of agricultural production and, above all else, on the government being willing to offer an incentive price for grain that was higher than procurement and market prices. Subramaniam urged a government policy of building government reserves by purchases of grain in the open market at incentive prices.[113]

Shastri and Subramaniam moved to incentive prices for grain production based on the recommendations of the Foodgrains Prices Committee (1964), which also recommended the formation of a permanent Agricultural Prices Commission. The commission, established in 1965 along with the Food Corporation of India, became the vehicle for promoting growth in agricultural production through price incentives. The commission's first report in 1965 noted, "Price support policy contributes to growth by inducing the farmer to adopt improved technology without fear of an excessive price fall."[114] Shastri accepted the commission's advice, and prices of wheat rose from 37.5 rupees per quintal in 1964 to 48 rupees per quintal in January 1965, to 50 rupees per quintal in November 1965. These higher prices enabled the Indian government to obtain much larger amounts of grain.[115] Although later critics argued that truly adequate incentive prices were never instituted,[116] the Agricultural Prices Commission paved the way for prices sufficiently high that farmers did respond and Indian grain production began a steady climb after 1967.

Despite these changes, in the view of political scientist Francine Frankel the new government was not highly successful either in keeping food prices reasonable or in quickly getting higher levels of production. Moreover, the possibility that famine had returned to India, in Kerela, emerged at the same time that the Communist Party of India split into two factions, one of which seemed poised to bring outbreaks of violence in rural areas as the poor struggled to survive. Congress Party sessions brought forth great criticism of the Shastri-Subramaniam plans for food security and economic development. World Bank critics added to the fire by threatening to withhold further aid to India unless public sector outlays came down, agricultural production went up, private enterprise was given a larger role to play in Indian development, and the rupee was devalued. Further bad news for Congress came when Marxists won the largest number of votes in an election in Kerela in March 1965. Pakistan's troop movements into the Rann of Kutch in March 1965 were simply the final blow in a long and complex series of events that led India to make changes that could not be easily reversed.[117]

Shastri, preoccupied with the foreign policy problems with Pakistan, passed leadership on the food front to Subramaniam, who issued a new plan, "Agricultural Production in the Fourth Five Year Plan: Strategy and Programme," in August 1965. This plan was the end of the community development type of approach that had endured as official government policy since 1952. In its place was to go a program of state support for agricultural entrepreneurs. Especially important was the decision seriously to embrace the Intensive Agricultural District Program that had emerged from the Ford Foundation report of five years earlier but had languished, with tepid support, ever since. Explicitly, Subramaniam's plan picked up on the fact that Rockefeller Foundation work in India, ongoing since 1956, had by this time developed successful high-yielding maize varieties. Indian agricultural scientists were also aware of the Rockefeller Foundation's work in Mexico that had produced successful high-yielding varieties of wheat. Similar reports from the Philippines and Taiwan indicated high-yielding varieties of rice were also available.[118]

According to Francine Frankel, it was the embrace of science, and its promise of significantly increased yields, that allowed Subramaniam to win approval for his plan in the cabinet, the Planning Commission, and the Parliament. Adoption of intensive practices on 12 percent of the farmland could bring over 80 percent of the increased grain production that was needed by 1970–71.[119]

Despite winning on the issues of incentive prices and more science, Subramaniam still had massive problems to face caused by the weather. By the end of 1965, he knew that the monsoon rains from June to November had been desperately low. Thus the harvest that should have arrived by December–January was less than was needed, portending scarcity, higher prices, hunger, and further erosion of India's foreign exchange position in 1966.[120]

Not only did drought hit in 1965–66; the monsoon rains failed on a massive scale again in 1966–67, thus throwing India into the most prolonged and dangerous food crisis since the Bengal famine of 1943. Production dropped from 89 million tons in 1964–65 to 72 million tons in 1965–66 and to slightly less than 75 million tons in 1966–67.[121] Food grain imports jumped to over 10 million tons in 1965–66,[122] the only defense against a return to the awful horror that India had suffered over twenty years earlier. As if famine were not enough of a tragedy, the drought also exacerbated the other long-standing problems of agriculture in India. India's need for grain and aid gave donor countries, especially the United States, an opportunity to pressure India on such issues as devaluation of the rupee, which Americans advocated. When India did devalue the rupee in June 1966, stalled American aid resumed within ten days.[123]

Subramaniam's embrace of agricultural science and technology occurred in a time of crisis, but it was based on an infrastructure that had been building since before the beginning of the Indian Republic in 1951. What is important to note here is that the new wheat varieties were neither created nor disseminated in a vacuum. India had explored, sometimes with great ambivalence, the prospects of agricultural science since the early 1950s. No doubt existed that the government, first under Nehru and then under Shastri, wanted to increase yields, better the lot of India's poor, and turn India into a modern country. Nehru tried to make the transformation without allocating a great deal of capital to the rural areas and hoping that social reform would raise the production levels. Shastri, although he did not live to see the full development of the intensification program, set the wheels in motion that allowed Subramaniam to transform the development project and embrace science, not social reform, as the engine of change. After Shastri's death in 1966, Indira Gandhi, Nehru's daughter, succeeded to the prime ministership. Despite her sympathies with the social reform of her father, she did not attempt to reverse the commitment to capital-intensive agriculture made by the Shastri government.

In the United States, the problems of India continued to be seen in offical circles as largely a problem of too much population and the ensuing threat to American security if India's poor became rebellious. In India, for the social reformers the intensive agricultural program was a sad capitulation to the evils of the wealthy landowners of India and to the imperialist ambitions of the Americans. For the latent capitalists of India, embrace of intensive agriculture was seen as Indian emergence from the stifling hand of a huge and awkward government bureaucracy that was incapable of keeping prices down and getting people fed. Whatever the vantage point and views of the commentator, it was clear that the engagement with the promises of the wheat breeders was both a political and a scientific expedition. Chapter 10 tells the story of wheat breeding in India and the transformation in wheat production launched by Shastri and Subramaniam.

9

Wheat Breeding and
the Reconstruction of
Postimperial Britain,
1935–1954

Most discussions about the green revolution focus on it as a package of technologies used to increase cereals production in the less industrialized world. Adoption of this package is generally asserted to stem from a crisis of overpopulation. The argument of chapters 6 through 8 was that a better understanding of the green revolution comes from recognition of deeper political questions, particularly those centered around national security issues. The United States promoted high-yielding agriculture in countries such as Mexico and India for reasons of its own perception of national security. Mexico and India, similarly, were driven to adopt the technologies in order to foster their own processes of industrialization and preservation of national security.

These last two chapters bring forth yet another dimension: adoption of high-yielding practices was by no means confined to the less industrialized countries. Both Britain and the United States also had a green revolution. Moreover, industrialized countries adopted the high-yielding practices for reasons related to questions of national security.

Revitalization of the British countryside, especially in its ability to produce wheat, had been a fond dream of many agriculturalists and agricultural scientists since the 1870s. Although the food emergency of the First World War temporarily rekindled an interest in British wheat production, its effects did not last after 1920. In contrast, the events during and after the Second World War sparked a rebirth of production and the construction of an integrated agricultural research system. Wheat breeding

was a key component of the postwar agricultural reconstruction in the United Kingdom. This chapter focuses on why and how Britain decided to adopt the technology
package of the green revolution on a timetable that paralleled events in India.

Agricultural Renaissance in Britain, 1936–1947

"Dismal" is probably the adjective best suited to describe agriculture and agricultural science in the 1930s, at least among those British who supported more domestic production of foodstuffs. Price guarantees in the Wheat Act of 1932 encouraged
British wheat farmers to increase their production, but the increase was not enough
to affect imports significantly. In 1939, for example, wheat was grown on 1.76 million acres, up from about 1.2 million acres in 1931.[1] Livestock production prevailed,
and grassland occupied 20.9 million acres at the outbreak of war in 1939.[2] Imports
accounted for about 70 percent of British calories consumed.[3] British policy remained
oriented toward the goal of providing cheap food for the urban working class, largely
through imports.[4]

Demoralization and a sense of futility may also have begun to pervade the small
network of agricultural scientists in Britain. Despite the start-up of the Agricultural
Research Council in 1931 and the council's establishment of its own research
center on animal diseases in 1937, in general the 1930s were not prosperous times
for research scientists. Moreover, the scientists on the council continued to be constrained by their partial subordination to the Ministry of Agriculture.[5]

Livestock agriculture was the only area of British agriculture that had a modicum
of prosperity in the years between the wars. Arable agriculture, particularly wheat
production, languished as foreign imports continually battered domestic producers,
a situation that surely did not improve the morale of wheat breeders at the Plant
Breeding Institute (PBI) in Cambridge. Moreover, Rowland Biffen's retirement as
director of PBI in 1936 brought to light the fact that the institute was beginning to
drift and no longer had a clear sense of a useful research agenda. A self-evaluation
report prepared for the visit of scientists of the Agricultural Research Council in April
1936 even raised the possibility that cereal crops would not be the prime focus of
PBI's future work. Ministry of Agriculture officials were surprised to learn how
strapped the institute was for financial resources and qualified researchers; in addition, PBI looked to the Agricultural Research Council for guidance.[6]

Although the mid-1930s were a nadir for the fortunes of British wheat breeding,
within a decade this science enjoyed a remarkable renaissance that vastly increased
its funding, personnel, research facilities, and respect from those in political power.
One person, Frank Leonard Engledow (1890–1985) (Figure 9.1), played an important role in this rebirth. Engledow provided quiet leadership both in wheat breeding
and in many other dimensions of British and world agriculture. In contrast to Daniel
Hall and Christopher Addison, who before 1939 had been stalwarts of a revitalized
agriculture, Engledow never occupied a major office in the cabinet or civil service.
Rather, he served primarily as a researcher, teacher, and government adviser from
Cambridge University from the 1920s through the 1980s. He clearly had a vision for
a British agriculture that was far more productive and robust than Britain had known
since the 1870s. Much of his thinking was driven by considerations of food supply

Figure 9.1   Frank Leonard Engledow. Courtesy of Ruth Engledow Stekete.

and national security rather than agriculture simply as an industry managed for economic reasons alone.

Frank Leonard Engledow[7] was a child of the working and lower middle classes, as his father was a police officer and his mother worked as a maid. Nevertheless, the Engledow parents wanted their children to rise through education, and Frank, their fifth child, fulfilled their every hope by his eventual appointment as a professor in Cambridge University.

He began his work in higher education at University College London, where he studied mathematics and physics, an indication of quantitative interests that would persist throughout his career. At the encouragement of his teachers, he entered St. John's College, Cambridge, in 1910. It was here that Engledow switched his interests from math to applied biology and earned a first in part I of the natural sciences tripos in 1912. In the fall of that year, he began work toward his diploma in agriculture with Rowland Biffen in the newly founded PBI. He emphasized the merger of two new fields, genetics and statistics, in his study.

Just before the outbreak of war in 1914, Engledow enlisted in the army and served in India and Mesopotamia. He was awarded the Croix de Guerre in 1918 and then served a brief stint as director of agriculture for Iraq[8] before returning to Cambridge in 1919. While he was away, his two companion books were Bateson's *Principles of Heredity* and Udney Yule's *Statistics*. Upon his return to Cambridge, he submitted a thesis, read by Bateson, and was admitted as a Foundress Fellow to St. John's College in November 1919, at the age of twenty-nine. There he began what was to be a lifetime association with Biffen, PBI, and Cambridge University.

Biffen had been the first to demonstrate that a physiological characteristic in wheat—resistance to a rust—was inherited as a Mendelian trait. Engledow's interests derived from Biffen's early work, and he spent the 1920s in a series of investigations on how a more complex physiological trait—yield—was also genetic. Engledow's work united the concerns of the geneticist, the plant anatomist, the field expert, and the farmer. Measurement of defined characters and statistical analysis provided the glue that held these disparate concerns together. Engledow also had broad interests in such fields as agricultural meteorology, plant variety nomenclature, the baking qualities of wheat, and the conditions of tropical agriculture in Britain's far-flung empire.

Engledow's work as a plant breeder resulted in the selection of four varieties of wheat that enjoyed a significant though not dominant role in British farming from the late 1930s through the 1950s.[9] He also bred a pea variety that is still used as a forage crop. In the mid-1920s, Engledow's interests began a slow shift away from straight research at PBI to more involvement in teaching and in service as an expert adviser on matters of colonial agriculture.[10] He began formal teaching as a lecturer in agricultural botany in 1925, and in 1930 he became the Drapers Professor and head of the School of Agriculture. This appointment was one of the most prestigious in agricultural science in the United Kingdom.[11] Financially it seems to have enabled Engledow, now forty, and his wife, Mildred Emmeline Roper, to build a new residence they called Hadleigh, which offered a large garden and room for three young daughters, soon to be joined by a fourth.

As an academic, Engledow began a series of consultations on tropical agriculture that were to expand his vision of the subject considerably beyond that of a statistically inclined geneticist, botanist, and plant breeder. In 1927 and 1929 he made trips to West Africa (Nigeria and Gold Coast) and to the British West Indies to advise on agricultural practices and research in the tropical colonies. In the 1930s, he traveled to Malaya to study rubber (1933), to India and Ceylon (now Sri Lanka) to study tea (1935–36), and again to the British West Indies (1938–39). After the war he went for a third time to the British West Indies (1945), to East Africa (1946), and to Southern Rhodesia (now Zimbabwe) and South Africa (1948). Additional trips occupied him in the 1950s.

One of Engledow's students was particularly important in wheat breeding. Benjamin Perry Pal (see Chapter 8) came as an Indian student to Cambridge in 1929, where he worked first with Biffen and then finished his doctorate with Engledow. In

1933 Pal returned to India, where he became the Second Economic Botanist of the Imperial Agricultural Research Institute in Pusa.[12] This post had been held previously by Gabrielle L. C. Howard, who with her husband, Albert, had been the main focus for wheat breeding in India (see chapter 4); later he advanced to major leadership posts in independent India's agricultural research. This connection between Cambridge and India solidified the spread of Mendelian genetics and scientific agriculture in India.

During the war years, Engledow was intensively involved in Britain's wartime food production issues, a series of tasks that his family remembers as very consuming of his energies.[13] He was appointed to the Agricultural Research Council in 1942[14] and also served on the Agricultural Improvement Council of England and Wales, a body to advise the minister of agriculture on issues of science and production. In 1943 Engledow represented the United Kingdom at the organizational meeting of the United Nation's Food and Agricultural Organization. Perhaps most important, Engledow also served on the Joint Advisory Committee on Higher Agricultural Education, appointed in 1944 to plan for the comprehensive restructuring of the educational and scientific infrastructure for British agriculture. On this committee Engledow was the clear visionary who saw a potential for British agriculture that had not been seen in the century since enactment of the Corn Laws.[15] He argued that agriculture should be governed by its ability to supply food for the nation, not simply as an industry with social and political constituents.[16]

For his scientific and policy achievements, Engledow became a knight bachelor in 1944 and was elected to fellowship of the Royal Society in 1946. Although earlier service in agricultural science had resulted in knighthoods and/or elections to the Royal Society (e.g., for Hall, Biffen, and Middletown), Engledow's recognition came at a time when the government of the United Kingdom was making a fundamental shift toward a highly productive agricultural industry, a shift in which he was a major promoter.

In 1943 Engledow became a founder trustee of the Nuffield Foundation,[17] an appointment that brought him into contact with a wide range of issues connected to science and research, as the Nuffield Foundation developed a program of support that was modeled partially on that of the Rockefeller Foundation in the United States. When in the United States for the organizational meeting of the Food and Agricultural Organization in 1943,[18] Engledow had visited the Rockefeller Foundation and talked extensively with Harrar, director of the Mexican Agricultural Program. He was impressed with the Rockefeller Foundation's work, which, in turn, perhaps inspired him to have the Nuffield Foundation prepare a study on the principles that should govern British agricultural policy. Begun in 1948, for a variety of reasons the report, *Principles for British Agricultural Policy*, was not issued until 1960. This report made the general argument that the need to maintain strategic preparedness for the possibility of war was the major justification for keeping British agriculture larger and more productive than it would be under laissez-faire market conditions.[19] A more productive British agriculture became government policy in 1947, before the Nuffield report was even started, so it is clear that forces other than the foundation were at work in altering the basic stance of the United Kingdom toward agricultural production and the science that underlies it.

His experiences in two world wars plus his work in the empire's far-flung colonies had made Engledow intimately familiar with the role of agriculture and food supply in the national security of Britain. More than most people, Engledow knew that food supply was crucial to military strength and that to achieve food security might require state intervention. Indeed, this theme dominated not only his contributions but virtually all reform efforts in British agriculture from 1939 into the 1950s.

War and the Climate for Change, 1935–1944

Engledow played a pivotal role in the reconstruction of British agriculture, especially after 1942. Prior to that time, British politics were not amenable to the political compromises needed to change agriculture, no matter how persuasively people like Engledow argued. The Liberals and the Labour Party were driven largely by a devotion to free trade in agricultural goods, the only mechanism they fully trusted to keep the price of food low for their constituents. Reformers on the Left, such as Christopher Addison and Daniel Hall, advocated science to make agriculture more productive, and they coupled their vision for science to land nationalization. The tiny Communist Party similarly shared a vision of a technically sophisticated agriculture operating on land owned by the state.

Conservatives, in contrast, tended either to line up with the free-trade position, because their positions as capitalist entrepreneurs were served by low-cost food for their workers, or to fear government interference and land nationalization because they were landowners. The Conservatives were not necessarily opposed to the practices of scientific agriculture, but as a party they did not want either to disrupt the economic patterns based on cheap (imported) food or to discuss state intervention in the farm economy when a clamor from the Left would argue for far more intervention than the Conservatives wanted. In addition, some Conservatives undoubtedly still saw the countryside as a quaint, nostalgic, green refuge from the dismal gray of industrial Britain. These romantics may have derived their incomes from the industrial bleakness, but they wanted to take their holidays and retreats in a place that scientific rationalism had not penetrated.

Without the outbreak of the Second World War, this stalemate might have lasted for many years. Plant breeding, as developed and advocated by Engledow and his colleagues, would have continued to play only a limited role in British agriculture, much as it had for the twenty-one years since the end of the World War I. Gathering war clouds over Europe, however, triggered preliminary movements by the government to improve the production levels of staple foods in Britain.

First, in 1935 the minister of agriculture appointed a committee to study the problems of food production in wartime. From this and subsequent work came the Agriculture Act of 1937, which provided further subsidies of wheat, barley, and oats. These inducements to higher production were expanded yet again in May 1939. In 1936 came the preparations to reform the War Agricultural Executive Committees, composed of progressive farmers in each county plus technical advisers. These committees, when activated in 1939, had the power to inspect the technical operations of each farm in the county, to recommend improved practices for those farmers found lacking, and, if necessary, to remove a farmer and turn the land's operation over to another farmer who would manage it with more intensive inputs.[20]

War Agricultural Executive Committees, which continued to operate for the duration of the conflict, played a significant part in raising the output of wheat harvested in 1939 (1.64 million tons) to the wartime high of 3.44 million tons in 1943. Similarly, potato production went from 4.35 million tons in 1939 to a wartime high of 8.70 million tons in 1945.[21] As advocated at the start of the war by Sir John Boyd Orr, production of staples such as wheat and potatoes was a critically important policy to pursue,[22] one that required each domestic farm to use the most intensive methods.

Britain had used War Agricultural Executive Committees in the last stages of World War I (see chapter 4), so the concept of stringent land-use regulation in the pursuit of increased yields was not new. These committees, however, represented a potentially draconian intervention of the state into the affairs of private landholders and tenant farmers. No political party seriously pursued such measures until war arrived in 1939, but it was not until 1940 that the full fury of the war was upon Britain; 1940 was also the first full year in which the programs of the War Agricultural Executive Committees could come into effect.

One other prewar step that was taken was in the field of agricultural education. In response to an appeal in 1938 by the Bedfordshire County Council for a small grant (£200) for a demonstration farm, officials of the Ministry of Agriculture argued that the ministry was ill equipped to provide advice or help on educational matters. Furthermore, the officials noted that a Consultative Committee on Agricultural Education was to be created, which would provide advice on how a coherent system of demonstration farms could be supported. In addition, ministry officials planned to launch a small committee that would conduct a full inquiry into all aspects of agricultural education.[23]

Serious reconstruction of Britain's system of agricultural education was needed to remove a fundamental constraint, already identified in the 1930s: Britain could not get the highest return from its agricultural research (i.e., higher national yields) because farmers, farmworkers, and their advisers were neither numerous enough (advisers) nor fully trained (farmers and farmworkers) to employ the potential of new developments. The problems were extensive: insufficient numbers of university students sought agricultural degrees; many of these had to emigrate for lack of work in the United Kingdom; domestic farmers were underserved by technically proficient advisers; and opportunities for students to get technical training at the advanced secondary level were inadequate.[24]

Potential disruption to food imports due to war suddenly made the undeveloped state of agricultural education easily visible. Research institutes often had little connection to universities and thus to the education of postbaccalaureate students; technical advisers of the Ministry of Agriculture had no necessary connection to research institutes or universities; demonstration farms were few; and support for education among those eighteen years old and younger was fragmented and poorly supported. Whether agricultural education was a responsibility of the Ministry of Education or the Ministry of Agriculture was not clear. Also unclear was whether farmers needed a university-level (i.e., baccalaureate degree) education, or whether they should confine their studies to schools lower than university. Was agriculture even a fit subject for universities to offer?

These questions all suggested that, despite major British contributions to scientific research in agriculture, the country had never embraced an industrialized system of agriculture. Without such a system, however, an industrialized country would probably be forced to rely on imports, because working people would find more lucrative employment in industry and thus abandon farmwork. Only highly productive, mechanized agriculture could expect a level of labor productivity that would enable agricultural labor to earn a wage comparable to that possible in a factory. Only products that were subsidized, not imported because of perishability, or prohibited from import by tariff or quota could hope to survive in an industrial country without a significant commitment to high-intensity agriculture.

These, then, were the conditions that characterized Britain in 1939. Most of its laboring people had left the farms for the cities. Given the commitment to free trade in food there was no need, short of war, for an educational system to support ever more complex technologies in agriculture. Without the educational system to generate a culture of commitment to high-intensity agriculture, it was not even possible to generate the necessary technologies in a form that was applicable to the specific requirements of British farms. Disarray in agricultural education, therefore, constituted a serious, long-term impediment to increased domestic food production.

It was this problem that the Ministry of Agriculture was attempting to come to grips with in late 1938 and early 1939. Because of unsettled conditions in Europe, ministry officials were not able to convene a working committee until March 1940. Thomas Loveday, vice chancellor of the University of Bristol, agreed to chair a committee whose charge was "to review the position of agricultural education of University Departments of Agriculture and Agricultural Colleges, as affected by war conditions, and to recommend from time to time such financial assistance from the Exchequer as may be considered appropriate, bearing in mind also the possible demand for higher agricultural education in the immediate post-war period."[25]

The Loveday Committee, as it came to be known, worked diligently through July 1944, with its major task the allocation of money to the fifteen colleges and universities that had agricultural programs. For the most part, the committee kept allocations at a steady level during these war years. Some changes in funding patterns were necessary for programs that lost most of their students, suffered bombing damage, or had a special onetime need for a piece of equipment.[26] Thus the Loveday Committee did not do more than inventory existing programs and allocate limited funds in the way it thought would best preserve them. In short, until 1944, the committee did not delve into a serious examination of the philosophy that should guide agricultural education, nor did it direct its attention to any programs other than the higher education programs in colleges and universities.

Not only did the Loveday Committee not get into the business of serious reform of Britain's agricultural education; it appeared at one time to have a rival. In July 1941, one year after the Loveday Committee first met, the Board of Education and the Ministry of Agriculture appointed a separate committee "to examine the present system of Agricultural Education and to make recommendations for improving and developing it after the war." This committee, known as the Luxmoore Committee after its chair, Lord Justice Luxmoore, was explicitly asked to go beyond higher education and to include formal programs for elementary and secondary students and informal programs for landowners, bailiffs, farmworkers, land agents, agricultural engineers, teachers, and advisers.[27]

The charge to the Luxmoore Committee clearly overlapped that of the Loveday Committee, especially on the mandate to chart postwar policy for agricultural education. Unfortunately, the surviving records of both committees do not clarify why the second committee was appointed, nor why the charges to the two groups were not more distinct. Fear of being usurped by a rival committee, however, led one of the Loveday Committee members, John S. B. Stopford, vice chancellor of the University of Manchester, to resign in disgust; he believed the Loveday Committee had been supplanted in the important task of reforming an important part of British education.[28]

Although on the surface the Luxmoore Committee's charge appeared to give it the enviable task of designing a postwar agricultural education system of the broadest scope, without the time-consuming tasks of allocating money to existing programs, in the long run the Luxmoore Committee did not prevail in its 1943 report.[29] Two factors probably accounted for its ultimate lack of influence. First, the Luxmoore Committee was split, not unanimous. One member, who thought agricultural education, at least at this level, should remain integrated with other educational programs, objected to having elementary- and secondary-level programs taken out of the hands of local education authorities. Second, the Luxmoore Committee was small (eight members), worked only two years, and lacked the breadth of membership required to revamp an industry as complex and large as agriculture.

Reconstruction Begins 1944–1945

By 1944, little doubt remained about the outcome of the war. To be sure, a great deal of uncertainty still attended the exact course and time schedule that would ultimately result in the collapse of Germany, Italy, Japan, and their allies. That Britain would not be forced to capitulate to Hitler, however, was not in doubt. In such a climate it was possible to begin a serious reform of the agricultural industry of the United Kingdom. Given the seriousness of the food supply situation that had begun in 1939, essentially no argument existed on whether Britain should substantially increase the intensity of its farming sector.

The minister of agriculture, R. S. Hudson, spoke for the government in Parliament on 20 January 1944 about the recently released Luxmoore report. He announced the government's intention to accept the report's recommendation for a more centralized national advisory service for farmers (in the United States this was called the Cooperative Extension Service), but he explicitly rejected the notion of centralized control from London over formal education programs. Instead, the counties would still be responsible for technical education,[30] and universities would remain in charge of higher education and training for research. Hudson wrote in May 1944: "The Luxmoore Committee furnished us with valuable advice on the main principles which should govern this branch of agricultural education, but I am satisfied that a good deal of further thought and planning is needed before we attempt to determine the shape of higher agricultural education for the years to come."[31]

In early 1944, Ministry of Agriculture officials began the work of assembling an extraordinarily broad-based committee to advise on how to create an educational infrastructure for the support of British agriculture. Two committees emerged, one for higher education issues and one for matters of secondary education. Thomas Loveday, then nearing seventy and wanting to retire from the vice chancellorship of the University of Bristol, agreed to chair the new ventures, and his committees were called, respectively, the Loveday Committee Reconstituted and the Junior Loveday Committee.[32]

Jointly with the Board (later Ministry) of Education, the Ministry of Agriculture recognized the need to include representatives from secondary and higher education. Educational members, however, were not sufficient. Farmers, representatives of the National Farmers Union, directors of research institutes, members of War Agricultural Executive Committees, representatives of the National Federation of Women's Institutes and of the Women's Land Army, and a representative of the Transport and General Workers' Union[33] served on the two committees. Thus, simply in terms of viewpoints represented, the twenty-three members of the new Loveday committees had vastly more strength than the earlier Luxmoore Committee. Moreover, the Ministry of Agriculture envisioned that shortly after the committee began work, Loveday would spend half of his time on it for five years and receive a salary of £1,000 per year (a generous amount for an academic at the time).[34] In short, revamping an entire system for knowledge generation for a country's agricultural industry was finally seen as the complex task it truly was.

The Loveday Committee Reconstituted began its work in July 1944 with a broad charge: "to consider the character and extent of the need for higher agricultural education in England and Wales and to make recommendations as to the facilities which should be provided to meet the need." Its review of the current situation indicated that in all of England and Wales, only 1,387 students were pursuing degrees at the university level in agriculture, down from over 2,000 in 1938–39.[35]

As the committees began their work, it became clear that one of the members of the Higher Agricultural Education Committee, Sir Frank L. Engledow, professor at Cambridge University, would play the role of chief visionary for the potential of British agriculture. Engledow had not been on the Luxmoore Committee, but he had prepared a series of papers for it, which were brought forward to the Loveday Committee Reconstituted. Their scope was broad and reflected Engledow's interests in production agriculture on a wide front.

In the first paper[36] Engledow outlined the problems of British agricultural education and sought objectives that the country should pursue. He argued that the existing agricultural education was inadequate in scope, unsuitable in form, without sufficient central supervision, and without clear objectives. Education, research, and advising functions were lumped with each other in a confusing, haphazard manner. In order to create a new system for higher agricultural education, Engledow proposed three objectives:

1. Promote general intellectual development.
2. Develop the student's understanding of the physical, biological, and economic principles that determined forms and practices of agriculture; this objective was particularly important for helping the adoption of new technology.

3. Promote technical efficiency, both in production agriculture and in agricultural research.

Engledow went on in this paper to identify two types of students that would attend the universities. Some would be headed for work in management, others for academic and research work. Engledow strongly believed that, for the latter students, universities should combine work, for example, teaching with research or teaching with advising.

Engledow's promotion of a better educated cadre of workers who would be capable of developing and using more productive technology was reinforced by a paper prepared for the committee by the Agricultural Research Council. The council estimated that only about 500 total workers were assigned to its own research stations and the Ministry of Agriculture stations. Britain needed at least an additional 300 to 400 workers, and that number should increase to about 1,000 within fifteen years. Universities therefore needed to move their production of research workers from the approximately 15 per year to about 30 per year. After the war, biological science would quickly replace physical science as the training most needed by the country.[37]

Work on the new Loveday committees reflected the postwar consensus about British agriculture. No clear opposition to efforts to increase the technical efficiency and intensity of British agriculture was to be found. On the contrary, a high degree of overlap emerged among groups that did not necessarily have common interests: the National Farmers' Union, the Transport and General Workers' Union, and the National Union of Agricultural Workers. The National Farmers' Union represented 150,000 farmers—the backbone of the middle-class managerial cadre of the country-side. These were the managers who hired the landless labor needed to run British agriculture and as such would be expected to resist the demands for better wages, working conditions, and living accommodations put forth by the Transport and General Workers' Union. All three groups, however, endorsed the notion that education, science, and technology were important aids to a better life.

For their members, the National Farmers' Union wanted to have education to degrees for research workers, teachers, advisory officers, technical inspectors for the Ministry of Agriculture, and technical officers of marketing boards. For the farm institutes, which provided technical training to farmers and farmworkers, the union did not think a degree was essential, but in any case the training offered should be directed more to the business management side of agriculture rather than the scientific. This position to the Loveday Committee Reconstituted was very similar to the position the National Farmers' Union had delivered to the Luxmore Committee in 1943. There it explicitly rejected land nationalization and favored a more highly developed educational and research system that could produce and disseminate new technical practices.[38]

Similarly, the Transport and General Workers' Union supported increased educational resources if British farmers were to compete with imports from other countries. Educational resources also had to be seen as educational opportunities for children, and the union wanted rural children to have opportunities equal to those in urban areas. In addition, the union wanted more flexibility from the universities in accepting students from rural areas, who often had lesser educations than their urban counterparts. Similar themes, such as the need to provide educational opportunities for farmworker children, came from the National Union of Agricultural Workers. This union was particularly keen on having more scholarship assistance and more places at university directed to children of farmworkers, not just to rural children in general.[39]

One of the only statements suggesting a more tepid interest in augmenting technical knowledge in British agriculture came from the Central Landowners' Association, founded in 1907 and representing over 10,000 land and estate owners. The association did not oppose the intrusion of technical expertise into the British countryside, but its position paper spoke to need for an estate owner to continue the traditional practice of being involved in "public duties" at both the local and national level. For this sort of work, education at the university level was important, but it should not become too vocational too early. Liberal studies, such as history, would be useful for the landowner who might engage in "public duties," but degree work in agriculture was certainly appropriate for the sons of large landowners.[40] It is quite probable the Central Landowners' Association was simply representing the traditional view of the remnants of the landed upper class, who occupied positions of power and prestige in London and the rural areas. In some ways these landowners may have feared or resented the intrusion of technically intensive farming into a lifestyle built on the wealth of extensive holdings. Even so, however, the association's position was one of mild support and certainly not opposition to intensification.

In its report at the end of 1945,[41] the Loveday Committee Reconstituted fully embraced the need for more science and technology in British agriculture. It defined agriculture as the science and practice of producing food from the land and marketing it. In the committee's view, agricultural education was to develop the men and women who would perform tasks intelligently and effectively to promote the welfare of agriculture. Higher education had to produce the leaders who would make British agriculture more productive. Graduates of higher education would work in a wide range of occupations, both in the United Kingdom and in the colonial empire. These included production agriculture, estate management, advisory officers, teachers, researchers, economists, engineers, and in various commercial activities dependent upon agriculture. Education in support of research was especially important because "research is the foundation of technical efficiency and progress." Fundamentally, the committee believed that "The welfare of agriculture depends on the constant application of new knowledge in improved practice."[42]

In its final report, the committee also devoted a great deal of effort to providing detailed estimates of needs: the number of trained professionals and their disciplinary backgrounds, the number of student positions at specific universities, and the finances to support the students and their professors. While these details are impressive in documenting the thoroughness of the tactical work of the Loveday Committee Reconstituted, the most important dimension of the study remained the consensus achieved among landless labor, production farmers (tenants and owners), and large estate owners about the importance of science and technology. Given the strength of this consensus across powerful class divisions, it was clear that Britain's postwar planning in agriculture would head in the directions outlined by Sir Frank Engledow.

Consensus for Reconstruction of British Agriculture, 1944–1947

As 1944 and 1945 progressed, it became increasingly clear that major changes would forever alter the face and character of British agriculture, long after the emergency of war had ended. First, in 1944 the Agriculture (Miscellaneous Provisions) Act brought into existence the National Agricultural Advisory Service, which for the first time gave Britain a comprehensive advisory system. This service, headed by J. A. Scott Watson, was the one recommendation of the Luxmoore Committee that found its way into policy. Watson, Sibthorpian Professor of Agriculture in Oxford University, was at the time of his appointment also serving on the Loveday Committee Reconstituted. Once a national commitment to provide expert technical advice to farmers was in place, it was unlikely that Britain would ever again let its agriculture languish as it had for the previous century.

In April and May 1944, further evidence that the restructuring was under way was provided by an extraordinary conference organized by the Royal Agricultural Society of England. Long a stronghold for those of the landed aristocracy who were also interested in technological progress in agriculture, the Royal Agricultural Society of England usually did not pay much attention to the hopes and aspirations of small-scale tenant farmers and the agricultural working class. This conference, however, convened twelve very different organizations, which issued a joint statement at the conclusion of their deliberations. A mere list of the participants indicates the breadth of the spectrum, from landowners to tenants to working laborers to the professional experts who managed British farms:

> Royal Agricultural Society of England
> National Farmers' Union
> Group of Peers
> Council of Agriculture for England
> Council of Agriculture for Wales
> Central Landowners' Association
> National Union of Agricultural Workers
> Transport and General Workers' Union
> Land Union
> Chartered Surveyors' Institution
> Land Agents' Society
> Land Settlement Association

The joint statement, endorsed unanimously, showed a similar breadth of issues involved in the restructuring of agriculture:

> The fundamental purposes of long-term policy should be the proper use and management of the agricultural land of this country for the production of the foodstuffs which it is best fitted to provide and which are most required to satisfy nutritional needs while maintaining the fertility of the soil, the raising of the standards of rural life and the increase in the rural population.

It is essential on national grounds that British agriculture should be maintained in a healthy condition, sufficiently prosperous to ensure a stable level of prices which will yield a reasonable return to the producer and on the capital employed in the industry and a scale of wages sufficient to secure a standard of living comparable to that of urban workers. . . .

In return for a guaranteed price level, all owners and occupiers of rural land must accept an obligation to maintain a reasonable standard of good husbandry and good estate management, and submit to the necessary measure of direction and guidance, subject to provisions for appeal to an impartial tribunal.

Agricultural education and opportunities for advancement within the industry should be expanded and research developed.[43]

Thus all of the interests most directly concerned with the fate of the countryside participated, and their combined statement represented an agreement on three critical points: (1) owners, managers, and labor all had a right to earn a decent (but unequal) living from the land, which would remain in private ownership; (2) owners and managers were willing to accept state intervention on the level and quality of their managerial practices in return for guaranteed prices; and (3) agricultural education and research were seen as advantageous to all.

Politically, consensus on the many issues affecting agriculture was more difficult to obtain in 1944. Prices of agricultural products, for example, were the key to prosperity among the rural classes and to the abilities of consumers to have a high-quality diet. Although the conference at the Royal Agricultural Society of England was able to agree that everyone should prosper and be well fed, the major political parties had very different attitudes toward subsidies, levies, tariffs, the rights of unions, and wages. Similarly, the major political parties had different ideas about ownership of the land. The Communist Party, for example, adamantly favored nationalization. Labour favored the state's right to take land but thought it was inadvisable to take control of all of the land too rapidly. The Liberals and the Conservatives were more reticent about land nationalization and probably had doubts that it was wise. They, too, however, accepted the legitimacy of state intervention to control the use of agricultural land, a stance that erodes a commitment to private ownership.[44] It was perhaps this underlying, potentially explosive debate over prices and landownership that made Churchill's wartime government reluctant to announce a long-term policy for agriculture during late 1944 and early 1945, despite persistent questions about the government's intentions in the House of Commons.[45]

In stark contrast to the debates over prices and landownership, the legitimacy of science and technology was unquestioned. All political parties, and indeed every other group that studied agricultural policy, favored more intensive education and research for postwar British agriculture.[46] Science, technology, and education were seen to offer the hope of advancement, prosperity, and national security to labor, tenant farmers, and landlords. So long as the majority of Britons, who were urban, could be assured of adequate supplies of good, nutritious food at reasonable prices, the way was clear for a substantial increase in British capacity for agricultural science and research.

War ceased in Europe in May 1945, and within weeks Churchill called for elections, to be held in July. Labour won an overwhelming victory: 47.6 percent of the popular vote (a plurality to the 39.6 percent of the Conservatives) and 393 out of 640 seats in the House of Commons (61 percent).[47] Clement Attlee became prime minister in late July.

Attlee's government was unexpectedly plunged into a financial and food crisis by the rapid end of the war against Japan and the precipitous ending of the American lend-lease program. Without the American aid, the United Kingdom, still dependent upon imports, had no immediate way of paying for those imports on its own account. Rebuilding of export markets shattered by the war would be necessary before the United Kingdom had enough foreign exchange to purchase its accustomed supplies of imported wheat and other food supplies. To compound these difficulties, which were primarily a domestic problem, within six months the Attlee government knew that wheat supply was a widespread problem in many parts of Europe and Asia (see Chapter 6).

Collaboration between the United States and Britain brought a resolution to the 1946 world food crisis without the eruption of a major famine. Nevertheless, even if a wartime consensus about the need for a more productive agriculture had not been formed in 1944, the events of 1945–46 made it abundantly clear that the reconstruction already under way must continue. In addition to continuing the development of the National Agricultural Advisory Service, the new government made two other initiatives that cemented the reforms in place. First, Attlee's government cut through the debate over prices for domestic agricultural goods by forging a compromise with farmers in the Agricultural Act of 1947. Second, the government made a decision to increase substantially the amount of support to all agricultural research as well as incorporating the researchers and research stations into a unified national system.

The Right Honorable Thomas Williams, minister of Agriculture through all of the Attlee administration (Figure 9.2), had the prime responsibility for creating the Agricultural Act of 1947. Williams, born in 1888, was elected a Labour MP from the western district of Yorkshire in 1922, a seat he held until his retirement in 1959. He began his experience in government as a parliamentary private secretary to Noel Buxton, minister of agriculture in the first Labour government in 1924. When Labour again came to power in 1929, Williams served as parliamentary private secretary for the minister of labour until 1931. During a stint out of government, Williams wrote a book, *Labour's Way to Use the Land*, that enhanced his position as a major Labour spokesperson on agriculture.[48] During World War II he served as the parliamentary secretary of the Ministry of Agriculture and Fisheries. Thus when Labour was elevated to power in 1945, he clearly had the appropriate experience to become minister of agriculture and fisheries, a post he occupied from July 1945 to October 1951. Williams served in the House of Lords from 1961 until his death in 1967 at the age of seventy-nine.[49]

Williams and the Labour government faced a horribly complex and daunting challenge when they came to prepare the centerpiece of their agricultural legislation, the Agriculture Act of 1947. Undoubtedly the primary driver for the Labour government was its desire, indeed its political need, to rebuild and improve the material conditions of life for working- and middle-class Britain. The Labour Party, a creation of and accountable to the trade unions, drew many of its members and leaders, much of its financial support, and virtually all of its votes at the polls from

Figure 9.2   Tom Williams, minister of agriculture, 1945–51. © Crown copyright, U.K. Ministry of Agriculture Fisheries and Food; photograph supplied through the courtesy of Lesley Holland.

the working class and the unions representing them. To the end of supporting a more abundant material life, the hallmark of the Labour program was to create conditions of full employment, a comprehensive scheme of social services, and a labor law that gave unions a stronger hand in dealing with the managers and owners of British industry. In terms of the empire and the larger world, Labour wanted to maintain Britain as a world power.

Agricultural policy had to be structured in ways that supported the overall goals of the Labour government. Parties internal to agriculture (the landowners, farmers, and farmworkers) perhaps could have been satisfied with a modest program that generated a larger demand for British agricultural goods at prices that were better than those obtained in the 1930s. Mechanisms to obtain such a program were varied and included price supports; income subsidies; and quotas, tariffs, and levies on imported goods. Farmworkers also would have demanded programs for better housing, higher wages, and educational programs to offer advancement for themselves and their children.

Such simplicity, however, was not to come to Williams. Extraordinarily powerful competing demands had to be accommodated before the government would have a workable agricultural policy. First and foremost was the simple fact that most Britons were not farmers or farmworkers; they were urban consumers who wanted food to be inexpensive. If foreign food was cheaper than domestic, then any move to more domestic production was surely to be difficult. At the same time, Britain ended the war with a much diminished capacity for paying for imported food. Prior to the war, British exports had earned enough money to enable Britain to pay for imported food and other products. After the war, the Labour government inherited a chronic problem of inability to earn enough foreign exchange, especially American dollars, to pay for the imports. Forgoing the imported food, however, created a danger of food shortage in the short term, which in turn would have further diminished Britain's capacity to export, which in turn would have necessitated further curtailment of imported food. Thus a vicious circle was waiting to descend on any government of the United Kingdom, a circle of disintegration that could, in fact, have led to the dissolution of the nation.

Agriculture was one of several "saviors" that the Labour government expected to use to solve the very real foreign exchange crisis. Increased domestic production would alleviate the need for imported food, and politically the costs of domestic production would be justified on the grounds that foreign exchange simply was not available to pay for imported food. Moreover, by reducing the foreign exchange requirements of Britain, agriculture could play an important part in keeping the United Kingdom at the financial, economic, and political center of the Commonwealth, a position that demanded a strong, stable, and convertible pound sterling.

Williams outlined the government's plan in a speech to the House of Commons on 15 November 1945. His cornerstone was a system of assured markets and guaranteed prices for milk, livestock, eggs, cereals, potatoes, and sugar beets. The minister of agriculture would discuss proposed price supports annually with the farmers, a practice that had begun in February 1945, even before the war ended. By establishing a guaranteed price sufficiently far in advance for key products, Williams argued that it would be possible to have the maximum extent of home production (thus saving foreign exchange) and prices agreeable to both farmers and consumers (thus avoiding resentment at the polls from disgruntled urban and rural voters). Williams went on to note that the government intended to continue to have the power to make sure that agricultural land was properly managed and equipped and that the National Agricultural Advisory Service delivered the best technical advice possible. County committees similar to the War Agricultural Executive Committees would continue to be the vehicle by which national policy was delivered to the local areas. As an afterthought, Williams also noted that world food supplies were seriously short, a brief hint of the world food crisis that would erupt in 1946 (see Chapter 6).[50]

Williams's speech in the House of Commons was focused largely on the domestic implications of agricultural policy, but internal studies made by the Ministry of Agriculture and Fisheries and the Ministry of Food in the winter of 1944–45 (when Williams was the parliamentary secretary of the minister of agriculture) fully developed the foreign exchange issue that shortly blossomed to its full effect. This study projected that domestic agriculture would replace £80 million per year of food imports at prewar prices by 1950–51. The report went on to note that severe rationing of food could yield the equivalent of £116 million per year at prewar prices and that the government did not want to contemplate such action. In order to have agriculture play its intended role in foreign exchange savings, the ministries recommended policies to move capital investment into agriculture, promote building of farm structures, ensure that labor supplies were adequate, control the technical quality of farming, and assure farmers of stable markets with adequate prices.[51] This study was not published until 1947, by which time the issue of foreign exchange was the most prominent crisis faced by the Labour government.

The Agriculture Act of 1947 worked its way through Parliament during 1946 and was given the royal assent on 6 August 1947.[52] It contained no real surprises, because Williams and the Labour government had indicated for nearly two years that Labour policy on agriculture would be based on the management of agriculture to meet the needs of the people as a whole, not just the landowners and farmers. To this end, the act was based on the incentives to farmers formed by guaranteed prices, reviewed annually with farmers, that were set far enough in advance for farmers to plan their production with full knowledge of the minimum price they would receive. For its part, the government had the power to raise and lower the guaranteed prices in a way that adjusted domestic production to the projected needs. It was this latter capacity that would allow the government to fine-tune the tension between the farmers who wanted higher prices and the consumer/taxpayer/voter who wanted lower costs. The major products of both animal and plant agriculture were covered, including wheat. Foreign exchange considerations formed the backdrop against which this adjustment would be made.

The act also contained punitive provisions that allowed the minister of agriculture to direct the technical practices that farmers must use and to dispossess farmers or owners of their lands if the directions were not followed. Comments in the House of Lords suggested that this was the end of a proud tradition of country gentry managing the land, but in fact the only new power of the government was its control over land use in peacetime as well as wartime.

The Agriculture Act of 1947 did not itself address the question of science and technology except in that it gave the minister of agriculture the power to decree the technical practices that would be used to serve the national purposes. This provision, taken together with the formation of the National Agricultural Advisory Service in 1946 and the movement under way at both the Ministry of Agriculture and the Agricultural Research Council, made it clear that national policy in Britain was now driven by the need to intensify the technology used in domestic production. British farmers would be guaranteed good prices, provided with new technology and advice on how to use it, and, if need be, told to use the new technology or else lose their land. In its internal deliberations the National Executive Committee of the Labour Party was explicitly attuned to the uses it had for the agricultural industries:

> The Government has launched a great long-term programme for growing more food at home. As we grow more food, we shall become less dependent on imported supplies. British agriculture will be one of the greatest dollar savers of all. British farmworkers and farmers will literally be "digging for dollars" on British soil.
>
> The Government's plan is, as announced by Mr. Tom Williams . . . on 22 August, 1947 to increase home food production by as much as £100,000,000 by 1951–52. This represents an increase of 50% over pre-war, 15% over the war-time peak of 1943–44, and 20% over 1946–47. . . .
>
> Prices for farm produce, acreage payments and subsidies are being increased. . . . The purpose of the increased financial assistance is threefold: to give farmers enough money to buy the necessary additional machinery, livestock and so forth; to cover the increase in the farmworkers' minimum wage and in other costs; and to create confidence in the future by establishing higher prices up to 1951–52. . . .
>
> This is the greatest programme for the development of British agriculture ever launched in peace-time. The challenge is a massive one, for without the utmost

efforts of the agricultural industry, either undernourishment or unemployment may have to be faced. The Government is confident that all concerned with agriculture will answer the nation's call.[53]

Perhaps only one issue in Labour's postwar policy was a drastic shift from the general thinking about agriculture that had begun in the 1930s. By 1947, Labour had abandoned all thoughts of a complete land nationalization, as had earlier been advocated by Christopher Addison, Sir Daniel Hall, and Tom Williams. The 1947 act provided the minister of agriculture with the statutory authority to purchase land compulsorily and to dictate land uses, but the party apparently concluded that nationalization was not the way to proceed. It is possible that it feared the disruption that would surely attend such a move and might have destroyed the land's abilities to help get the United Kingdom out of its foreign exchange crisis. Alternatively, it perhaps concluded that strength at the polls would not be significantly enhanced by land nationalization. Whatever the reason, after the passage of the 1947 act it was clear that Britain fully intended to have a much more productive agriculture but that the increases would come through price programs, technical advising, and substantial investment in new agricultural research, not the radical reform of land nationalization.

Expansion of Agricultural Research and Wheat Breeding, 1942–1954

Support for increased research was inevitably present from the scientists concerned, and agricultural scientists had called for increased resources since before the turn of the century. What was new, therefore, in the years after World War II was the growing acceptance of their ideas by Parliament.

War emergency prompted the government to increase funding for research from about £170,000 per year in the late 1930s[54] to about £570,000 for the year ending 31 March 1942.[55] For the remainder of 1942 and 1943, the Agricultural Research Council debated the means by which research could be restructured. Frank Engledow prepared a memo on needed changes in Britain's research capacities after he joined the council in 1942. His memo, plus that of the council's secretary, W. W. G. Topley, were the basis of extended discussions on 23 November 1943. Topley summarized the proceedings in terms of two major themes: the relationship of the agricultural research institutes to universities, and the problem of recruiting suitably trained and motivated personnel. Topley concluded that the five research institutes at Cambridge University, including the Plant Breeding Institute, would be better able to accomplish their missions if they moved to management by autonomous governing boards rather than governance by departments at the university. He noted that a committee was currently reviewing the PBI and plant breeding as a whole for recommendations on restructuring. He further concluded that substantial demand for experienced researchers might necessitate recruitment abroad in the United States, Canada, and Australia.[56]

Between 1944 and 1946, the council developed a comprehensive and highly specific plan for postwar growth of the agricultural research establishment. Ten survey groups, each composed of Britain's best known agricultural scientists, provided a set of recommendations for different research specialties in a report called *Post-War Programme*. Their scope included the adequacy and organization of existing research stations and the need for new stations, new personnel, and new capital expenditures. Overall, they recommended that research expenditures should move from the existing £737,000 to £1,441,000; capital expenditures on the order of £2,535,000 were considered necessary to bring existing stations up to an enhanced capacity for research and to initiate entirely new facilities. A time line of ten to fifteen years was recommended by the council as appropriate for the amount of growth envisioned.[57]

Engledow chaired the group studying the needs of plant breeding, and the group examined the major plant-breeding research stations: the Plant Breeding Institute (University of Cambridge, cereals), the John Innes Horticultural Institution (independent, basic genetics), the Welsh Plant Breeding Station (University of Aberystwyth, grasses), and the Scottish Society for Research in Plant Breeding (independent, potatoes and other crops of importance to Scotland). It also turned its attention to issues in breeding of fruits and vegetables. All existing major research stations were recommended for enhancement, and the group also advocated a new plant-breeding station for vegetables.[58]

Annual expenditures for wheat and other cereal breeding at the PBI, suggested Engledow's group, should go up by about fivefold, from £5,150 per year to £24,000 per year. Capital expenditures of £75,000 were needed to augment the institute's facilities. For many years it had performed its experiments on a tiny facility of 21 acres, and postwar planning envisioned an expansion to 380 acres and substantial additions to the staff.[59] In a move supported by all political parties, the Labour government set in motion the expenditures needed to foster this massive expansion, not only of the PBI but of the whole of agricultural science research in Britain. John Fryer, secretary of the Agricultural Research Council from 1944 to 1949, noted after the release of the memorandum on *Post-War Programme* that "there is every reason to suppose that this plan has the full sympathy of the Government,"[60] which indeed was the case.

Expansion on such a scale at the PBI demanded new leadership. Howard Hunter had moved into the directorship of the institute in 1936, when Rowland Biffen retired. Hunter retired in 1946 after having spent the war years directing both the PBI and the National Institute of Agricultural Botany. G. Douglas H. Bell became acting director on Hunter's retirement and full director in 1948. For the following twenty-three years, until 1971, Bell directed the PBI in an expansion so large as to clearly demarcate the new institute from the small research station established in 1912 by Biffen.

Even as a boy, Bell (born 18 October 1905 in Swansea, Wales) was interested in plants. He obtained a degree in agricultural botany at the University of North Wales and in 1928 began two years of study at Cambridge with Engledow and Biffen, followed by a trip to the United States to study barley. When he returned to England in 1931, Bell immediately went to work for Biffen at the PBI, with whom he earned his doctorate. Three years later Engledow also asked Bell to take over some of the lecturing in pure and agricultural botany at Cambridge, a post he held for fourteen years, until he became the full director of PBI under its new arrangements.

Surviving records do not completely explain the difficulties in launching the growth of the PBI into a major research institute. Nevertheless, Bell was left in a state of uncertainty both by the process of his hiring and by the ways in which the government expected the PBI to expand. First, the Agricultural Research Council agreed internally to offer the position to Bell in early 1947, then decided to look at an additional candidate, and finally made its decision in January 1948, with effect from 1 October 1947.[61] Given Bell's long association with Engledow and Engledow's prominence in Agricultural Research Council affairs, it is simply unclear why it took about a year to complete the appointment. Second, Bell recounted years later with some surprise, that while he was waiting for the Agricultural Research Council and the Ministry of Agriculture to decide on his appointment, he traveled to London to meet with the secretary of the council. He wanted to find out what the council had in mind for the expansion but was told that it was his job to tell the council what the PBI would become![62] Bell, however, had substantial experience and ideas, and at the age of forty-two he began to preside over the PBI's transformation.

Bell wanted the PBI to be approved by the scientific community, the farming industry, the industries dependent upon farming, such as milling and baking, and the government's officers at the Agricultural Research Council and the Ministry of Agriculture. With so many masters to please, it is perhaps not surprising that the governance of the PBI appeared complex, even convoluted. When Bell first became director, he reported to the Agricultural Research Council's secretary. Shortly after coming on as director, he also began to report to the Ministry of Agriculture. Cambridge University ceased all responsibilities for management of the PBI in 1950. Finally, in 1952, the PBI began operating as a company controlled by its own governing body, chaired by Sir Frank Leonard Engledow of Cambridge University. Seven of the nine members of the board were scientists, thus symbolizing that the PBI was first and foremost a scientific research organization. Funds to run the institute continued to come through the Agricultural Research Council from the Ministry of Agriculture.[63] Engledow's continued involvement suggested that the PBI was not entirely divorced from the university, but the university had no formal control over the institute's work.

Legal organization, however, was probably the easiest problem to solve. Far more difficult were the efforts to acquire an expanded land base and new buildings. Bell wanted the PBI to remain next to the town of Cambridge as an incentive for staff: proximity to the town meant that staff children could easily attend good schools, but a more remote location would not offer such an advantage. Yet obtaining a large block of land near a prosperous university town turned out to be a long process. As early as 1948, Bell noted that a 380-acre farm, Ansty Hall, in Trumpington, a village near Cambridge, was under negotiation for purchase. Acquisition was completed in 1950, after the death of the elderly tenant farmer.[64]

Building on the Ansty Hall site commenced in 1951, and the first staff occupation of the new facilities began in late 1952. Work on cereals and administration of the PBI remained at the School of Agriculture at Cambridge University until late 1954, however. Formal dedication of the completed site took place on 15 July 1955, eight years after Bell had begun as the PBI's acting director.[65]

Despite the difficulties and delays of purchasing land and building facilities in the midst of postwar shortages, Bell realized that staff and the research program were the most important dimensions of his directorship. He laid out the PBI's first priority as the maintenance of existing breeding and varietal selection work. Retirements and interruptions from the war had left the PBI in the immediate postwar years with only a tiny operation. Bell wanted to expand the hybridization work as soon as possible, and his appointments quadrupled the size of the scientific staff from three to twelve between 1947 and 1951. Overall, the staff of the PBI went from twelve to thirty-five in this period.[66]

One of the first hires made by Bell was Francis G. H. Lupton, a young man who had started his study of the natural sciences, especially botany, at Cambridge before the war. After five years in the Royal Navy, Lupton resumed his studies at Cambridge in 1946. Although most of his contemporaries in botany had only disinterest or disdain for agricultural subjects, Lupton was influenced by F. T. Brooks to move from basic botany to the agriculture diploma and an applied study of botany, zoology, and physiology. Brooks was also on the Agricultural Research Council (from 1941 until his death in 1952), which put him in a good place to recommend Lupton to Bell in 1948.[67] He became head of the cereals department at the PBI in 1970, as Bell was retiring from both head of the department and from directorship of the institute.[68] Lupton worked at the PBI for thirty-five years until his retirement in 1983.[69] A few years later, in 1954, Bell hired John Bingham into the cereals department. Before coming to the institute, Bingham had earned a baccalaureate degree in agricultural botany at the University of Reading. Like Bell and Lupton, Bingham, too, had a long career at the PBI.[70]

Bell, Lupton, and Bingham received numerous honors, both scientific and civil, for their work. Bell was elected to fellowship in the Royal Society in 1965, as was Bingham in 1978.[71] Bell and Bingham were also elected as honorary fellows of the Royal Agricultural Society of England. Bell was named a CBE (Commander of the British Empire) and Lupton an OBE (Order of the British Empire) in recognition of service deemed meritorious by the British government. The honors for these three scientists reflected their substantial achievements in breeding new high-yielding varieties of wheat suitable for Britain and for other contributions to agriculture.

Concluding Remarks

From 1939 until the late 1980s, a period of five decades, the United Kingdom reversed the agricultural decline that had begun a century earlier with the repeal of the Corn Laws. All governments, both Labour and Conservative, remained committed to minimum price guarantees and public support of research as the focus of public policy. Results of the research program established at the PBI by Bell in the late 1940s took nearly two decades before they began to yield new varieties of wheat, but after 1970 the PBI's products dominated British wheat production, and yields were at unprecedented highs for Britain (see Chapter 10).

Between 1945 and 1973, a number of institutional changes were initiated by the government that altered the situation for agriculture and agricultural science at the margin. Most prominent was the passage of the Plant Varieties and Seeds Act (1964), the movement of the Agricultural Research Council (by then called the Agricultural and Food Research Council) into the Department of Science and Education (1965), the establishment of the National Seed Distribution Organization (1967), and, most significantly, the entry of the United Kingdom into the European Community in 1972. While each of these events altered the specific context in which wheat breeding was performed, none in any way reversed the fundamental commitment of the government to the concept that Britain would produce more of its food from its own soil.

Britain had come close to famine in two world wars, was stripped of its empire by national liberation struggles in India and elsewhere, and was forced by World War II and the loss of India to revamp its economy[72] in ways that substituted domestic food production for imported foods. Britain produced only 23 percent of its wheat in 1936–39 but 67 percent in 1974–75 and 77 percent in 1980–81.[73] Similar gains were to be found in other commodities. Transformation of British farming and agricultural science required a profound reworking of the relationships between British culture and the country's environmental resources. It truly remade the English countryside and reconstructed Britain's relationship to its environment. At the very least, a proper loaf of bread could finally be made from wheat grown in English soil. It is also important to understand that Britain's agricultural transformation was its version of the green revolution. Just as Mexico and India intensified their agriculture for serious national security purposes, so, too, did Britain.

Science and the
Green Revolution
1945–1975

Higher Yields from Science

In the years after the end of World War II, farmers, agricultural scientists, and policy makers in many countries all knew, or learned, that higher yields of wheat were what they wanted, and they were successful in achieving them. Their specific motivations were different, but their objectives were not. Not only were the objectives clear, but a central method by which the higher yields were to be achieved was plant breeding. Plant breeding itself was an applied science that had to be nested within organizations that supported it and its allies in the agricultural, biological, and engineering sciences.

By 1950 wheat breeders believed that the number of factors governing yield was small, which meant that the research avenues likely to be fruitful were also few in number. The amount of water available and the responsiveness to soil fertility, especially nitrogen, were in most cases the key ingredients for higher yields. For wheat, the ability of the plant to resist invasion by fungal pathogens was almost as important as water and soil fertility. Water and fertility were needed in every crop year, but damage from fungal pathogens varied with weather. Thus plant disease was not necessarily a destructive factor every year. Control of water, soil fertility, and plant disease was therefore at the center of research programs in wheat breeding. A wheat breeder would find success if his or her program produced new varieties that gave higher yields within the context of water, soil fertility, and plant disease existing in the area.

Ancillary questions also existed and in some cases matched the major factors in importance. Weed control was always a problem, so high-yielding wheat had to have some capacity to resist competition from weeds. Similarly, in some areas and some years, insects could cause damage. Wheat varieties therefore had to be able to withstand them somehow.

Other factors of importance to wheat breeders were habit of growth and the color and quality of the grain. Winter wheats were useful in climates that had winters mild enough to allow planting in the fall and thus higher yields the next summer. Spring wheats were mandatory where climatic conditions were more severe. Consumers were used to red or white grains, and they resisted purchasing an unfamiliar color. Similarly, wheat seeds varied enormously in their physical-chemical properties, which in turn affected the uses to which they could best be put. Some served well to make leavened breads while others were more suitable for pastries and confections. A bread wheat did not make a good doughnut, and a cookie wheat made a poor loaf of leavened bread. Still others were best suited for pasta or for animal feed. Because consumers resisted buying a grain not suited for its intended purpose, wheat breeders knew that their programs had to produce varieties that matched expectations of the consumers as well as performed well under the conditions of water, soil fertility, and plant disease.

Scientific expectations for wheat breeding were thus reasonably clear after 1945. Unfortunately, other than for a few private breeders, research programs existed only in theory unless governments made a conscious decision to provide systematic support over a period of many years. Moreover, support had to include efforts to educate research workers and technicians, finance the research programs in appropriate research stations, and provide the advice and financial climate needed by farmers to acquire the new varieties and other new, related technology. It was sometimes difficult for governments to make these decisions because the political economic and cultural context of wheat production were complex.

Decisions were made, however, and research programs were launched to find new ways to get more wheat from the soil (see chapters 5–9). In the thirty years between 1945 and 1975, the United States, Mexico, India, and the United Kingdom each traversed a path that took it to a remarkably higher level of wheat yields (per hectare per year). Wheat breeding was a core component of each country's abilities to produce the grain. Although the ingredients needed to produce the higher yields were numerous, a key step everywhere was the recognition of the semidwarf genes that permitted a wheat plant to be highly responsive to nitrogen fertilizer. The genetic material containing the semidwarf genes was taken from Japan to the United States and then to Mexico, and then, by different pathways, to India and the United Kingdom. These genes gave rise to a green revolution on a global basis.

Semidwarfing Genes and the Exploitation of Cheap Nitrogen

People, other animals, and plants all have nitrogen as an important constituent of their bodies. Nitrogen is included in proteins, nucleic acids, vitamins, and other materials. Quite possibly one reason for the ubiquity of nitrogen in living creatures is that the earth is enveloped in massive amounts of the material, which constitutes about 78 percent of the volume of the atmosphere. Despite the overwhelming abundance of nitrogen, living organisms nearly starve for lack of nitrogen in the right chemical form.[1] Atmospheric nitrogen gas is a molecule consisting of two atoms of elemental nitrogen bound tightly together. In fact, those two atoms are so tightly bound that under most conditions neither atom will leave the other to combine with any other material. For this reason, atmospheric nitrogen is essentially "inert," that is, it generally does not react with anything and cannot be used directly to sustain life.

Under certain specific conditions, however, the nitrogen molecule in air will react chemically. Its constituent nitrogen atoms will combine with other materials and become "fixed" as ammonia or as one of several oxides of nitrogen. Ammonia is used directly or is "nitrified" to nitrites and nitrates, which are then taken up by green plants to support vital processes. Unfortunately for human beings, the conditions necessary to fix nitrogen into usable forms are extremely limited. Lightning, photochemical reactions, and the combustion of fuels fix nitrogen. In addition, a complex array of bacteria have an enzyme, nitrogenase, that permits them to fix nitrogen to ammonia. Some of these bacteria live symbiotically with legumes and other types of plants, and some live freely in soil or water. Outside of these limited physical and biological pathways of nitrogen fixation, however, nitrogen gas in the air around us remains inaccessible, with one major exception: the manufacture of synthetic nitrogen fertilizer, which will be discussed shortly.[2]

Energy is the key to fixation of atmospheric nitrogen. The enormous electrical energy of a lightning burst, for example, can break the bonds that hold nitrogen atoms together, and the fixed nitrogen can enter the soil to be absorbed by plants. More subtle are the biological processes involving nitrogen fixation. Symbiotic bacteria in legumes and other plants, for example, take sugars made in photosynthesis in the plants and convert this chemical energy into other high-energy compounds, which are used to fix atmospheric nitrogen. Plants with symbiotic nitrogen-fixing bacteria thus pay a "tax" in the form of sugars lost from the plant to the bacteria. In return, of course, the plant receives essential amounts of nitrogen from the bacteria.[3]

Similarly, cyanobacteria (formerly known as blue-green algae), for example, can photosynthesize both sugars and the high-energy compounds needed to break the powerful bonds holding nitrogen atoms together. These microscopic plants can thus do for themselves what more complex plants like legumes must do in consort with their symbiotic bacteria. Cyanobacteria occur in moist soils and in aquatic environments. Agricultural systems, especially rice, may receive some nitrogen through such organisms.[4]

Once fixed, a nitrogen atom becomes part of one of the most important of the ecological biogeochemical cycles. In the nitrogen cycle, fixed nitrogen moves from plants to consuming animals to decomposers to eventual return to the atmosphere as nitrogen gas, where it can again be fixed. People derive immense benefits from the "free fertilizer" provided by the nitrogen cycle. For most of the time in which humans have been on earth, the natural pathways of the nitrogen cycle were generally adequate to support the plants and animals upon which we depended for survival. We thus took our part in the nitrogen cycle and had no need to concern ourselves with its mechanisms or even its existence.

Three factors operating over thousands of years began slowly to encourage a more active human involvement with the nitrogen cycle. First were processes involved with agriculture. With the Neolithic revolution, humans began their ongoing codependency with wheat and the hard labor of tilling the soil (see chapter 2). Sometimes the harvest was appropriated entirely by the tillers, but after the development of more "complex" societies it was taken from the tillers by a ruling class. In either case, more crop for the same hard labor was easily seen as an advantage. Materials such as manure to fertilize the crop thus became valuable resources. Thousands of years after the Neolithic, modern chemistry finally explained why manures were valuable and identified nitrogen, phosphorus, and potassium as the most important constituents of a manure. The search for ways to provide nitrogen was well developed by the end of the eighteenth century in Britain and had spread or developed independently elsewhere.

Second were processes involved with warfare. People originally did not need science or sophisticated weapons for their interminable, bloody squabbles. With the advent of gunpowder in China and Europe, however, military strategists came to appreciate the vastly more efficient mayhem that explosives could provide. Gunpowder, learned the chemists, had fixed nitrogen in the form of nitrate salts. Nitrates as oxidizers in intimate contact with carbon sources provided the extraordinarily rapid combustion that produced an explosion. Thus to the hunger of agriculturalists seeking new manures was added the generals' desire for ever cheaper and more powerful bombs and firepower. By the nineteenth century, chemists who understood nitrogen became the front line for those seeking increased production in agriculture and more efficient engines of destruction.

Third and finally came a factor that became important in the eighteenth and nineteenth centuries. Increased food supplies, better hygiene, and medicine reduced the human death rate, first in Europe, then everywhere. As a result, the number of people on earth began to increase at a faster pace. Until the twentieth century, people were supported entirely by naturally fixed nitrogen. In the twentieth century, however, natural rates of nitrogen fixation became inadequate, at least given existing land-use practices. An alarm about "overpopulation" was easy to raise, even though a more careful analysis indicated most such alarms were simplistic. Despite these complexities, the notion that larger populations could not be supported at natural rates of nitrogen fixation had an important element of truth. As the human population has continued to grow past 5 billion individuals, the magnitude of and necessity for human intervention in the nitrogen cycle has also grown.

Although three factors have affected the search for ways to fix nitrogen, the quests for manures and explosives clearly were the most influential. Science provided what may have been its first successes in agriculture in eighteenth-century Britain. The Norfolk four-course rotation allowed a crop to be grown on every piece of land every year, a marked advance over the medieval tradition of a three-year rotation, one of which was a fallow year. During one of the four years, the crop was legume clover, an innovation that led to a greater amount of fixed nitrogen reaching the soil than when the land was fallowed one year out of three. A cereal such as wheat followed the clover, and in the third year a root crop such as turnips provided animal fodder. The root crop allowed heavier stocking of animals per hectare, which in turn produced more manure. In the fourth year, another crop of cereal was grown. Results of this crop rotation, coupled with enclosure of common land, enabled yields and income to increase substantially.[5]

European expansion to the New World resulted in the large-scale exploitation of the guano deposits of Peru and the sodium nitrate mines of what is now Chile. The existence of these huge deposits of natural bird dung and the highly soluble sodium nitrates were due in part to the extremely arid climates of these parts of South America. Guano, of course, was a result of natural processes of nitrogen fixation, but the ability to move it from one place to another was a deliberate human manipulation of the natural processes. Both products were exported to the United States and elsewhere in large quantities during the nineteenth century and represented major enhancements of natural nitrogen flows to the importers. Chilean sodium nitrate provided fixed nitrogen to the United States from 1830 onward, and in the first half of the twentieth century sometimes supplied nearly one-third of the country's annual consumption of fixed nitrogen. Other sources of naturally fixed nitrogen that were developed commercially included wastes from fish- and meatpacking plants, sewage sludge, cottonseed meal, and other waste products.[6]

Expansion of the iron and steel industries during the nineteenth century created a second source of new fixed nitrogen in agriculture. In this case, however, the nitrogen was not "natural," that is, it was strictly a product of human activity and thus for the first time represented an enhancement of nitrogen supplies outside the natural pathways of the nitrogen cycle. In 1893 the Solvay Process Company of Syracuse, New York, built twelve coke ovens that could capture the gases coming from the heated coal, including ammonia. By 1930 over 100,000 tons per year of nitrogen in fertilizer (mostly as ammonium sulfate) were produced in by-product coking, about one-third of American consumption of fertilizer nitrogen at the time. This route to nitrogen fixation was tied directly to iron and steel manufacture, so by-product ammonium sulfate was not a product manufactured for itself.[7]

Markets for guano, sodium nitrate, and by-products of coking developed slowly in the nineteenth and early twentieth centuries. Many farmers in the United States had neither the means nor the motivation to invest in fertilizers.[8] Before 1939, British farmers used about 60,000 tons per year of nitrogen fertilizer, a figure that rose rapidly to 182,000 tons in 1944 under the pressure of maximum production (see Chapter 9).[9] Indian farmers used low amounts of nitrogen fertilizers before 1939, especially on the staple cereal crops.[10]

A series of inventions starting in the early twentieth century set in motion a dramatic, steady decrease in the price of synthetically fixed nitrogen. By the 1950s, synthetic nitrogen fertilizers were sufficiently inexpensive that many growers, even those who grew cereals as well as those who raised the more lucrative cotton, tobacco, fruits, and vegetables, were increasingly inclined to use as much of the new fertilizers as possible. Economic return was the driver in decision making. Decades before the 1950s, technologically progressive farmers had begun to seek methods that enabled them to use nitrogen's abilities to magnify yields.

These inventions were a series of efforts to attack directly the recalcitrant nitrogen molecule of the air. Developments of hydroelectric facilities, such as at Niagara Falls in the late nineteenth century, probably triggered research to put to work the mammoth amounts of energy now available. Some of the methods to destroy the bonds between the two nitrogen atoms included electric arcs to oxidize atmospheric nitrogen to nitrate, a material directly utilizable by plants as food. Over a period of about fifty years, numerous variants of this strategy were tried, none terribly successful.[11]

A second, more indirect, attack on the stable nitrogen molecule was somewhat more successful by the 1950s. Limestone (calcium carbonate) was burned to form lime (calcium oxide), which in turn was placed with coke in an electric furnace at 2200°C to form calcium carbide. Nitrogen gas mixed with the calcium carbide at a high temperature (1100°C) yielded calcium cyanamide, which could be used directly as a fertilizer. Unfortunately, the huge amounts of electricity needed to produce the high temperatures, plus the use of about two tons of coal for each ton of fixed nitrogen, made this route to fixed nitrogen expensive.[12]

Two German chemists, Fritz Haber and Carl Bosch, sponsored by the Badische Anilin- und Soda-Fabrik, put together a workable method of combining nitrogen with hydrogen to form ammonia (NH^3). By 1909 Haber had found a number of catalysts for the reaction, which permitted the chemist to form ammonia at relatively low temperatures and pressures. Still, however, his machine could produce only about eighty grams per hour, a remarkable achievement for research but far from adequate for large-scale production. Bosch and his colleagues engineered a plant that by 1913 could fix twenty or more tons per day of nitrogen, which they quickly expanded to thirty tons per day at the Ludwigshafen-Oppau plant of the Badische Aniline- und Soda-Fabrik.[13]

World War I led the U.S. government to attempt to build a synthetic ammonia plant in Alabama, but successful production never occurred there. After the war, a number of companies built plants, and synthetic ammonia became a strong contender in the American fertilizer business. By 1940 the United States was making over 100,000 tons per year of synthetic nitrogen (over 25 percent of the total nitrogen fertilizer), much of it ammonia produced by many variations and refinements of the original Haber-Bosch process. This figure had risen to 587,000 tons of nitrogen fertilizer by 1949, about 64 percent of the total. This method of fixing nitrogen from the air surpassed the coke by-product production in 1941 for the first time. After 1945, direct fixation from the air vastly exceeded the production from coke. Twenty-seven plants were operating in the United States and Canada by 1950, with a capacity in excess of 1.5 million tons per year.[14]

Human ability to radically augment the nitrogen cycle was thus born in the period 1900 to 1950. The significance of the Haber-Bosch process was recognized with Nobel Prizes, in 1918 for Haber and in 1931 for Bosch. Important leaps in ammonia production came with both world wars, but agriculture was the major market for fixed nitrogen over the long term. From 4,000 tons of nitrogen fixed in 1913–14, the amount of synthetic nitrogen produced exceeded 11.8 million tons in 1959–60,[15] and the energy cost of synthesizing the materials dropped substantially.[16] More recently, ecologists have noted that the amount of nitrogen fixed synthetically is larger than that fixed by natural processes. This development could have profound, negative environmental repercussions.[17] At the same time, however, it is important to note that the large human population is now dependent upon the ability to fix such large quantities of nitrogen.

Agriculture was profoundly affected by the advent of inexpensive nitrogen that could be applied almost at will. Farmers had known for years that manures were good for yields, but these synthetic manures took time to move into the role of a stable production tool. They cost money to buy, which hindered growers who were worried about unstable and uncertain prices. They were unfamiliar to use and could damage crops if not handled properly, so growers and their advisers had to learn to handle the new technology before it could be seen as a reliable tool. Rather quickly after the Haber-Bosch process achieved commercial success, complaints from organic enthusiasts focused on their long-term potential to harm the soil, and allegations of inferior produce ricocheted around the new fertilizers. Albert Howard, the illustrious English wheat breeder in India, achieved more name recognition by leading this movement for organic agriculture than he ever had as a wheat breeder.

Nevertheless, the fixed nitrogen was so cheap and so effective that its eventual hold on the agricultural producer was unshakable. The question for growers became not whether to use the materials but when, how much, how applied, and for how much money. "Progressive" farmers were those who moved first to exploit the new sources of nitrogen. In turn, these advocates of heavier fertilization with nitrogen created the conditions for plant breeders: new varieties of wheat and other crops would have to be bred with heavy nitrogen fertilization as the assumed backdrop.

Use of manures on wheat, natural at first and later with synthetic nitrogen, began in the nineteenth century, notably in Japan. With the desire to increase yields by adding nitrogen came the recognition of a prominent "flaw" in the wheat plant as it had been grown for millennia: wheat stems were spindly and tall, as much as 200 centimeters (2 meters). Wheat actually produced two crops, grain and straw. These traditional varieties produced grain and straw in proportions that were considered appropriate for the most part. Grain for food was the prize of the crop, but the straw had vital functions as animal bedding and feed and, in some areas, for construction purposes such as roofing.

Traditional varieties may have been well suited to producing quantities of grain and straw in amounts that were useful. However, farmers who wanted to increase their yields of grain found that increasing the amount of manure on wheat was often useless. What mattered was the fate of the straw. With heavy manuring, the plant produced a straw that could not support the grain head. As a result, the plant fell over ("lodged") before harvest. Not only was any increase in grain yield lost, but also manuring often resulted in yields that were even smaller than those that relied on natural levels of nitrogen in the soil. In short, manuring the traditional wheats was at best unhelpful and at worst counterproductive.

Despite the tendency for the world to be awash in too much wheat in the nineteenth century (see chapter 2), at least one country, Japan, may have lacked sufficient supplies and accordingly began to seek wheat varieties that would not lodge with manuring. Japan had limited agricultural land and worked to expand production of grain, partly through military expansionism in the early 1900s into Formosa and Korea. The second method to enhance total production of grain, begun in the late nineteenth century, was to find plant varieties that would yield more per hectare. Fertilization was assumed.

Success in raising yields per hectare was visible to the visiting American commissioner of agriculture, Horace Capron, who headed an advisory delegation to Japan in 1873. Capron recorded that the Japanese had found wheat varieties, sometimes as short as 0.5 meters, that responded to manuring by producing a larger grain head and would not lodge. These plants produced more grain relative to straw than did the traditional varieties. Even before Capron's visit, some of these short Japanese wheats were sent to France, and in 1911 other Japanese wheat seeds were distributed to wheat breeders in Italy.[18] Thus even before synthetic ammonia became available, some countries were seeking wheat types that would usefully respond to manures and alter the traditional balance between grain and straw.

Japan continued its efforts in wheat breeding up to World War II, and the Ministry of Agriculture released several improved varieties that had been selected from crosses of wheats of very diverse origin—some Japanese, some from the United States, and possibly some from Korea. A few of these varieties served as the sources of "semidwarfing" genes after 1945.[19] All of them had straw that was shorter than considered "normal," but wide ranges of height occurred in both the semidwarfs and the traditional varieties. The most important characteristic of the semidwarfs was their ability to respond to nitrogen fertilizer without lodging. By no means were the Japanese semidwarfs the only source of short-strawed wheat, but they were preeminent in breeding efforts after 1945.

Samuel Cecil Salmon[20] of the USDA was the critical link in moving the genetic material of the semidwarf Japanese wheats into the international networks of wheat breeders. Salmon, born in a sod house on the frontier in South Dakota in 1885, received his baccalaureate at South Dakota State College of Agriculture, later joined the faculty of Kansas State College of Agriculture and earned a master's degree there, and received his doctorate with Professor H. K. Hayes at the University of Minnesota in 1932.

Hayes, meanwhile, was himself in the midst of a research group seeking resistance in spring wheat to stem rust (*Puccinia graminis tritici*), a project also involving E. C. Stakman (see chapter 6). This exercise ultimately led to the release of the variety Thatcher in 1934, a major achievement for the spring wheat area of the United States.[21] Salmon, therefore, was trained in one of the most important centers for wheat breeding. Almost immediately upon finishing his doctorate, Salmon moved to become the USDA project leader on wheat. Here he had the responsibility of promoting the improvement of all wheat in the United States.

At his new job in Washington, D.C., Salmon was largely confined to a desk and the work of coordinating the vast research network among the federal and state wheat-breeding stations. Perhaps it is not surprising, therefore, that his major direct contribution to wheat improvement came quite serendipitously. When Douglas MacArthur established a government of Japan by the American Army of Occupation, he created a Natural Resources section. Postwar shortages of food were potentially quite severe in Japan, and the Natural Resources section, headed by Major Warren H. Leonard, requested assistance on wheat affairs from USDA. Salmon left Washington, D.C., for Japan in December 1945 and remained until July 1946.[22]

Most of Salmon's work was to advise on the organization of an effective national research network in wheat. To this end, he visited many stations and ultimately recommended that many of them be consolidated into larger units, that work between them be coordinated, and that a national advisory service be created. Among the many policy-type sessions he held, however, he also was able to observe many test plots of the 1946 wheat crop. Several varieties caught his eye, one of which was known by the Japanese as Norin 10.[23] Even Salmon, however, did not know just how remarkable it would turn out to be: "I did not visualize the future extensive use and value of Norin 10, but I do think it fair to say that I recognized it as different from any variety in the U.S. and potentially valuable."[24]

Salmon's many years of experience with wheat were at the right place at the right time. He arranged that samples of several Norin varieties be sent to wheat breeders in the United States.[25] Salmon's direct involvement with the Japanese wheats ended once he arranged for their transmission overseas. Nevertheless, his appreciation for a wheat that could respond to nitrogen led to the introduction of new genetic material to a research network in the United States that was seeking the properties exhibited by the Norin wheats.

Orville Arthur Vogel: Wheat Breeding in the "Inland Empire"

Samples of four Norin varieties arrived in the United States between June and August 1946. They were first grown in an isolated nursery in Arizona during 1946–47 to check for possible new disease organisms. In the following year, seven breeding stations received samples of these Japanese wheats. Further samples of wheat from Japan and Korea arrived in subsequent years, but the four Norin varieties quickly entered into ongoing breeding work. One of the research stations receiving the Norin wheats in 1947 was the joint program of USDA and Washington State College of Agriculture (now Washington State University) at Pullman, Washington.[26]

In retrospect, it is easy to see that the genes contained within the Norin wheats, especially Norin 10, were spectacularly successful when introduced into wheats grown in Washington, Oregon, and Idaho, often dubbed by local enthusiasts as the "Inland Empire." In 1961, fourteen years after first receiving the Norin samples, the Pullman station released the new variety Gaines, a soft white winter wheat that gave from 5 percent to 50 percent higher yields than the varieties then in use.[27] By itself, producing Gaines was a significant achievement in plant breeding for yields. Too close a focus on the achievements of Gaines alone, however, obscures two other broad sets of questions. First, why were the workers at the Pullman station so receptive to the Norin materials? Second, what were the follow-up ramifications for subsequent wheat breeding? Answers to these inquiries are needed to understand why the scientific work to produce Gaines occurred at Pullman and what difference it made.

Orville Arthur Vogel (1907–1991) (Figure 10.1) played the lead role in first making use of the Norin genes in the United States. Vogel, one of seven children, was born on a farm near Pilger, Nebraska. His grandparents had immigrated from Germany, and his parents still spoke German. Vogel's first language was German, and he did not learn English until he was five. His parents separated when he was in the seventh grade, which led to moving to Pilger, where he completed high school.[28]

Figure 10.1   Orville Arthur Vogel (*center*) describing a new eight-row planter for wheat-breeding experiments. Courtesy of U.S. Department of Agriculture, Pullman, Washington.

Vogel's timing on entering the world could not have been worse for the job he may have preferred: farming. American agriculture suffered from chronic low prices all through the 1920s, and the onset of the Great Depression in 1929 lowered prices to catastrophic levels. Under any circumstances it would have been difficult for a young person to enter farming, and Vogel instead pursued education. He took two years of work at Yankton College in South Dakota, taught high school in Wynot, Nebraska, for two years, then finished a bachelor of science degree in agriculture at the University of Nebraska in 1929. He continued at Nebraska for a master's degree in 1931. Immediately upon finishing his graduate work, he received an offer to be a junior agronomist with the Bureau of Plant Industry, USDA, stationed in Pullman, Washington. Vogel and his wife, Bertha, from Lincoln, Nebraska, headed for Washington State, where he began his duties on 16 February 1931 as a young man of twenty-three.[29]

Vogel thus entered a world that, like his native Nebraska, was not far removed from initial Euro-American settlement. Washington had become a state in 1889, just forty-two years before Vogel arrived. Settlement of the Oregon Territory had begun in the 1840s, but when Vogel arrived the state of Washington was still sparsely settled. Forest products and fishing dominated the economy of the western part of the state, and wheat and apples were the big industries of the eastern areas.

Physically and economically, Washington, Oregon, and Idaho were not easy areas for commercial wheat production. The deep, fertile loess soils that covered parts of Washington, known as the Palouse, were attractive for wheat production and usually had enough rainfall to make a reasonable crop. Much of eastern Washington, however, lacked either good soil or sufficient rainfall or both. More important was the remoteness of the area from the main population centers of the United States, which made movement of any agricultural crop to market expensive. In fact, no economic transport of a bulky, low-value crop like wheat was even imaginable until railroads connected Washington and Oregon to the eastern United States. Based on the state's geography and demography, therefore, wheat in Washington from the beginning had to be an "export industry" in a highly competitive market. Large-scale wheat production was never feasible for the local market of the Inland Empire. Wheat cultivation in Washington was part of the "excess wheat acres" in the Neo-Europes (see chapter 2), and reduction of production costs was always a paramount goal of both farmers and research workers.

Perhaps the remoteness of Washington State, and consequently the need for reduction of production costs, helped encourage a tradition of wheat breeding to begin soon after Washington entered the Union. William Jasper Spillman served as professor of agronomy at Washington State College from 1894 to 1901 and initiated a program of breeding for winter wheats in 1899. He moved to USDA in 1902. This effort aimed to capture the higher yields possible with fall-sown wheats than with the spring-sown varieties that dominated wheat production at that time. Spillman was successful in finding winter hardiness suited to the area, which considerably increased the production of winter-grown wheats and therefore the total wheat production of Washington. Spillman's varieties were club wheats, an ancient form adapted to hot, dry conditions and with short straw. A particular disease, loose smut (*Ustilago tritici*, a seedborne disease), however, increasingly plagued Washington wheat growers, and Spillman's new winter varieties were thus limited in their yield potential.[30]

After Spillman, Edward Franklin Gaines began a program of breeding for loose smut resistance in 1915. Gaines had earned his bachelor's and master's degrees at Washington State College in 1911 and 1913, respectively. He earned a doctor of science degree at Harvard in 1921. Gaines was successful in breeding a variety, Ridit, that was released in 1924 and resisted all of the then recognized races of loose smut.[31] Loose smut, however, continued to form new races, and in any case Washington wheat continued to be subject to enormous downward pressures on prices throughout the 1920s because of overproduction of wheat on a global basis. Thus Gaines' successes always occurred in a context of persistent severe problems. The extreme competitive pressures battering American farmers who were in the heart of their transition to full-scale industrial models allowed no resting on laurels for even a moment.

Congress appropriated further funds to establish the Western Regional Cooperative Wheat Improvement Program in 1930. This money enabled the Washington experiment station to add two additional USDA researchers to the staff, one of whom was Vogel. Vogel's assignment was to learn varietal testing and breeding from Gaines. At the time of his hiring, research objectives were quite clear: resistance to loose smut and lodging. In addition, Vogel was to work on finding resistance to "shattering"—the habit of grains breaking out of the grain head before threshing—in white winter wheats.[32] A variety that shattered was difficult to harvest without extensive losses of grain. On the other side, of course, farmers did not want varieties in which the grain was held too securely to the grain head because that trait would make threshing difficult.

From his very first day of work, therefore, Vogel knew that his job was to help find wheat varieties, particularly winter wheats, that would yield more, be resistant to loose smut and other diseases such as bunt, and not shatter. Problems with loose smut remained particularly important and in some areas and years of the 1930s could take up to 90 percent of the crop in the western states.[33] The idea of higher yields was also tied to resistance to lodging, especially because some of Washington's soils were quite fertile and even without heavy manuring often supported grain crops that lodged.[34] Vogel was quite explicit in 1937 about the types of wheat varieties considered valuable: "A number of the winter wheat hybrids appeared especially promising because of their short, stiff straw, high resistance to bunt and reasonably good head and kernal types."[35] Vogel thus began his work in a setting that (1) valued high yields, (2) saw short plants resistant to lodging and disease as key to achieving those yields, and (3) expected to create those plants by crossing currently used varieties with other germ plasm that appeared promising.

By the 1940s Vogel had a number of accomplishments. While working at the Washington station, he earned his doctorate in 1939 with a thesis based on a study of factors that affected shattering.[36] Vogel also proved extremely inventive in the design and construction of planting and threshing machinery for plant-breeding research.[37] He also served from 1947 to 1955 as the coordinator of Cooperative Wheat Improvement for Western States, an administrative job that he was happy to leave in order to return to his first love, the practical breeding of new wheat varieties.[38]

In the 1930s Vogel set in motion the hybridization experiments that led to a steady stream of new varieties, which over three decades dominated the wheat industry of the northwestern states (Washington, Oregon, and Idaho). When Vogel began his wheat breeding, Elgin was the variety in Washington State that gave the highest yields and thus was the standard against which potential new releases were measured. Vogel's first two releases were Orfed in 1943, followed shortly thereafter by Marfed in 1947.[39] In 1949 he and his colleagues released two new varieties of winter wheat, Brevor and Elmar. Brevor, he noted, was a soft white wheat with straw that was short to medium in height and highly resistant to lodging. Compared with Elgin, Brevor gave 5 to 12 percent greater yields in many places, and average yields for Brevor were nearly 63 bushels per acre in contrast to Elgin's 59.3 bushels per acre.[40]

Elgin and Elmar (both short-strawed) and Brevor (very short-strawed) were all well accepted in the Inland Empire. They could be planted in the fall on land with high fertility, and their short straw made them resistant to lodging. They were also resistant to bunt.[41] As much as forty pounds of nitrogen per acre could be put on Brevor without causing lodging.[42] Use of fertilizer to get high yield may have been the most important characteristic, but it was not the only one sought by farmers.

Erosion control, especially on some of the steep and highly erodible soils of the Palouse, was increasingly important to both growers and natural resource managers. Erosion control was aided by fall-sown winter wheats, but early sowing in the fall was better than late fall. Elgin, Elmar, and Brevor, unfortunately, were not good for early fall sowing. They tended to grow too much in the following spring, which led even these short-strawed varieties to make too much straw, lodge, and suffer yield loss. For this reason, Vogel and his colleagues were interested in finding yet shorter varieties than Brevor, which tended to reach a height of 95 to 113 centimeters. Elmar reached 100 to 120 centimeters.[43]

About a hundred Norin 10 seeds, sent to Vogel by Burton B. Bayles and Samuel C. Salmon of USDA, arrived in Pullman in time to be planted in spring 1948. When no grain heads were visible by mid-July, Vogel knew he had winter-habit seeds.[44] In fall 1948 Vogel had a graduate student, Dick Nagamitso, make crosses between Norin 10 and Brevor to produce F_1 seeds.[45] Second-generation plants (the F_2) showed the segregation typical of Mendelian genes, and some were of the very short stature of Norin 10. Norin 10 itself tended to be male-sterile, which meant that its progeny with this characteristic tended to outbreed with other wheats rather than be self-fertile and thus true-breeding. Vogel and his colleagues searched among the F_4 progeny to find varieties that were male-fertile as well as short and high-yielding. Two of the selections entered varietal tests in 1954, where they both yielded more than Elmar and Brevor, especially when sown early. Increases in yields were as much as 20 percent (67.8 bushels per acre compared with 56.5 bushels per acre). These new plants were short like Norin 10 — only about 65 centimeters tall, compared with the approximately 100 centimeters or more of Brevor and Elmar.[46]

Vogel and his colleagues realized that they had significantly new material, even though they had not yet brought the Norin 10 genes into a commercial variety for release:

> On the basis of present data it appears that the semidwarf growth habit represents a highly important development toward the breeding of winter wheats better suited to highly productive soil and climatic conditions, and especially for use in very early fall seedings for controlling soil erosion in the Pacific Northwest. These results warrant a re-evaluation of previously accepted limitations in winter wheat production.[47]

Subsequent selection work and further crosses finally gave stable plants of acceptable total characteristics by 1958, when Vogel began to consider how to release the new semidwarfs as a commercial variety. Grain from 1,000 plants was planted in 1959 to yield about 75 bushels in 1960. About 50 bushels were used by commercial seed growers to give 6,800 bushels in 1961, which were sold to the Washington State Crop Improvement Association. By 1962, half a million bushels were available, enough to plant about one-fourth of Washington's wheat.[48] Gaines was the new variety, the first semidwarf wheat plant developed within the United States and a harbinger of substantially elevated wheat yields not just within Washington State but ultimately globally. Orville Vogel had substantially increased the ability of farmers to make use of the now cheap supplies of fertilizer nitrogen, and thus of people to harvest solar energy.[49]

Vogel was tied into an extensive network of cooperating wheat breeders, and, just as the Japanese breeders had generously shared the Norin varieties with the USDA, Vogel shared the material with a wide range of other breeders. Of greatest consequence was his sending in 1953 of F_2 seeds from the cross of Norin 10 and Brevor to Norman E. Borlaug of the Rockefeller Foundation program in Mexico. Borlaug started using the material by 1954.[50]

What is important to note here, however, is that the conceptual framework for recognizing and developing Norin 10's genes was well in place for at least a decade before the Japanese varieties came to the United States. Vogel knew that farmers wanted higher yields, which ultimately meant increased use of nitrogen. The breeder's job was to find varieties that could respond to higher nitrogen levels without lodging. Plants also had to have appropriate disease resistance, milling and baking qualities, and other attributes suitable to climates, soils, and markets. Furthermore, Vogel knew long before Norin 10 came his way that crosses planned within a framework of Mendelian genetics were the primary way of constructing suitable new varieties. He also had techniques to distinguish the differences caused by genetic differences rather than the environmental influences of climate, soils, and growing conditions.

Exploitation of Norin 10, therefore, did not require any new objectives or methodologies. What Norin 10 provided was evidence that the existing frameworks of analysis were effective in finding the types of wheat plants that farmers wanted.

Norman E. Borlaug: Wheat Breeding to Transform Mexico

Norman E. Borlaug, born in 1914 on a small farm in northeastern Iowa near the town of Cresco, became in the 1970s the most famous of the scientists who created the green revolution (Figure 10.2). He was awarded the Nobel Peace Prize in 1970 for his work in breeding semidwarf wheats,[51] a project begun with seeds from Orville

Figure 10.2   Norman E. Borlaug. Courtesy of University of Minnesota Archives.

Vogel of the F_2 of Norin 10 × Brevor. By that time Borlaug had a career that included plant pathology, plant breeding with an emphasis on wheat, the role of high-yielding varieties in national development, and the relationships of demography to agricultural change. He was a crusader who took the word about higher yields to anyone who would listen. Borlaug was equally effective with farmers in their fields, scientists in their experiment stations, and heads of state in their palaces. Perhaps Borlaug was at his best when it came time to meld the biological results of breeding experiments with changes in policy needed to move national average yields to higher plateaus. He was completely convinced that the growth of the human population was a serious problem and that the achievement of higher yields was simply a way to buy time while other efforts sought a reduction in the population growth rate.

For all his achievements later in life, Borlaug's early career did not immediately show this promise. Somewhat on a lark, he left Cresco for the University of Minnesota in 1933. Once in Minneapolis, he was told he must attend University College to do more preparatory work before matriculating at the university. At least some of his advisers raised the question of whether he was really cut out for university work. Borlaug, however, if nothing else, had a stubborn streak plus a competitive flair that kept him going in his studies. He successfully gained full entrance to the university and graduated with degrees in plant pathology—a B.S. in 1937, an M.S. in 1940, and a Ph.D. in 1942.

Elvin Charles Stakman, the well-known plant pathologist who led the Survey Commission on Mexican Agriculture for the Rockefeller Foundation in 1941 (see chapter 5), was Borlaug's primary mentor; he persuaded Borlaug to change from a specialization in forestry to plant pathology. Borlaug was also influenced by Herbert Kendall Hayes, one of the major consolidators of the science of plant breeding, a successful breeder of rust-resistant spring wheats, and mentor to S. C. Salmon. Borlaug's professional training thus took place in what was probably the most active center of wheat improvement in the world at the time. Minnesota's preeminence may have derived from its status as a major milling and baking center for wheat grown in many parts of the Great Plains, as well as a major wheat producer in its own right.

The United States was at war when Borlaug finished his doctorate, so his first work was dictated by the Selective Service System and the manpower allocation schemes of the War Production Board. As a trained mycologist, he joined the ranks of staff scientists at the DuPont Company in Delaware, where he worked on such topics as mildew resistance in the rayon fabrics used by the military. Stakman, however, had other intentions for Borlaug's talents. In 1942, at the annual meetings of the American Phytopathological Society, Stakman and J. George Harrar began to recruit Borlaug to join the newly formed Mexican Agricultural Program financed by the Rockefeller Foundation. Harrar had also earned his doctorate with Stakman, in 1935, and had accepted the directorship of the Office of Special Studies in the Ministry of Agriculture in 1943 (see chapter 5). Borlaug was released from manpower regulations on 1 July 1944 and prepared to join the staff of the Rockefeller Foundation.[52]

Borlaug arrived in Mexico in October 1944, to be followed shortly thereafter by his wife of eight years, Margaret Gibson Borlaug, and young daughter, Norma Jean. He faced a number of obstacles. First, he had never worked on wheat, maize, or beans, the crops of most interest to the Rockefeller program. Second, he had never before been outside of the United States and knew not one word of Spanish. Third, the Rockefeller program was still in its very early stages, without a clear sense of which research directions were most likely to be successful. Fourth, the Rockefeller team was without much in the way of well-developed facilities. On a more personal level, the Borlaug family was grieving their child, who died shortly after birth in 1945.

Borlaug's primary duties were to work with Harrar on the wheat improvement program. Between 1943 and 1945 Harrar had conducted three annual surveys of wheat areas in which he compiled information on methods of cultivation, stand health, varieties used, and diseases endured. His general conclusions were that Mexican wheat production was amenable to many improvements.[53] Based on his initial work, Harrar proceeded to outline what was by then a standard, multifaceted research program based on collection of local varieties, selection from the local plants of varieties that looked promising, importation of varieties from elsewhere, a uniform testing protocol for yield and resistance to stem rust and other diseases, breeding of new varieties, and the distribution of the best new varieties. By 1945 he had collected and analyzed 277 varieties of local wheat varieties grown in Mexico and made 42 crosses of local varieties for testing.[54] Within a few years Harrar found that 5 out of 300 introductions had done well.[55] In spring 1945 Stakman made a visit to Mexico and noted that it was essential to work at multiple research stations, a fact that Harrar had noted also. Stakman further noted that wheat in Sonora was important but that no work was under way there at the time.[56]

Although both Harrar and Borlaug were trained as plant pathologists, not wheat geneticists, the crossover of skills in the two disciplines was high. Resistance of wheat to the many pathogens that attack it was a prime objective of wheat breeding, so a thorough background in plant pathology was in fact quite essential to wheat breeders. Mexican wheat, in fact, was plagued with severe disease problems, mostly from stem rust (*Puccinia graminis tritici*). Moreover, since Stakman's elucidation of plant pathogen evolution and the importance of races in wheat disease control (see chapter 4), plant pathology itself had been almost as concerned with genetics as was plant breeding. Thus both Borlaug and Harrar brought useful knowledge to the tasks of wheat improvement, despite the fact that both also had to acquire new skills and adopt new approaches.

By the time Borlaug arrived, Harrar had planted one experimental crop of wheat. In spring 1945 Harrar and Borlaug began the first crosses of wheat for plant breeding. At this time Harrar concluded that his overall administrative duties were too much for him to continue to also direct the wheat improvement program. Therefore, he turned the program over to Borlaug's leadership.[57]

Subsequent years demonstrated that Harrar's talents in administration and diplomacy were outstanding, which took him successively to the New York office of the Rockefeller Foundation as deputy director for agriculture (1951–55); director of agriculture (1955–59); vice president for medical, natural, and agricultural sciences (1959–61); and president and trustee (1961–72). His involvement with the administrative dimensions of the foundation's programs included the substantial expansion of the agriculture program and, in the 1960s, the organization of the international agricultural research centers.[58]

Harrar's preoccupation with the administrative functioning of the Office of Special Studies left wheat improvement in Borlaug's hands. Borlaug recalled forty-four years later, in 1989, that he knew what his major responsibility was: to get higher levels of wheat production in Mexico as soon as possible.[59] This was, of course, the major rationale behind the Mexican government's desire to have the Rockefeller Foundation program in Mexico at all (see chapter 5). Exactly how to achieve this well-established objective, however, was not at all clear or uncontroversial.

Borlaug knew that soil fertility was a major problem with Mexican wheat yields, but it was not the only problem. In the 1940s wheat was grown in two major areas of Mexico: El Bajío, the highland areas of central Mexico, near the capital district of Mexico City, and the newly irrigated deserts of the northwest in the states of Sonora and Sinoloa (Figure 10.3). Farms in El Bajío were small, operated by impoverished peasants, some Indian and some mestizo. Northwestern farms were large, well capitalized, and owned either by Mexicans who fully identified with the European aspects of Mexican culture or by North Americans from the United States.

Borlaug's colleague Edwin John Wellhausen ran the maize improvement project largely in El Bajío and was convinced that solutions had to be created for the majority of farmers in the highlands. Borlaug, however, came to see the situation for wheat quite differently. By 1948 he was convinced that raising Mexican national levels of wheat production must involve the large, irrigated farms of the northwest, as well as the highland areas.[60]

Two factors probably led him to this conclusion. First, use of nitrogen fertilizers and other capital inputs would be essential if wheat yields were to rise. The farmers of the northwest were likely to be more receptive to and more capable of adopting reliable schemes of wheat production based on higher fertility levels than the smaller farmers of the highlands. Second, for Mexican wheat growers in the mid-1940s, disease and the resulting instability of yields were at least as great a concern as the absolute level of yields. Thus Borlaug knew that obtaining resistance to stem rust, the most devastating of the diseases, was the first task. Varieties with better yield potential could follow once rust resistance was achieved.

Borlaug thus planned his research program around the idea that he would work in both the highlands and in the northwest. Experimental plots in both regions immediately opened the door to an important innovation in the breeding methodologies: "shuttle breeding," or the production of two generations a year by using the different growing seasons of two distinct regions. The orthodox view at the time, learned from H. K. Hayes, was that the plant breeder grew all plants in the area in which they were to be used. The idea behind this theory was that the plant had to be adapted to the soils and climate in order to be successful. Only if all selections were made exactly in the area of production could the breeder be confident that selected new varieties would be well adapted.

Borlaug began, probably between 1946 and 1948, to violate this orthodoxy and to move breeding experiments back and forth between his plots in Toluca in the highlands and Sonora in the northwest, separated by 2000 kilometers. Thus he had to cross the best advice of his teachers, who were not shy about pointing out that he seemed not to have learned his lessons very well.[61]

Why did Borlaug launch this challenge to orthodoxy, especially since Harrar was not keen about it for reasons of travel expenses? Two explanations seem most likely. First, development of experimental plots in Sonora was the prime vehicle for bring-

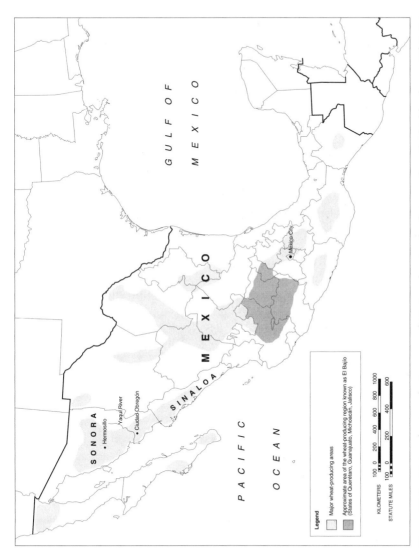

Figure 10.3 Major wheat-producing areas of Mexico, showing the distances from the highland areas (El Bajío) to the northwestern areas (Sonora and Sinoloa). Line drawing by Tim F. Knight. Outline of map adapted from U.S. Department of the Interior, Geological Survey, *North America 1982* (1:10,000,000) (Reston, Va.: Geological Survey, 1982), 1 p. Wheat-growing areas adapted from U.S. Department of Agriculture, *Major World Crop Areas and Climatic Profiles* (Washington, D.C.: U.S. Department of Agriculture, 1987), Agriculture Handbook no. 664.

ing the Rockefeller work to the attention of growers in the northwest, an essential ingredient, Borlaug felt, in raising national production levels of wheat. Second, rust problems were severe in all parts of Mexico but especially in the northwest. Borlaug thus had to solve the rust problem of the northwest, and he wanted to do so as quickly as possible. Traditionally plant breeders counted on ten generations of crosses to get new varieties, which, at one generation per year, meant a total of ten years to get a commercial release. Borlaug hoped that, at two generations per year, he could reduce the time needed to get a new variety to about five years or maybe fewer.

Borlaug was so convinced of the necessity to conduct his research program in both Sonora and El Bajío that he initiated work when the physical facilities in Sonora were, to put it mildly, extremely difficult. The Yaqui Valley of Sonora was reachable only by poor roads, or by going through the United States. A plot of land was available, on an experiment station established by Don Rodolfo Calles, governor of Sonora, in the 1930s, but support for research in the area had been poor and the station had gone to ruin. Borlaug endured sleeping in haylofts and cooking his own food. He had either to borrow machinery from neighboring farmers or do the work by hand.[62] Years later, he remembered these expeditions to Sonora as bleak, to the point that he even wondered whether he ever should have left his promising career with the DuPont Company.[63] Nevertheless, he so believed that Sonoran wheat was crucial that he stubbornly refused to abandon shuttle breeding. When Harrar, backed by Stakman, told Borlaug in 1948 that he should not go to Sonora anymore, Borlaug actually resigned. Within a day, Stakman and Harrar backed down and refused to accept the resignation. So Borlaug continued his work in both El Bajío and Sonora.[64]

Years after the fact, Borlaug was not entirely sure why Harrar and Stakman had changed the decision about the program in the northwest. He surmised, however, that a letter from Don Aureliano Campoy, a farmer whose land adjoined the experiment station in Sonora, may have been the deciding factor. Campoy's letter, which arrived the very day Borlaug resigned, extolled the virtues of Borlaug's team under conditions of inadequate support from the head office. Borlaug, probably unknown to Harrar, also received a copy of the letter from Campoy. In addition, Borlaug's work was well known and respected by Don Rodolfo Calles, former governor of Sonora and son of former president Elias Calles.[65] Whatever the reasons, the patronage of influential persons in the northwest was advantageous to building local support for wheat research in the newly irrigated Sonoran deserts.

This support from Calles represented a substantial success for Borlaug's program. Sonoran wheat farmers had suffered disastrous outbreaks of rust in 1939, 1940, and 1941, all before the Rockefeller Foundation began its work. Their fields were so devastated by the disease, and the response of agricultural scientists was so inadequate at the time, that prominent growers in the area disdained the promises of science.[66] Borlaug believed it was his preliminary work in Sonora that began to turn the situation around to the point that the growers supported his work.

Inclusion of Sonora in Borlaug's wheat-breeding program had at least three consequences, two of which Borlaug had anticipated and one of which was serendipitous. First, breeding two generations per year did in fact reduce the time for finding useful rust-resistant varieties. By 1950 Borlaug and his colleagues had released at least eight new varieties, all of which had some advantage over the multitude of wheats previously grown in Mexico.[67] Second, Borlaug's program made the northwest growers enthusiastic about achieving higher production levels on their own lands, which in turn raised total Mexican production levels. In addition, wheat selection, plus policy changes in the Ministry of Agriculture to increase the production of wheat at the expense of barley, led to increased production of wheat in the high elevations of central Mexico by 1950.[68]

The third and serendipitous consequence was that the breeding scheme designed by Borlaug had the unanticipated effect of making any varieties he selected quite well adapted to multiple soil and climatic conditions. Perhaps most important, he later learned that most wheats are sensitive to photoperiod, that is, they will not flower and set seed unless the light conditions are proper. Generally this means that wheat must experience lengthening daylengths in order to flower.[69] Borlaug, however, planted his experimental seeds in Sonora in November, when daylengths were shortening. He then harvested in April, moved the new generation to Toluca, and planted in May, when daylengths were getting longer. These plants were harvested in October, just in time to move them to Sonora for a November planting. As it turned out, this regime of selection eliminated varieties that had strong requirements for lengthening days in which to flower. Only daylength-insensitive cultivars thrived and were carried on in Borlaug's program. In later years this unanticipated daylength insensitivity was highly important in the success the plants had in many other parts of the world.

Based on the improvements made to wheat varieties, the expanded area given to wheat, and especially the increased production from the northwest, total Mexican wheat production went from 365,000 metric tons (average yields of 750 kg/ha) in 1945 to 1.2 million metric tons (average yields of 1370 kg/ha) in 1956.[70] By 1950 Harrar's overall assessment of the wheat improvement effort in Mexico was that the yield potential had been reached and that further work would be aimed at getting better seed stocks and improving the cultivation practices of Mexican growers. Additional increases in agricultural production would also come from similar work in beans, maize, barley, and rice.[71] It seemed from Harrar's report almost as if the scientific knowledge then available had already contributed as much as possible to the question of Mexican yields.

Borlaug's wheat improvement program between 1948 and 1954 might best be characterized as in a period of refinement and consolidation. The initial jumps in yield that came from a vigorous, systematic breeding and selection program were essentially in place by 1948. Farmers who aggressively sought the higher yield potential of the new varieties adopted them, and extension work promoted the more widespread use of the improved varieties among farmers less inclined to take risks with new practices. Increased use of nitrogen fertilizer made the higher yields possible. Even an outbreak of new stem rust race, 15B, which caused much damage in Canada and the United States, had relatively little impact on Mexico because the new variety Kentana, released in 1948, had good resistance to it. Kentana accordingly came to be grown even more widely. It did well in El Bajío, the northwest, and other areas. Lerma Rojo, a new variety released in 1954, was likewise highly successful because it grew well in soils of low and high fertility and had stable, broad resistance to many diseases.[72]

Thus within ten years of Borlaug's arrival in Mexico, he could point to a number of results that had transformed the wheat production industry of Mexico. As with Vogel's work in Washington State, Borlaug, Harrar, and their colleagues had not made radically new conceptual developments, aside from the introduction of shuttle breeding and the abandonment of carefully breeding new varieties to fit a precise ecological setting. For the last six years of this first decade, Borlaug's publication pattern turned increasingly to extension brochures to advise farmers how to get the highest yields and to announcements and descriptions of the varieties released.[73] Unless something dramatically new came to their attention, it looked in the early 1950s as if Borlaug and his colleagues had reached a plateau of achievement that would not soon be raised.

It was not that Borlaug was satisfied with the yields that he and his colleagues were able to reach. He recalled many years later that from the mid-1940s he and Joe Rupert had kept their eyes open for shorter strawed wheats that might not be susceptible to lodging under conditions of high fertility. Although they found many such plants, none ever turned out to be useful in breeding for short stature. For the most part, these plants were probably aneuploids, that is, plants missing a chromosome and some of its genes. Such plants would not breed true and were short only because they were deficient in their genetic makeup.[74] In 1952–53, Borlaug also made a concerted search of the USDA World Wheat Collection for wheats with shorter straw, to no avail.[75]

Burton B. Bayles of USDA provided the crucial information that moved the semidwarfing genes from Vogel's to Borlaug's breeding program. Bayles visited the Rockefeller program several times in the late 1940s and early 1950s as part of his coordination work on wheat improvement in the western states. In 1952 he told Borlaug about Vogel's success with the Norin 10 varieties, and Borlaug wrote Vogel to ask for samples. In 1953 Vogel sent several selections of F_2 seeds from Norin 10 × Brevor and Norin 10 × Baart, and Bayles sent samples of Norin 10 to Borlaug. Borlaug attempted to get the semidwarf genes into good Mexican varieties, but the first attempts were plagued with problems. Norin 10 was a late flowerer, which meant it had to be used as the female parent and bear the seeds of the next generation. Unfortunately, Norin 10 was also highly susceptible to the rusts in the area, and all of the first crosses died. Borlaug also used the seeds from Vogel to cross with Mexican varieties.

By 1955 several crosses were successful, and Borlaug realized that he had a remarkably new type of wheat that could potentially surpass the yields of anything known up to that time. Mexican wheats, which were spring varieties that matured within six months, could never match the yields of Vogel's winter wheats that had nine or ten months to grow. Nevertheless, under Mexican conditions the semidwarfing genes allowed farmers to use high levels of nitrogen fertilizer without causing lodging, and this was the key to higher yields.[76]

Borlaug and his associates released two varieties of semidwarf wheats in 1962: Penjamo 62 and Pitic 62. Both had been derived from the Norin 10 × Brevor cross sent by Vogel.[77] Release of these two new varieties came before Borlaug and his colleagues felt they were ready. In fact, Borlaug would have preferred never to release either one in order to continue the breeding cycle long enough to remove certain defects. The necessity to release them came from the fact that curious farmers who visited the experiment stations were taking samples of breeding stock. Borlaug noted that many commercial fields were showing a complex mixture of the new short semi-dwarfs and the older tall varieties like Lerma Rojo. It was not only that Penjamo 62 and Pitic 62 were better than these haphazard mixtures, but Borlaug realized that the stations would have to make the release soon or lose credit for their development.[78]

In the subsequent five years (to 1967), the Rockefeller Foundation scientists put out a steady stream of new releases with the semidwarf genes of Norin 10. Sonora 64 and Lerma Rojo 64 both came out in 1964 (the number in the varietal name gives the approximate year of introduction). They were followed by Jaral 66, Tobari 66, INIA 66, Noreste 66, Norteno 67, and CIANO 67.[79] Just as the introduction of the new varieties between 1948 and 1954 had facilitated a substantial increase both in yield per hectare and in total national production, these new semidwarfs provided yet another sharp boost to yield potential. Improved varieties without the semidwarfing genes could yield between 4000 and 4500 kilograms per hectare, but the new types with the Norin 10 genes could yield 6000 to 6500 kilograms per hectare.[80]

Elimination of wheat imports as a way to conserve foreign exchange was a primary motivation for the Mexican government's desire to have the Rockefeller Foundation begin its program (see chapter 5). By 1963 the American embassy in Mexico City reported that on this ground the Rockefeller Foundation program was more than a complete success. In the 1940s the Mexicans had imported an average of between 196,000 and 278,000 tons of wheat per year. By 1955–59 this annual importation had dropped to about 23,000 tons per year, approximately 90 percent lower than importations in the late 1940s. In the early 1960s the importations dropped to below 500 tons per year and were usually fewer than 10 tons per year. In 1962 Mexican exports of wheat were about 1,000 tons, and in 1963 this figure rose to 72,000 tons.[81] Mexico not only was conserving foreign exchange, improved wheat production was earning foreign exchange. Without the semidwarfs, self-sufficiency was possible; with the semidwarfs, Mexico moved into being surplus in wheat.

In 1960 the Rockefeller Foundation concluded that its work in the Office of Special Studies was finished. On its own account, the Mexican government was inclined to absorb completely the work of the foundation. Indeed, as a full partner, the Mexican government worked with the Rockefeller and Ford Foundations to transform the Office of Special Studies into the second of the international agricultural research centers. The Centro Internacional de Mejoramiento de Maíz y Trigo (International Center for the Improvement of Maize and Wheat, or CIMMYT) opened its doors near the village of Texcoco outside of Mexico City.

For the first time in sixteen years, Norman Borlaug believed he might be leaving Mexico, which by this time was truly his adopted home. However, under Harrar's leadership the Rockefeller Foundation was moving toward a new and comprehensive program to increase grain production in all parts of the less developed world. As part of this effort, Borlaug drew the assignment of remaining as a Rockefeller Foundation consultant to CIMMYT with the task of promoting increased wheat production in any country that wanted assistance. In this capacity he became the chief courier of both the Norin 10 genes and the gospel of making the wheat plant an efficient tap into the now globally cheap sources of nitrogen fertilizer. These duties soon brought him to India and contact with M. S. Swaminathan and other Indian scientists.

Monkombu Sambasivan Swaminathan and Indian Wheat Breeding

By the early 1960s, word of the remarkable properties of the Norin 10 genes was rapidly making the rounds of the network of the world's wheat breeders. In contrast to the situation Vogel faced in the late 1940s, no doubt remained that the semidwarfing genes could be moved from the Norin strains and put into varieties that worked in different climates, soils, cropping patterns, and political economic circumstances. Movement of the semidwarfing genes thus became a matter of information and of samples reaching a new location that was prepared to develop the varieties needed by local farmers and to demonstrate how to make the new seeds work.

India by the early 1960s was fully outfitted with a research and education network that was seeking genetic material with properties like those of the Norin 10 genes. The policy of the Indian government was ostensibly favorable to higher yields, but the pathway to implementation was tortuous (see chapter 8). As it turned out, movement of the Norin 10 genes into Indian agriculture was possibly not entirely necessary and certainly not sufficient to raise India's aggregate national yields of wheat to levels of self-sufficiency. Political economic policies also had to be adjusted to stimulate a sufficient number of farmers to obtain higher yields. Natural and political calamities threatened the integrity of the Indian state in the mid-1960s, which pushed India into a new policy framework. With the semidwarfing genes, India's wheat production quickly rose to levels almost unimaginable a few years earlier. Thus was born the green revolution. Many were involved, but M. S. Swaminathan and Norman E. Borlaug played pivotal roles.

Monkombu Sambasivan Swaminathan was born on 7 August 1925 in the town of Kumbakonam, known for its many temples, in what is now the southeastern state of Tamil Nadu (Figure 10.4). His father, a medical doctor, was M. K. Sambasivan, from the village of Monkombu in Travancore, now part of Kerala. His mother, Shrimati Thangammal, was from Pudukottai, at the time a princely state in what is now Tamil Nadu. Swaminathan's father died when he was only eleven years old, and he, his mother, two brothers, and one sister were assisted by various brothers of his father and mother.[82]

Although Swaminathan's father died when the boy was quite young, he later recalled at least two features of his father's work that had substantial impact on his own thinking. First, Dr. Sambasivan became quite prominent in the town of Kumbakonam for his leadership in campaigns against mosquitoes that transmitted filariasis. The effort involved mobilizing of the community to do such things as fill in holes that held water and thus bred mosquitoes. Sambasivan became a leader in town politics through his medical work. Second, Dr. Sambasivan was an avid participant in the growing struggle, led by Mahatma Gandhi, for Indian national independence, embodied in *swaraj* (self-rule) and *swadeshi* (self-reliance).[83] This effort also included the struggle for reform of the caste system and the opening of temples to Harijans

Figure 10.4   J. George Harrar (*left*), H. K. Jain (*second from left*) and Monkombu Sambasivan Swaminathan (*second from right*), inspecting improved varieties of seeds in Delhi, 1971. Courtesy of Rockefeller Archive Center, North Tarrytown, New York.

(outcastes or untouchables), a reform that led the high-caste priests to refuse to attend Dr. Sambasivan on his death.[84] In later years Swaminathan's work in agricultural research drew upon his early exposure to notions of community involvement, extending new opportunities to the very poor, and, most important, the idea that self-reliance in food production was essential for Indian national dignity.

Swaminathan earned his first baccalaureate degree at the age of nineteen from Travancore University in 1944. He then studied agriculture at Coimbatore Agricultural College in Madras, where he received a second baccalaureate degree in 1947. Swaminathan at that time considered going into farming on family-owned lands, but an uncle urged him to continue his studies. Accordingly, in the year of Indian independence (1947), he entered the postgraduate program in genetics and plant breeding with B. P. Pal at the Indian Agricultural Research Institute (IARI) in New Delhi, where he received a diploma in 1949. During this work, however, he still remained somewhat uncertain about his future in agricultural science. Accordingly, in 1948 he sat for the exam to enter the civil service and was offered a position with the Indian Police Service. Swaminathan declined the offer and in 1949 won a UNESCO fellowship to study genetics at the Netherlands Agricultural University in Wageningen.[85]

After one year in Wageningen, Swaminathan moved to Cambridge, England, where he began work on potatoes with H. W. Howard at the Plant Breeding Institute (PBI). Thus Swaminathan reinitiated the pathway that B. P. Pal had begun in 1929 when he left India to study with Biffen and Engledow at PBI. Swaminathan earned his doctorate under Howard in 1952 and then set off for a year of research work at the University of Wisconsin in the United States (1952–53). During these first years of his professional work, therefore, Swaminathan published entirely on the potato (through 1955) with a focus on cytogenetics, cytology, and plant breeding. He returned to India in January 1954 after declining a position at the University of Wisconsin.[86]

Return to India at first meant a temporary post in rice breeding at the Central Rice Research Institute in Cuttack. From April through September 1954, Swaminathan worked on crossing *japonica* and *indica* varieties of rice. Significantly, this work brought Swaminathan for the first time into contact with the genetic basis of yield. The *japonica* varieties could yield five to six tons per hectare, but the *indica* varieties could not go beyond two tons per hectare. At fertilizer rates of higher than twenty kilograms of nitrogen per hectare, the *indica* varieties lodged. Although Swaminathan remained at Cuttack for only six months and did not publish results of his work there, some of the segregants from his original crosses reached Malaysia, where they were still grown three decades later.[87]

In October 1954, Swaminathan obtained a permanent post as assistant cytogeneticist at IARI in New Delhi, where he remained as a teacher, researcher, and administrator for the next eighteen years. During these years at IARI, he maintained a vigorous research program with students, which resulted in over fifty postbaccalaureate degrees awarded and over 200 publications by 1970. Swaminathan also switched the focus of his work to wheat when he joined IARI. He became especially interested in the prospects of using radiation to induce useful mutations in economic crops. Swaminathan became one of the most prominent agricultural scientists of India and contributed greatly to the development of IARI as a major research station.[88]

In the 1960s and 1970s, he received many prestigious awards. These included the Mendel Memorial Medal by the Czechoslovak Academy of Sciences (1965), selection as an honorary member of the Swedish Seed Association (1971), the Silver Jubilee Commemoration Medal of the Indian National Science Academy (1973), election to the Royal Society (1973), election as a foreign associate of the U.S. National Academy of Sciences (1977), and election as a foreign member of the V. I. Lenin All-Union Academy of Agricultural Sciences of the USSR (1978).[89]

At least some of these awards might have come to Swaminathan for his extensive work in cytology, cytogenetics, and studies in mutagenesis. Swaminathan, however, played a catalytic role in bringing the high-yielding wheats to India and in promoting the policies needed to stimulate higher aggregate national yields. In 1961 IARI began the All India Coordinated Wheat Improvement Project, located in its Division of Genetics.[90] Although Swaminathan did not directly coordinate the All India effort, his movement into the headship of the botany department brought him out of his laboratory and into the very practical, field-based breeding program.

Swaminathan recalled many years later that it was in 1959 that he first became aware of Vogel's work on the semidwarf wheats and their ability to respond positively and without lodging to high levels of nitrogen fertilizer. He wrote to Vogel and asked for seeds, but Vogel's high-yielding wheats were in winter varieties, which required a period of cold to flower and were suited to conditions that permitted the crop to be in the ground for seven to nine months before harvest. Indian wheat is generally planted in November or December, and it must be ready for harvest by April, before the fierce summer heat arrives in May. Thus Vogel suggested that Swaminathan might want to contact Borlaug in Mexico in order to obtain some of the Mexican varieties that were of the spring habit, designed for a short season of about five months.[91] Swaminathan's interest in the semidwarfing genes thus developed at exactly the same time that the Ford Foundation recommended more intensive inputs into Indian agriculture and prodded Nehru's government into organization of the Intensive Agricultural Districts Program (see chapter 8).

While correspondence was going on between Swaminathan and Vogel, S. P. Kohli of IARI's botany department initiated crosses with some Italian dwarfs, but these short wheat plants also had an ear that was short and thus did not yield much grain. Swaminathan knew by 1962 that short, stiff-strawed wheats that had normal ears and resistance to numerous races of black, brown, and yellow rusts were likely to be the kinds of plant that would allow Indian farmers to take advantage of nitrogen fertilizers and more irrigation later in the season. In February 1962 he noticed some experimental plots at IARI that had plants with the requisite short stems and normal ears. He traced the records of these plants and found that they originated from Borlaug's program in Mexico. After informal discussions with Pal and Ralph Cummings, field director of the Rockefeller Foundation in India, Swaminathan prepared a formal proposal to bring Borlaug to India during the next flowering and harvest period for wheat, February through April 1963.[92]

Pal and Cummings were both supportive of Swaminathan's proposal, and Borlaug arranged to be in India during March 1963. Swaminathan prepared an extensive tour of the varied areas growing wheat, each of which had special characteristics of climate, soil, and local wheat varieties.[93] Borlaug traveled almost constantly during his month in India, frequently with members of the IARI staff, including Swaminathan, Kohli, and others. By 1963 Borlaug knew well from his experience with the Norin 10 genes in Mexican wheats that at least in theory it should be possible to develop comparable wheat plants in India. In his summary of findings from his trip, therefore, he focused on the major factors that he judged had to be accommodated in order to increase Indian wheat production: disease resistance, soil fertility, water, and breeding for high-yielding potential under adequate systems of fertilization and irrigation.[94]

Borlaug especially linked the problems of soil fertility, the prospects for higher uses of fertilizer, and the appropriate objectives of a wheat-breeding program:

Most of the Indian soils where wheat is cultivated, are extremely deficient in nitrogen. . . . In order to appreciably increase the production of wheat in India, it will be necessary to bring about a great change in the breeding, agronomic and soil fertility research, as well as in the application of the plant pathology information. . . . In the past ten years it has been standard practice to search for small increases in yield. . . . Now, when there is every hope of a revolution in . . . the use of fertilizer, it is necessary to . . . change . . . the plant itself. . . . This modification in the breeding program . . . should discard the idea of searching for a variety which will . . . increase only a few pounds per acre the potential yield. . . . There is an immediate need for a new agronomic type of wheat which will catalyze . . . the use of fertilizer and improvements in irrigation. . . . The first and most clearcut, obvious factor limiting yield when cultural practices are modified, is lodging. New varieties must be developed which willbe [*sic*] resistant to lodging when heavy rates of fertilization are applied.[95]

Borlaug went on to recommend that wheat improvement needed a full-time coordinator in order to bring true, national coordination to what he saw as a seriously fragmented exercise. He pointed out that even within IARI, two different and largely uncoordinated wheat-breeding programs were under way. He also recommended initiation of a summer breeding station so that two generations per year could be grown and thus the time to obtain a new variety reduced. Borlaug's orientation was clearly on the project of raising national aggregate production figures as quickly as possible, that is, a transfer to India of the very success he had in Mexico.

At the conclusion of his Indian visit, Borlaug promised to prepare a selection of Mexican varieties and get them to India in time for sowing in the fall of 1963. The shipment from Mexico arrived in November and contained 100 kilograms each of Lerma Rojo 64A, Sonora 63, Sonora 64, and Mayo 64. These varieties had all been released for commercial production in Mexico. In addition, Borlaug sent smaller samples of over 600 additional lines in advanced stages of selection but not yet released in Mexico.[96]

Sufficient quantities of the four released varieties allowed the IARI scientists to divide the samples among four different research stations. Yields the following spring immediately indicated that the Mexican wheats were remarkable. Indian improved wheats without the semidwarfing genes generally did not yield above 2700 kilograms per hectare, and in good years most of them tended to yield in the range of 2000 to 2400 kilograms per hectare. In contrast, all of the Mexican varieties indicated in the first year that they could yield generally in the range of 2900 to 3700 kilograms per hectare. Some yields came in at nearly 4600 kilograms per hectare.[97]

In comparison with yields being obtained on farmers' fields, the Indian-improved varieties grown on research plots were thus generally about three times more productive. All India average wheat yields in the period 1956–57 to 1960–61 were 759 kilograms per hectare. The Mexican varieties demonstrated immediately that yield limits might be as much as five to six times as high as Indian farmers were then obtaining.[98]

Even given the spectacular yield potential of the Mexican wheats under Indian conditions in the first year of experience with them, Borlaug was convinced that, with adequate fertilization, varieties such as Sonora 64 and Sonora 63 should yield more, as much as 5500 to 6000 kilograms per hectare,[99] a figure that was nearly eight times as high as Indian farmers were then obtaining. From Borlaug's point of view, these figures were sufficient to justify an immediate effort to push the Mexican wheats into the mainstream of Indian wheat production:

> *These varieties or advanced lines should be multiplied rapidly and used for widescale testing and demonstration purposes. If results continue favorable after another year of extensive testing they should be used commercially without hesitation. Such varieties possessing high yield potential and resistance to lodging, when heavily fertilized and properly irrigated will serve as a catalyst toward revolutionizing Indian wheat production if used vigorously as an extension and demonstration tool in the Village "package developing programs."* (emphasis in the original)[100]

Additional excitement was also generated by enthusiastic press coverage of Borlaug's second visit to India. On 15 March 1964 the *Times of India*, the *Sunday Statesman*, and the *Sunday Standard* ran articles that trumpeted the high yields obtained in the first trials of the Mexican wheats at IARI.[101] Borlaug's enthusiasm may also have been stirred by a formal movement by the board of trustees of the Rockefeller Foundation, under the leadership of J. George Harrar and John D. Rockefeller III, in September 1963, to designate five areas of concentration, the first two of which were "Toward the Conquest of Hunger" and "the Population Problem."[102] Positive press coverage and a formal designation of higher yields as the top priority for the foundation were powerful incentives to push for rapid adoption of the Mexican wheats.

Borlaug's enthusiasm notwithstanding, a number of scientific and political factors stood as barriers to further promotion of the Mexican wheats. Although no argument attended the notion that the Mexican wheats provided genetic material that could be highly valuable under Indian conditions, Pal and Kohli were not keen on the idea of quickly releasing the Mexican varieties as they were. They cited as matters of concern the shriveled grains and susceptibility of the Sonoras in Punjab to yellow rust. They argued for caution and for crossing the Mexican wheats with Indian wheats in order to get new genetic varieties that would combine the best traits of the Mexican wheats with the best of the Indian strains. Kohli, who was the coordinator of the All India Coordinated Wheat Improvement Project, wanted Borlaug to send 200 to 300 kilograms each of eight Mexican varieties and 20 kilograms each of an additional twenty-one Mexican strains. Guy Baird of the Rockefeller Foundation staff in India echoed the concerns of Pal and Kohli and urged caution. He pointed out that premature release of U.S. hybrid maize in India had created problems.[103] All parties also realized that the Mexican wheats were not absolutely essential to increasing wheat production in India. More fertilizer and irrigation on Indian wheats, provided the amounts were not so high as to provoke lodging, could also produce higher yields.

Not only did scientific questions surround the idea of pushing the new wheats into rapid adoption; the very notion that wheat itself was worthy of high attention was possibly controversial. Rice was the major grain of India, which may have created questions of how much money should be devoted to research on wheat. In addition, India and the Rockefeller Foundation had invested as much or more in maize improvement since 1957, which may have promoted work on that grain as perhaps a better use of future research funds.

Despite these reservations, Swaminathan, with the support of Pal and the Indian Council of Agricultural Research, moved in March 1964 to propose that Borlaug send about twenty tons each of Sonora 63 and Sonora 64 to be used by IARI and other research institutions for about 1,000 acres of demonstration plots and to increase the supply of the Mexican seeds. In addition, the Ford Foundation staff were interested in moving the Sonora 63 and Sonora 64 into seed farms in the Intensive Agricultural Districts they were supporting. These plantings, too, were proposed more for demonstration and for seed increase than for full commercial production.[104] On its own account, IARI took the seeds produced in the spring of 1964 and multiplied them over the summer at its station in Wellington, in south India. Those seeds entered the All India Coordinated Wheat Improvement Program, to be harvested in spring 1965.[105]

On the scientific front, therefore, 1964 began with great excitement at the yields of the Mexican wheats, but the results prompted an intense desire for more research and further initiation of breeding work to combine Mexican and Indian wheat strains. Administratively, the Rockefeller Foundation added R. Glenn Anderson (1924–81) to its New Delhi staff as cocoordinator with S. P. Kohli of the All India Coordinated Wheat Improvement Project. Anderson had come from his native Canada to Mexico to work with Borlaug before accepting the assignment in India. Anderson's appointment represented a substantial increase in the Rockefeller Foundation program in India and a direct infusion of experience from Mexico to India. The foundation's increased commitment to wheat was also reflected in a substantial overall budget increase for its Indian Agricultural Program.[106] On the Indian side, the Education Commission of Indian government appointed B. P. Pal, director of IARI, to chair a task force on agricultural education, which met into 1965. Ralph Cummings, field director of the Rockefeller Foundation program in India, was added to the task force in May 1965 and proceeded to play a highly influential role in drafting the final report.[107] The year 1964 therefore marked a substantial deepening of both India's and the foundation's engagement with wheat as a vehicle for reform of Indian agriculture.

If scientific considerations alone had not been sufficient to result in a slow embrace of the Mexican wheats, political factors soon rose to paramount importance shortly after the wheat harvest of March and April 1964. Of most significance was the death of Prime Minister Jawaharlal Nehru in May 1964, after his debilitating stroke of the previous January. Lal Bahadur Shastri was elected to the leadership of the Congress Party, and thus to the prime ministership, on 9 June 1964.[108] Thus at the very time the Mexican wheats were first showing their prowess for yields under heavy nitrogen fertilizer, the top levels of the Indian government underwent a major transition, the first in the seventeen years of independence. Uncertainties stemming from new leadership may have dampened enthusiasm among administrators like Pal, who headed the IARI. When a new boss is coming in, the last thing an experienced civil servant wants is to be responsible for starting a bold new venture that fails or is not in line with new priorities.

It quickly became apparent, however, that the promise for higher yields would not be outside the goals of the new leadership. Shastri's first alteration of the cabinet was to appoint C. Subramaniam, then the minister of steel, to be minister of food and agriculture, an arena of government policy that was in crisis and disarray. Shastri and Subramaniam both believed that India needed to raise its total agricultural production. In June, during his first month in office, Shastri appointed a prices committee, which recommended price incentives for commercial farmers and larger investments in production inputs. Subramaniam's new Food Corporation of India, announced in July 1964, was to be the major conduit for procuring grain at prices attractive to farmers, although for its first year of operation it would purchase mostly rice, not wheat, from domestic growers. In July Shastri's government also began negotiations with the United States to provide about 4 million tons per year of wheat imports for the following five years. Thus at the time the new Mexican wheats first showed promise, the new government clearly remained on a trajectory to rely on imports of wheat to meet domestic demand.

Subramaniam continued the development of his thoughts in a speech to the National Development Council, Committee on Agriculture and Irrigation, on 1 January 1965. The National Development Council was Shastri's vehicle for opening the debate about India's planning to political leaders of the states, who had had little influence with the Planning Commission under Nehru's rule.[109] Subramaniam called for a reorganization of central government efforts in order to bring better cooperation between the Ministries of Agriculture, Irrigation, and Community Development and Cooperation. He was particularly forceful in arguing for a wider use of science in reforming Indian agriculture, including the use of better seeds, more and better use of fertilizer, and more efficient use of irrigation. Prices to reward the farmer and credit to enable the farmer to invest, he argued, were also needed.[110] With this speech, about seven months into his tenure as minister of agriculture and with no previous background in the field, Subramaniam clearly staked out a broad and comprehensive reform agenda based on the new technology provided by agricultural science.

Not only did Subramaniam prove to be a quick student in learning the substance of an industry he had previously not known; both Subramaniam (Figure 10.5) and Shastri (Figure 10.6) showed the personal interest to visit the experimental plots of Mexican wheats at IARI. Shastri's visit was at a field day for local farmers, and the

Figure 10.5   C. Subramaniam, minister of agriculture (*second from right*) inspecting a field of Sonora 63 with officials of the Indian Agricultural Research Institute and the government of India. Courtesy of Rockefeller Archive Center, North Tarrytown, New York.

Figure 10.6   Lal Bahadur Shastri, prime minister of India, 1964–66, being presented with a sheaf of grain by a village cultivator at a field day during the Diamond Jubilee Celebration of the Indian Agricultural Research Institute. Courtesy of Rockefeller Archive Center, North Tarrytown, New York.

prime minister delivered a speech in simple Hindi in which he said the modern wheats were to older wheats as airplanes were to bullock carts. Subramaniam relied on scientists like Swaminathan to prepare much of his material. Swaminathan recalled later that although Nehru was interested in science, the new leaders of India provided higher access to people interested in agriculture.[111]

More than mere interest in agriculture, however, was involved. As Subramaniam moved to embrace fully the promise of the high-yielding varieties of wheat, he was simultaneously rejecting the entire basis of India's development plans as they had been developed by Nehru and the Planning Commission since 1947. It is likely that only the near-calamitous political conditions in 1964 and 1965 permitted Shastri and Subramaniam to promote a policy that was so accepting of the new wheats at IARI. Possible outbreak of famine, eruption of political violence over shortages of food, and stern pressure from the World Bank all combined by August 1965 to complete the transition in the central government to a full embrace of the technology needed to get higher agricultural production (see chapter 8).

Subramaniam's proposal was to make real India's commitment to the Intensive Agricultural District Programme, which the Ford Foundation had recommended in 1961. Endorsement of the IADP carried with it a substantial reallocation of India's foreign exchange reserves. Where 191 crores of rupees (1 crore = 10 million) had been spent on imports for agriculture during the third five-Year plan (1961–66), Subramaniam's new proposals called for over a fivefold increase, to 1,114 crores of rupees in the fourth five-year plan. During the course of the plan, estimates of India's foreign exchange earnings were 5,100 crores of rupees, of which 5,300 crores of rupees were projected to meet existing imports and 1,350 crores of rupees were needed for foreign debt servicing. Thus without Subramaniam's agricultural projects, India was already projecting a deficit of foreign exchange of 1,550 crores of rupees.[112]

Financing development, including Subramaniam's plans, thus required foreign assistance of at least 2,650 crores of rupees, an amount larger than the 2,600 crores of rupees estimated to be received from 1961 to 1966. Such mammoth needs for inputs to agriculture, mostly irrigation equipment and nitrogen fertilizers, also raised the possibility that if foreign assistance were not forthcoming, India would have to accept private foreign investment to build manufacturing plants for these items but on terms of foreign ownership that up to that time India had found unacceptable. Opponents within Shastri's government focused on the inescapable needs for foreign assistance, the unpalatability of foreign capitalist penetration, plus the inegalitarian nature of the IADP scheme for rural development to stall Subramaniam's proposal.[113]

If 1964 was the year of cautious scientific optimism on the promise of the Mexican wheat and the turmoil of change in the political sphere, then 1965 was the year of consolidation of confidence on the technical capacities of the semidwarf wheats and a continuation of unsettledness in many dimensions of Indian political life. Most serious among Indian political worries were the worsening relations with Pakistan. Skirmishes in April over the Rann of Kutch turned into full-scale war in September, with major battles in the far north (Jammu and Kashmir) and in the Punjab between Delhi and Lahore.[114] To the disruptions caused by conflict came a year of drought starting with the monsoons of the summer months. The full effects of the water shortage were visible by early 1966.[115]

Wheat harvested in March and April 1965 was sufficient to approve Lerma Rojo 64A and Sonora 64 for release to commercial production on irrigated land. The Mexican wheats were especially good at resistance to lodging.[116] In early July the Indian government placed an order to Borlaug for 200 tons of Sonora 64 seed. The Rockefeller Foundation was to pay Mexico in U.S. dollars, and India was to reimburse the Rockefeller Foundation in New Delhi in rupees.[117] In late July India wanted to change the order to 100 tons of Lerma Rojo 64 and 100 tons of Sonora 64, if possible, because of Lerma Rojo's better rust resistance. Unfortunately, the switch was not possible, because the 200 tons of Sonora 64 were already en route when the revised order was placed. Accordingly, India ordered an additional 50 tons of Lerma Rojo 64.[118] Borlaug advised that Sonora 64 should not be grown in areas in which stripe rust was a problem.[119]

A shipment of 250 tons of seed could plant about 2900 hectares (about 7,100 acres) if seeded at 86 kilograms per hectare, which was the seeding rate used for the 1965–66 yield trials.[120] IARI planned to use the 250 tons for further testing, demonstrations, and distribution to over 5,000 farmers.[121] Thus by mid-1965 both the scientific and political consensus was that India was moving to disperse the Mexican wheats as they were, without waiting to breed new varieties that had useful Indian traits. Such breeding work could come later, of course, and ultimately replace the original Mexican types.

Strategically, this move to rapid adoption of the Mexican wheats was a major shift from the usual caution and slower pace of new variety adoption. It clearly reflected Borlaug's missionary sense that dramatic increases in yields, even when the varieties available were not perfect, was the way to proceed:

> Since Indian wheat production is entering a phase of dynamic change it will be necessary to use a dynamic approach to the multiplication and distribution of new varieties. . . . I would urge you to be aggressive in deciding to multiply and distribute promising lines. Don't look for the perfect variety—or for a variety that will last for 15 years commercially, for you will never find it. If your breeding program is dynamic and aggressive, the lines entering increase will already be obsolete. This is the way it should be.[122]

This dramatic shift in the speed at which Indian agriculture was being reformed was based as much on political considerations as on scientific considerations. Indian scientists had two years' worth of data, for 1964 and 1965, which indicated that the Mexican wheats could respond without lodging to nitrogen fertilizer and irrigation. Further extensive trials were under way for the 1965–66 growing season.[123] They were not perfect in terms of resistance to Indian rusts, especially Sonora 64, but with care they apparently could be grown safely in India with great success. Farmers had to learn new methods of seeding, irrigation, fertilizing, and controlling weeds, but such adjustments did not pose a major conceptual problem.

Politically, in the summer of 1965 Shastri's government could not have felt anything but beleaguered. In addition to the chronic concerns about food supplies and the uses of foreign exchange, India had fought, not very successfully, a skirmish with Pakistan on their southern border in the Rann of Kutch. India, with its much larger population, was militarily much more powerful than Pakistan, but both countries were heavily armed, largely by the United States, which worked to keep both countries in the Western camp against the Soviet Union and China. Dependence on the United States for both food and military equipment was not a welcome state of affairs in New Delhi.[124] Thus any avenue to relieve these political stresses must surely have looked attractive to Shastri's government.

To make matters worse, in the summer of 1965 the United States was in the process of revamping its programs of food aid, as part of a larger concern about global population growth. India specifically was informed by the State Department that future aid would depend on India's allocation of foreign exchange for fertilizer or on building fertilizer plants in India. In addition, in August 1965 the Johnson administration put India on virtually a month-to-month arrangement for food aid.[125] These explicit links between population, food aid, and agricultural policy were stimulated by a conference of demographers, policy makers, and others, which was held in July and organized by the Rockefeller Foundation.[126] Thus the foundation worked on the reform of Indian agriculture from two sides simultaneously: first as collaborators with Indian agricultural scientists on wheat yields and second as catalysts for policy making in the United States for ties between food aid and the actions of recipients of aid.

Almost immediately on the conclusion of the war with Pakistan (late September 1965), India proposed to the Rockefeller Foundation to import a large amount of Mexican wheat seed, to arrive in time for planting in the fall of 1966. At first the proposal was for 5,000 tons. This figure dropped to 2,000 tons in February 1966, then rose to 21,000 tons in April. Although the Rockefeller Foundation was delighted to act as a catalyst in arranging for this enormous shipment, Harrar, as president of the foundation, wanted no responsibility for the seed because of possible problems with poor germination. Harrar was willing, however, to provide India up to $100,000 to help pay for it.[127]

Subramaniam made the political importance of the Mexican wheats completely clear in statements published in October and November 1965. It is important to see in his own words how he linked agricultural science, the development of India, and India's national security:

> The inadequacy of our agricultural production has thrown a grave challenge to the nation engaged in the task of its economic growth and in the increasing responsibilities of its defence. It is a challenge to our will to live in prosperity and freedom. . . . Our men of science are called upon to provide the ideas and leadership for bringing into the field methods and techniques which will effect a breakthrough in our agriculture and sustain its dynamic growth. . . . Agriculture in this country should be regarded as a management problem and not merely a way of life, and I am sure, the productivity approach is going to help us in maximizing output.[128]

> [Indian and Pakistani armies have stopped, but we should not blunt our vigilance.] In fact, we must remind ourselves that there is an enemy lurking still—the enemy of ignorance, poverty and obscurantism. This enemy has to be fought and fought to a finish. Never has this need been so great as now; and never has our country been in a mood as ready to respond as today to new initiatives to sweep away the cobweb of delay, obstruction and reaction. . . . India cannot continue to remain in the same state as before. . . . [The current crisis] has focused the need to introduce in a maximum way science and technology into our administration, our economic activity and our defence. The scientific revolution has a great role to play in making our society strong and successful. Research has to be organised in a bold and purposive manner to this end.[129]

Prime Minister Shastri was even more explicit about the links between agricultural production and the nation's security. In a broadcast on All India Radio from Delhi on 10 October 1965, he stated:

> I consider self-sufficiency in food to be no less important than an impregnable defence system for the preservation of our freedom and independence. . . . Dependence on food imports is not only bad for the economic health of our country but it undermines our self-confidence and self-respect. We have to stand on our own legs and a beginning has got to be made right now towards self-sufficiency in food. The food front today is almost as vital as the military front. . . . The *jawan* is giving his blood, he is staking his life for the country. I am asking the *kisans* to give their toil and their sweat.[130]

Although Subramaniam resolved to move as quickly as possible to increase India's agricultural production, the Shastri government did not last long enough to complete the mammoth import of the Mexican wheats. Lal Bahadur Shastri died in

Tashkent, USSR, on 11 January 1966, the morning after concluding a peace agreement with Pakistan that had been brokered by Premier Aleksei Kosygin. Shastri's death reignited the battle within the Congress Party over the directions of Indian policy. Indira Gandhi, Nehru's daughter, emerged as the new leader and prime minister on 19 January 1966.

In the confusion of establishing a new administration, those within the Planning Commission who opposed Subramaniam's plans for the Mexican wheats may have sensed an opportunity to raise anew the objections to moving to capital-intensive agriculture. Staff of the Rockefeller Foundation heard the commission's objections that nitrogen fertilizer would not be available in 1966 and that in any case the Mexican wheats had not responded well to fertilizer. They launched a request to Borlaug in Mexico for solid information about the potential for nitrogen responsiveness in the Mexican wheats.[131] Although surviving records do not indicate exactly how Borlaug and the other foundation staff responded, the objections based on availability of nitrogen fertilizer very probably reflected the intense battle within the Indian government over how to allocate scarce foreign exchange.

Subramaniam's plans for Indian agriculture survived the transition from the Shastri government to the Gandhi administration. In fact, they became a major piece of discussion when Gandhi visited Washington in March 1966 as part of the new government's efforts to improve relations with the United States. Gandhi was obliged to meet the demands of the Johnson administration that India devalue the rupee, enhance its own ability to increase agricultural production, and in other ways show evidence of development that were consistent with what the world's largest capitalist country thought development should be. Gandhi was successful in impressing the American government with India's policies, even though she later faced intense criticism both within Congress and among opposition parties.[132] She also was successful in gaining a commitment from the Johnson administration for the increased food aid that would be needed in 1966 because of the failure of the monsoon rains of 1965–66.[133] Subramaniam's influence within the new administration was enhanced by his appointment to a seat on the Planning Commission,[134] which previously had been a source of opposition to his thinking.

Given the now firm directions of the new government, in 1966 India sent a team of three, including S. P. Kohli of IARI as the chief technical adviser, to Mexico to select and purchase a large quantity of Mexican wheat to seed a significant portion of the 1966 wheat crop of India. A total of 18,000 tons of Lerma Rojo 64 began loading on the Greek freighter *Phaedra* at the Sonoran port of Guyman on 18 July 1966. This was the largest single seed transaction ever in the developing world, and it was tremendously complicated. The Indian delegation had to deal with Mexican cooperatives and individual farmers, select fields, get rid of weeds that would contaminate the seed, and oversee the warehousing, cleaning, and sacking of the grain for shipment. For their efforts, Borlaug felt they got a good price and handled the whole affair very skillfully.[135] By mid-September the *Phaedra* was unloading its cargo at Khandla in Gujarat.[136]

With such a large shipment of grain, India was launched into a transformation of its wheat production. Borlaug estimated that 18,000 tons plus seed raised in 1965–66 in India would permit about 1 million acres to be planted with semidwarf varieties in the fall of 1966, to be harvested in the spring of 1967. With the new harvest, India would be fully set to use the high-yielding varieties in 1967–68.[137] At the time, India had about 33 million acres under wheat, of which about 10 million had some irrigation.[138] Thus the huge shipment of Mexican grain, plus that raised domestically, had the potential for seeding about 10 percent of the suitable (irrigated) wheat land.

In 1966–67, for the second year in a row, the monsoons failed in India. Drought in India, however, was always a problem for the areas without irrigation. If irrigation was available, then yields might suffer little or no ill effect from the drought; such was the case with the 1966–67 drought. Total grain production of India was down sharply for the second year in a row, and the drought of the previous year had left the country with no carryover stocks. Production of 89 million metric tons in 1964–65 had dropped to 72 million metric tons in 1965–66 and less than 75 million metric tons in 1966–67. Imports, including a record-setting 10 million metric tons in 1966, averted famine but not a loss of national pride.[139] In contrast to the decline in national yields, the state of Haryana, just west of Delhi, had extensive irrigation works. The index of its production of wheat in 1964–65 was 137.8, which dropped slightly to 135.1 in 1965–66 and rose to 157.8 in 1966–67 (1956–57 = 100). It is possible that Haryana's production of wheat in 1966–67 went up because of the first influx of the Mexican wheats on a large scale. Fertilizer consumption was increasing rapidly at the time: 1,081 tons of nitrogen were used in 1960–61, but this had risen to 12,626 tons in 1966–67.[140]

For India as a whole, the year 1966–67 marked a turning point in its national wheat production. Continuous pushing of the high-yielding variety program by the Indian government and the Rockefeller and Ford Foundations pushed the area planted to the new varieties from 504,000 hectares in 1966–67 to over 10 million hectares in 1972–73. Correspondingly, India's reliance on food grain imports began to drop, from 4.7 percent of total in 1960–61 to 0.8 percent in 1972–73.[141] Similar changes occurred with rice production as dwarf varieties that responded to fertilizer came into production. The green revolution did not end the question of agriculture and food production in India, but it shifted it onto an entirely new plain in which self-sufficiency was a realistic goal.

The people most involved went on to positions of greater responsibility, a factor that surely reflects the prestige they earned in the eyes of the Indian government for their efforts. Subramaniam appointed B. P. Pal, director of the Indian Agricultural Research Institute, to be director general of the Indian Council for Agricultural Research in June 1965. This was a move that had been recommended by a review team headed by Dr. Marion Parker of USDA and was welcomed by the Rockefeller Foundation staff in New Delhi. For the first time a scientist would head the council, and all scientific research units sponsored by the central government came under Pal's supervision.[142] For his general supervision of the green revolution, therefore, Pal became the chief agricultural scientist of India. He remained in that role until his retirement in January 1972.

M. S. Swaminathan, who had played such an important catalytic role in identifying the semidwarf strains as potentially interesting, remained at IARI until Pal's retirement in 1972 and then became Pal's replacement as director general of the

Indian Council for Agricultural Research. He remained the chief agricultural scientist of India until 1979, when he moved successively to be secretary in the Ministry of Agriculture and Irrigation (1979–80) and then member of the Planning Commission (1980–82). Swaminathan then left the service of the Indian government to become director general of the International Rice Research Institute in the Philippines. He retired from that position in 1988, returned to India, and continued to serve as president of the International Union for the Conservation of Nature and Natural Resources, a post he began in 1984.[143]

In India, Pal and Swaminathan brought the plant-breeding work that originated with Rowland Biffen and Frank Engledow in England together with the plant breeding that stemmed from the work of Orville Vogel and Norman Borlaug of the United States and Mexico. In their turn, the Indian scientists contributed further varieties of wheat that ultimately transformed the Indian agricultural landscape, just as Vogel and Borlaug had transformed those of the United States and Mexico. To complete the circle, we now turn to the work of Francis G. H. Lupton and John Bingham, who transformed the English agricultural landscape in a similar fashion.

Francis G. H. Lupton and John Bingham

Two English scientists, Francis G. H. Lupton and John Bingham, were at the heart of the efforts to breed new winter wheats at the Plant Breeding Institute after World War II. Lupton joined the PBI staff in 1948, at the start of the institute's postwar reconstruction (see chapter 9). Bingham began work on 1 December 1954.[144] The two scientists and their colleagues eventually reached about 1,000 new crosses per year, but it was not until 1964 that the PBI scientists released a new variety: Maris Widgeon. This was the first substantially new wheat variety created in Britain since Frank Engledow's Holdfast was released in 1935, twenty-nine years earlier. Private breeders in Britain released a few varieties of wheat, such as Redman in 1939 and Hybrid 46 in 1946,[145] but for the most part British-grown wheat tended to be varieties produced either by PBI or by foreign research stations. Thus British wheat breeding endured a spell of nearly three decades without many new varieties.

An absence of new releases for such a long time by no means indicated a lack of interest in wheat by PBI. Instead, the long hiatus represented, first, virtually a complete disruption of wheat-breeding efforts for nearly fifteen years by World War II and the reconstruction of agricultural research that followed it (see chapter 9). Second, the initially high postwar enthusiasm for increasing British food production, which could have been a powerful stimulus to introduce new wheat varieties, began to dissipate in the 1950s. Reforms in British agriculture initiated by the Labour government after 1945 elicited higher food production from farmers without remarkably new varieties.

Higher farm production in the 1950s occurred in a context of general economic recovery from the damages of the war. As a result of the more prosperous times, the sense of urgency about the British food supply and the foreign exchange reserves for food imports diminished through the 1950s and 1960s, but the period from 1952 to 1954 produced brief foreign exchange scares that were reminiscent of the sterling crises of 1947 and 1949. Engledow and a few others[146] continued to maintain that

Britain needed to achieve even higher levels of food self-sufficiency as a national security measure, but the mainstream sentiment became more complacent and thus happy with the higher levels that had already been achieved. Wartime austerity rules on wheat milling ended in 1953, and most consumers probably did not care if the bread was made with imported wheat or domestic grain, as long as the price was reasonable.

Not only did improving economic conditions during the 1950s reduce the sense of crisis about the need for British food self-sufficiency, but also the wheat breeders at PBI had one of the most technically and politically difficult jobs among wheat breeders in the entire world. Compare their situation with those elsewhere: in Mexico, Borlaug and his associates entered the scene when very little improvement of wheat varieties had been accomplished. A few judicious selections from the varieties already growing, some imports of varieties not found in Mexico, and some simple crosses, all combined with some advice about use of a little fertilizer, sufficed to make significant changes in the Mexican context. This comment is not to diminish the significance of what Borlaug's team achieved, but, in contrast, British wheat breeders had to produce varieties that were better than the already improved varieties being grown, with modest amounts of fertilizer, in order to be successful. Moreover, the Mexican government at the level of the president acutely wanted higher national wheat production, and Borlaug's unit reported directly to the minister of agriculture, which gave it a continuing high political visibility.

Similarly, Indian wheat breeders after 1947 already had a substantial backlog of improvement work. In contrast to Britain, however, India had until 1964 a government that did not match its words of enthusiasm for agriculture with policies that encouraged farmers to increase production. The Labour government's Agriculture Act of 1947 (see Chapter 9) established a variety of programs that remunerated farmers who moved to higher production levels, and the Conservative government that followed made changes that encouraged more intensive practices even more effectively. India's government did not move to real incentives for higher production until the mid-1960s. Thus PBI wheat breeders faced the ironic situation that government policies to encourage production were already so successful at improving yields of existing varieties that it was increasingly difficult to find new ones with higher yields.

Perhaps the British context for wheat breeding was most similar to the American one. American breeders, too, worked in a situation that already had substantially improved varieties, so they also had to search hard for new varieties that could outyield the old ones. In contrast to the United Kingdom, however, agriculture in general was a much more powerful industry in the United States. The government's attention never drifted far from agriculture because the political consequences could be severe. British farmers could not wield such a block of votes in Parliament, so there was a tendency for government in Britain to pay attention to agriculture most when the Treasury feared a collapse of the pound sterling and a foreign exchange crisis. PBI's breeders were thus caught up in the shifting levels of attention they could expect agriculture and their craft to enjoy from Parliament and the public.

Thus the program in wheat breeding at PBI after 1950 became one without a strong sense of crisis to push for rapid introduction of new varieties. Instead, Lupton, Bingham, and other PBI scientists pursued a breeding strategy linked to an extensive research program on (1) quality for bread making of wheat protein, (2) plant physiology, (3) plant pathology, (4) cytology, and (5) cytogenetics. Although the specifics of the research instituted after 1948 had all the trappings of the most current biological science, it is significant to note that the grand strategy for postwar PBI breeding programs actually dated to the work of Biffen and Engledow in the 1920s. They argued explicitly in 1926 that what was needed for British wheat were varieties that had shorter, stiffer straw that would enable farmers to put on more manure without causing the crop to lodge. In addition, they believed that British wheats could be constructed to produce strong flour for bread and to be resistant to disease.[147]

Partly as a result of this extensive research program, and partly as a matter of good fortune from newly imported genetic material, PBI ultimately produced new varieties that had astonishingly high yields, perhaps close to the theoretical maximum based on efficiencies of photosynthesis currently found in wheat. In addition, PBI wheats bred by Lupton and Bingham had protein quality suitable for use in British breads, which reduced the amount of wheat imported from North America to vanishingly small amounts. Britain, in fact, achieved the vision held by Rowland Biffen in 1905, self-sufficiency in wheat production, in the 1980s.[148]

The period 1948–55 was one major reorganization and shifts for both PBI and British agriculture as a whole, and it is important to understand these events in order to see how Lupton's and Bingham's research program emerged and succeeded. When Lupton joined PBI in 1948, he had just completed his baccalaureate degree from Cambridge with a tripos in botany, zoology, and physiology. He had not had any experience, practical or theoretical, in agriculture, but PBI Director G. D. H. Bell hired him because he was well versed in the biological sciences, which reflected Bell's sense that PBI staff should be good scientists first and learn their agriculture as they needed it. Lupton spent a substantial part of his first year attending the lectures in agriculture at Cambridge, and his first assignments were to assume responsibility for the wheat breeding previously conducted by Engledow.[149]

During the reconstruction of PBI, a substantial amount of time was spent organizing the new facilities, defining the breeding objectives, and devising the breeding methodology. A particularly important point was whether the program should seek varieties that made "strong" flour. British millers and bakers were accustomed to buying imported grain from North America, which had the kinds and amounts of protein necessary to make a strong flour, which in turn could make the type of bread loaf preferred by the British. English wheat tended to make a weak flour, which would not rise properly and gave a loaf of poor crumb structure. At the dedication of PBI's new buildings, an elderly miller went so far as to tell PBI scientists that they would be wasting their time to seek wheat varieties that could produce strong flour under British conditions. The miller's implications were that Britain would always be dependent on imported wheat, unless British consumers developed a preference for a different type of bread. Engledow, Bell, and Lupton, the three senior wheat scientists, however, disagreed and insisted that the quality of the grain should be an important objective of the PBI program.[150]

After the initial objectives of the work were clear, Lupton began research that focused on the resistance of potential new varieties to disease (yellow rust, *Puccinia glumarum*; powdery mildew, *Erysiphe graminis tritici*; and eyespot, *Cercosporella herpotrichoides*). In addition, Lupton delved deeply into the methods for conducting large-scale breeding experiments. Factors he considered were how to (1) select parents, (2) select progeny for further testing, (3) lay out the plots for testing the progeny, important for distinguishing the environmental effects of soil variation from genetic differences, (4) assess variation between seasons and thus further distinguish environmental from genetic effects, (5) select for disease resistance, and (6) select for grain quality.[151] Later, in the 1960s, Lupton added studies on photosynthesis in wheat and the mechanisms by which the sugars made in the green parts of the plant were moved to the grain for storage as starch.[152] Always, the goal was to find new wheat varieties that had higher yield, quality, and reliability for growers.[153]

Bingham joined the staff in 1954, six years after Lupton. He was born in Norfolk and had completed his baccalaureate in agricultural botany at Reading University before spending two years in the Royal Air Force.[154] When Bingham arrived in Cambridge, PBI had consolidated its postwar growth plans. The new facility at Ansty Hall Farm was just occupied, the formal dedication by the minister of agriculture (D. Heathcoat Amory) was coming up in mid-1955, and a batch of new scientists (including Bingham) had just been hired by Bell. Bell also had outlined a new five-year plan for the institute's research, which was endorsed by external reviewers just before Bingham came on board.[155]

Bingham's efforts focused on questions of protein quality for bread making, an objective that constituted about one-half the effort on wheat breeding at the time. For Bingham, the practical work of breeding was more important than the accompanying research in physiology and plant pathology, but he believed that such research was critical to the conduct of the breeding work. He, like Lupton, published a steady stream of papers on various aspects of the work at PBI.[156]

Lupton and Bingham made up a team that spent most of its efforts on winter wheat. Spring wheat was done by R. N. H. Whitehouse, who joined PBI with Lupton in 1948. Whitehouse also worked on winter oats while Lupton and Bingham worked on spring oats. Bell worked on barley, lectured at Cambridge, and had the overall administrative responsibilities for the institute. These four were the senior workers in the Cereals Division in the first decade after PBI's reorganization in 1948. As they worked during this first decade, however, the context to which they mainly related, production agriculture, was itself undergoing a profound shift. In retrospect, it is easy to see the changes in British farming that in some ways made it difficult for the scientists to produce new varieties that had a chance to compete for the farmers' attention. On the other hand, the new conditions included higher uses of nitrogen fertilizer, which made it imperative to find varieties that would withstand the higher fertilization without lodging. Furthermore, farmers' attention was keenly tuned to the advantages of any such varieties, should they be found.

In order to see the shifting structure of British agriculture as a whole, it is necessary to sketch the broad changes wrought by the Second World War. In 1939 Britain imported about 70 percent of its food and raised about 30 percent domestically (percentages are of wholesale value) for its 46 million people. Liquid milk and potatoes were almost entirely homegrown, meat was 50 percent domestic, and wheat, sugar, oils, and butter were largely imported. For wheat and flour, about 88 percent was imported.[157] A total of about 5 million acres (out of about 30 million acres of total farmland) was used for grain crops, about 45 percent of which was in wheat. About two-thirds of the British wheat went for confectionery products (biscuits, or "cookies" in America, and cakes). One-fourth of the crop went to feed animals. The rest was for seed and bread, but bread was primarily made from imported strong flours.[158]

During the war, improved prices, combined with strong government policies to produce more food locally, stimulated farmers to plow up grassland for wheat, barley, and potatoes. One-fourth of Britain's sheep were slaughtered because of the loss of grass pastures, but cattle remained at about constant numbers. Wheat and barley production rose by about two-thirds, and potato production went up 100 percent. Wartime restructuring therefore led British farmers to produce more domestic grain for bread and animal feedstuff.[159] For the consumer, this meant that body and soul were kept together with bread (made with more domestic wheat) and potatoes, supplemented by rationed high-protein products such as milk and beef, which was produced with more domestic feed grains (barley, oats, and wheat) than prior to the war. As noted earlier (chapter 9), wartime Britain still relied on many imports of food and feed from North America, but an increased proportion of its calories came from British sunshine.

At the conclusion of the war, no one wanted to see the prewar depression conditions return to British agriculture; in addition, increased domestic production was a prime tool adopted by the Labour government for alleviating Britain's foreign exchange crises (see chapter 9). Although it was not an explicit goal of the Agriculture Act of 1947, the government aimed to have national agricultural production 60 percent higher than 1939 by 1956. Deficiency payments provided in the act were the prime mechanism by which farmers were to be encouraged, albeit indirectly. In theory, farmers' higher profits were to go for investment in more intensive technologies, which in turn would create higher national production levels.[160] With higher production levels would come reduced demand for imports and thus important savings on limited foreign exchange reserves.

Experience with the first several years of the 1947 act indicated that farmers did not respond to the incentive payments in ways that policy makers in the Ministry of Agriculture intended. Rather than intensify their farming practices to increase their aggregate outputs, farmers purchased consumer items such as automobiles. National output rose a modest 8.5 percent between 1945–46 and 1950–51, not encouraging for policy managers who wanted to see national output up by 60 percent in 1956. Consequently, in 1951 the government moved to a program of grants to subsidize specific types of farming practices. Grants were available after 1951 for purchases of fertilizers, plowing up grasslands for grain production, and raising of beef calves.[161] This tactic began to show effects quite quickly. Wheat production rose from 2.4 million tons per year in 1950–52 to 2.7 million tons in 1955–57, even though acreage dropped from 2.3 to 2.1 million acres. Similarly, barley production rose from 2.0 to 2.9 million tons per year, and barley acreage expanded from 2.0 to 2.4 million acres.[162]

Change to a Conservative government when Winston Churchill led his party to victory in the 1951 elections did not significantly affect these policies to stimulate aggregate levels of output. Although the Conservatives paid homage to the notion of free markets and people doing without government handouts, the practice of production grants was continued and extended in the years after 1951. Grants for major capital improvements, hedgerow removal, and land drainage were significant stimuli to move British farmers from a pattern of mixed animal-arable farming to more highly specialized, mechanized farms that raised only grain or only animals, on a progressively larger and more intensive basis.

In one sense, the government's policies were a success: national output was up 56 percent by 1953–54, compared with the goal of output up by 60 percent. This "success," however, was tempered by the government's recognition that global agricultural output was also increasing and that the United Kingdom's agricultural policies were not only expensive to the Treasury but arguably unneeded in the context of general British economic recovery and cheap supplies of food imports. In 1954, for example, the United States adopted the P.L. 480 program as a way of disposing of its enormous supplies of surplus grain. Some raised questions about whether it might be wiser for Britain to revert to its old habits of reliance on imported foods.[163] Still, through the 1950s the government continued to use production grants to encourage the increase in homegrown feed grains, beef, and milk.[164]

Deliberate government action to encourage the intensification and specialization of British farms produced a moving target for PBI scientists, but they continually matched the selective conditions for identifying promising new varieties as closely as possible to the emerging practices of farmers. In both barley and wheat, the institute showed some successes during the 1950s. Bell, who had worked on barley since the 1930s, had an immense success when he released several new varieties of barley and field peas. The most significant of these was Proctor, a new barley released in 1953.

Proctor was important for two reasons. First, it gave the new PBI enormous credibility because of its incredible success. By 1960, within seven years of its official release, Proctor occupied 70 percent of the barley acreage in the United Kingdom.[165] Not only was Proctor the main barley of choice, but acreage devoted to barley also expanded tremendously, from 2.0 million acres in 1950–52 to 3.6 million in 1960–62 and to 5.4 million in 1970–72.[166] Expansion of barley and selection of Proctor as the barley of choice was not really foreseen by Bell and his PBI colleagues because Proctor was bred to be a good malting barley for the brewing industry. This was because at the time of the crosses leading to Proctor, a major use for barley was for making beer. About half of the crop went for this purpose, and the rest was used as livestock feed.[167] Government promotion of more cereal and livestock production generated the vastly increased interest in barley, which essentially replaced oats as a major British feed grain. Total British production of barley rose from 2.0 million tons to 4.9 million tons in 1960–62 and to 8.6 million tons in 1970–72.[168]

Second, Bell believed that the success of Proctor provided an important vindication for the overall breeding strategy at PBI. Barley, like wheat, can respond to nitrogen fertilizer, but lodging is a serious problem unless the variety has short, stiff straw. Proctor and the other new barleys introduced in the 1950s had the properties that made them all resistant to lodging. In addition, Proctor came from an extensive series of crosses, treated as one breeding program with a focused objective, and conducted over a long period (about fifteen years).[169] The design of the breeding program for barley was comparable to that for wheat, so Proctor's successes surely gave PBI scientists hope that they would eventually have similar successes with wheat.

PBI success with wheat was not as rapidly forthcoming as it was with barley. Perhaps a good deal of the time lag was due to the belief that it took many generations of selection to perfect a variety to the point that it could do well in yield tests and be useful to growers. Bell noted in 1947, at least one year before Lupton joined the staff, that PBI was interested in a number of new varieties from elsewhere in Europe, particularly France and Sweden. Breeding work in France had not ceased during the German occupation, and French breeders by the mid-1940s had some short, strong straw varieties that could withstand more nitrogen without lodging. These varieties were imported immediately after the war for testing in Britain, but Bell recommended that growers stay with already recommended varieties until the imports could be better characterized.[170] By 1949 the number of recommended wheat varieties for Britain was at twenty-two, all dating from before the war and representing a mix of British, French, Belgian, Dutch, and Swedish efforts. The new French varieties were not yet ready for recommendation.[171]

By mid-December 1952 the situation had changed considerably. At the Crop Conference of the National Institute of Agricultural Botany, the new French varieties Nord Desprez and Capelle Desprez were noted to be heavy-yielding, with short stiff straw and good resistance to yellow rust. At least one grower was eagerly anticipating the possibilities of higher yields of wheat with the new varieties and argued that it was time for the institute to take some of the older varieties off the list because they weren't much good and caused confusion among farmers. In addition, he recommended that the institute conduct yield tests at higher levels of soil fertility and that the Ministry of Agriculture should set targets of cereal yield increases at 30 percent higher in 1956, not 15 percent higher.[172]

Tests at the outlying experiment stations quickly confirmed the interest in the new French wheats, especially Capelle Desprez. Plots planted in fall 1946 at the Norfolk Agricultural Station and harvested in 1947 showed Nord Desprez as the "outstanding variety."[173] For the next several years, Nord Desprez continued to appear impressive, but from the test plots harvested in 1950 it was clear that Capelle Desprez had a significant advantage over Nord Desprez. Not only did Capelle Desprez show the high yields as Nord Desprez, but also it was significantly more resistant to eyespot disease and yellow rust.[174]

Capelle Desprez was recommended by the National Institute of Agricultural Botany in 1953 and became the most widely grown wheat in Britain from 1958 to 1968. Capelle Desprez was noticeably shorter than other varieties then in production. For example, Little Joss, Biffen's first variety released in 1908 and still grown on 2.3 percent of the wheat acreage in 1950, was 142 centimeters in height, compared with Capelle Deprez's 110 centimeters.[175] In terms of what Biffen and Engledow had envisioned in 1926, this short, stiff-strawed, disease-resistant variety was almost perfect for British conditions. Capelle Desprez's major fault was that its protein quantity and quality were not adequate to make a strong flour for British bread.

Given the original desire in 1947 to make Britain more independent of imported wheat, on the surface Capelle Desprez seems to have been a failure because during its reign Britain continued to import strong wheats to make the British loaf possible.

A more careful examination of the Ministry of Agriculture's program, however, indicates that the policies used to increase British agricultural production were in fact precisely suited to ensuring the success of Capelle Desprez. Deficiency payments, the first incentive adopted in 1947, merely gave the farmer a government check for the difference between the target price and the average price of wheat in the market. No premium was established to pay the farmer a higher amount for wheats that could make strong flour. In short, only yield helped a farmer's income, and Capelle Desprez was the best and most reliable yielder. By 1965–67, British wheat yields had risen to 3.8 million tons from 2.4 million acres, compared with 1.9 million tons from 2.2 million acres in 1945–47, a 100 percent increase in production on a 9 percent increase in land.[176]

In addition, the production grants instituted after 1951 for more use of fertilizer and more drainage had the effect of promoting intensive, heavily fertilized growing of wheat, which was increasingly produced with little or no crop rotation. The famous Norfolk Four Course System (clover, turnips, wheat, wheat; or one of its many variants) was thus firmly buried as a relic of the past. As a result of the intensive, continuous wheat production, the disease eyespot, known for many years, suddenly became a serious problem in English wheat production. Capelle Desprez was the variety of choice because it offered stronger resistance to eyespot than any of the other British wheat varieties. This French import was so good under the conditions that emerged in postwar British wheat farming that it is not surprising that it was difficult to find anything to outcompete it.

Wheat breeding at PBI by the late 1950s and early 1960s actually looked rather bleak. To be sure, Bell had been eager to import the French wheats immediately after the war because one aspect of varietal improvement was to identify useful cultivars created elsewhere for adoption and adaptation. Plant breeders, however, needed to create their own varieties, if only to maintain a sense of pride in their scientific skills. It was thus a time of embarrassment in 1961 when the National Institute of Botany removed the last PBI variety of wheat from the recommended list, the first time this had ever happened since the institute was organized in 1930.[177]

Bell, Lupton, and Bingham, however, never forgot their objective of trying to get a wheat that was high-yielding and also could make flour that worked well for the British loaf of bread. For a time, however, yield, regardless of quality, was the overriding concern. Nevertheless, PBI scientists, among their many crosses, mated Holdfast (introduced by Engledow in 1935) with Capelle Desprez in an effort to get the bread-making quality of Holdfast with the high yields of Capelle Desprez. By 1960, results were promising, and by 1964 Lupton had released a new variety, Maris Widgeon, which combined high yields with the ability to make strong flour. Yet the first indications of adoption were disappointing, a factor that led PBI scientist to wonder whether it was worth continuing the emphasis on grain quality.[178]

Ancillary developments, outside of the major work on breeding, however, combined to make a new context for Lupton's and Bingham's wheat-breeding efforts. First, in 1961 the research organization for the British bakers developed a new process of mixing the dough that made it possible to use a lower quality flour and still get a loaf of bread that was acceptable to British consumers. Essentially this Chorleywood Baking Process combined oxidizing agents with higher levels of physical work done on the dough in mixing to create a protein matrix that expanded properly when the yeast fermented to produce carbon dioxide. At PBI itself, work on micromethods of testing wheat for protein quality and quantity made it possible to select more efficiently for the exact kinds of protein that made a strong flour. Yet other work at PBI on the cytogenetics of wheat from the 1940s through the 1970s made it possible to incorporate disease resistance from wild relatives of wheat in the genus *Aegilops* into wheat, thus giving new strengths to genetic resistance to disease. These three developments, each in their own way, helped to dispel the feeling that developed in the 1960s that Capelle Desprez and Maris Widgeon might have represented the peak of wheat yields that was not to be surpassed.

Successes in the field also helped immensely. Bingham brought out Maris Ranger in 1968, a variety also based on Capelle Desprez but yielding even higher amounts of a grain suitable for making biscuits, a confectionery that needed weak flours. Enthusiasm for the possibility of new PBI varieties was also enhanced by that old ally of agricultural reform in the United Kingdom, worry about foreign exchange considerations and the ability of higher domestic yields to reduce food imports. Largely due to the Chorleywood Baking Process, yet another assist was given to the good reputation of Maris Widgeon. By 1967, some milling companies were offering a premium price for Maris Widgeon grain, so long as its overall protein content was 10 percent or more. At these levels of protein in Maris Widgeon, the Chorleywood Baking Process could use flour that was as high as 75 percent homegrown and still produce a loaf that would sell in the markets of London.[179]

Thus if the 1960s began with a deep pessimism about the potential of the PBI program, the decade ended with a revived sense that the course of the research would lead to new varieties that would outcompete the existing ones. Two developments in the 1970s combined with the existing context of English wheat production to take wheat production levels to the place of national self-sufficiency and even made Britain a minor exporter of wheat. First was an institutional change, the United Kingdom's entry into the European Economic Community and participation in its Common Agricultural Program (CAP) on 1 January 1973. In terms of technology, the CAP was nothing new. Its ability to elicit change from British farmers lay simply in the fact that it provided generous grants to farmers all over Europe to invest in more intensive practices, including machinery, drainage, and fertilizer use. The CAP also provided a guaranteed market for all grain produced, at a price that was highly attractive to British grain farmers. Thus after 1972, British wheat yields began to climb steadily as farmers continued to invest in new technology and plowed up every available hectare to cash in on the guaranteed market.[180]

Technically, the process encouraged by the financial policies of the CAP was vastly enhanced by the introduction to Britain of the Norin 10 semidwarfing genes. Entry of the Norin 10 genes to Britain was difficult only in the sense that the varieties in which the genes first appeared were not suitable for English conditions. Certainly no conceptual problems separated the attributes of plants carrying these genes from the objectives of PBI's program. After all, the institute had been looking for short, stiff-strawed varieties that could respond, without lodging, to nitrogen since 1926.

John A. Rupert, who had worked closely with Borlaug in Mexico for the Rockefeller Foundation, left Mexico and moved to Chile when the foundation began an agricultural program in that country. In Chile, at a latitude in the south comparable to Britain's in the north, and in a maritime climate, Rupert crossed lines containing the semidwarfing genes with Capelle Desprez and Marne Desprez. PBI obtained F_1 selections of these crosses in 1962.[181] At first, however, they were mostly troublesome because they had very high susceptibility to mildew and yellow rust, common British diseases that the lines had not been bred to resist.[182]

PBI scientists recognized the power of the new genetic material and knew of its immense successes in the United States, Mexico, and India. In 1977, fifteen years after the genes first came into the United Kingdom, PBI released Hobbit, the first semidwarf variety well suited to the British economic and physical environment. A string of other releases followed shortly thereafter, and British yields on a per hectare basis and on a national basis began to soar. In 1984, a particularly productive year, Britain produced wheat on 1.97 million hectares with an average yield of 7.6 tons per hectare and a total national yield of 15 million tons. The national market for wheat was only 10 million tons, so Britain was in the position of being a wheat exporter, a position the country had not occupied since the nineteenth century.[183]

Britain did not achieve these sorts of yields without igniting a complex series of criticisms of the technology involved. Environmental criticisms centered on the large amounts of fertilizer and pesticide used to achieve the higher yields. In addition, the CAP payments made it worthwhile for wheat and barley farmers to plow even quite marginal land and put it into production, thus destroying precious wildlife habitat in a very densely populated country that had already lost a great deal of its habitat. The CAP itself was criticized as a waste of taxpayers' money and as a generator of surpluses that were "dumped" at below production prices in the third world. The dumping provoked the same response in the third world as did the U.S. P.L. 480 plan: depression of yields in the recipient country because its farmers could not compete with the imported grain.[184]

Nevertheless, by the 1980s and 1990s the scare about starving to death had left the United Kingdom. No longer could agricultural science receive the hopes of the nation for preserving the ability to live in the face of an enemy that threatened to starve the country into submission. Even the farmers, who in World War II had won accolades for preserving Britain, became known simply as "barley barons" by their critics more interested in environmental conservation, ending unneeded drains on the taxpayer, and fostering food self-sufficiency in third world countries. Despite all the changes in attitude between 1945 and 1985, it must still be recognized that the PBI scientists, particularly Lupton and Bingham, had created a remarkable new technology that changed the very way the British people saw their land and its potential to sustain them. This was the science that truly made the green revolution a global phenomenon.

Epilogue

Implications of History
for the Future

A Brief Recapitulation

This book began as an effort to understand how and why the green revolution occurred in less industrialized countries between about 1960 and 1980. Specifically, it aimed to understand India and Mexico, both of which embraced new technology that yielded substantially more wheat than did their respective traditional practices. Clearly the "how" had its proximate causes in the adoption of new varieties, higher amounts of fertilizer, and more intensive use of water resources:

$$\text{Seeds} + \text{nitrogen} + \text{water} \longrightarrow \text{increased yields}$$

More distally, the means by which the new production practices were built rested on the development of Mendelian genetics, applied plant breeding, the ability to make inexpensive nitrogen fertilizer, and water development projects. Thus the "how" of the green revolution rested firmly on the ability to coordinate several component technologies, each of which was needed to produce the higher yields.

It was in the question of "why" that the story quickly became more complex. Starting from the end point and working backward leads to a series of linkages, or layers of explanation within layers of explanation:

- The green revolution happened because of new practices made available by plant breeding, soil fertility science, and hydrologic development.

- The new practices were part of a larger set of changes in agricultural technology that were adopted by farmers, each of whom saw the practices as a workable solution to their own problems. The farmers who adopted the new practices saw them as a way for themselves to better control and use a portion of the earth.
- The new practices themselves came from a complex of research initiatives. Nations embraced programs of plant-breeding research, plus the general education of farmers and scientists, because these programs were considered essential to each nation's security. At various times and places, these considerations dealt primarily with issues of direct military security and more indirect issues of protecting foreign exchange reserves.
- Considerations of national security were part of the larger issue that is fundamental to the existence of nation-states: How does a country maintain control over a portion of the earth? Just as farmers had to be concerned with their ability to control a small piece of land, nations had to worry about the collective project of controlling larger pieces of land. Military threats from outsiders and losing control of foreign exchange reserves both threatened the viability of nations. Ample supplies of grain were seen by leaders as a bulwark for defense of the realm and of the currency.
- Societies collectively sought control of large land areas in order to guarantee the survival of their own cultural identity. Efforts to preserve cultural identity, however, were linked to more fundamental concerns about methods of tapping the energy trapped in photosynthesis. Societies learned about 10,000 years ago of the ability of the wheat plant to capture light energy from the sun in a form that was convenient to harvest and store. These peoples thus formed a cultural identity around the use of wheat. Both wheat and these societies were irrevocably changed as they developed an inseparable codependency.

Seen in this light, we may ask again, Why did the green revolution occur? The answer would lead through a series of explanations, starting with the yield enhancement given new seeds by fertilizer and water, through the efforts of farmers to control their farmland and nations to control their national territory, and ending with the mutual codependency of people and wheat.

Does this mean that the codependency, evolved millennia ago in the Neolithic age, set in motion a train of events that led inexorably and inevitably to the green revolution? No. The chain of events linked here was not absolutely determined by the activities of our Neolithic ancestors seeking ways to capture photosynthetic energy.[1] Historical inquiry simply offers description and explanation after the fact.

Despite the limitations of history, this larger explanation of the green revolution sheds immediate light on a number of questions[2] one hears about it:

- *Was the green revolution in less industrialized countries the result of humanitarian aid from the more industrialized countries?* No need exists to doubt the genuine humanitarian impulses that often accompanied the aid programs, but humanitarian explanations alone are grossly deficient in capturing the full complexity of the green revolution. We must learn to see the embrace of the aid programs by both donors and recipients as reflections of their respective desires to ensure their nation's security.
- *Was the green revolution a strategy of multinational corporations to "hook" less industrialized countries into needing their products, such as fertilizers and pesticides?* Multinational corporations certainly found it in their interest to sell their products in the third world, but the evidence presented here indicates that the green revolu-

tion was not of their initiative. It seems most reasonable to say that the involvement of these companies came substantially after the decisions were made to embrace the promise of high-yielding agriculture and that their sales strategies had little or nothing to do with why the high-yielding systems were developed.

- *Was the green revolution confined just to less industrialized countries?* No, many industrialized and less industrialized countries adopted the new varieties, more fertilizer, and more water for their grain production in order to get better yields. The green revolution was a global phenomenon.

- *Is the green revolution now over?* If one sees the green revolution as a wave of new technology that substantially increased cereal yields, then in many countries the wave of technological change is now past. Grain farmers in those countries continue to use the technology, and most of the world's grain now comes from farmers who adopted the green revolution practices. The green revolution is "over" only in the sense that it is no longer new. Instead, it is the major standard practice.

- *Was the green revolution a failure or a success?* To answer this question it is imperative to establish the criteria for success. If success means an increase in the aggregate physical supply of grain, the green revolution was a success. If success means an end to hunger, then the green revolution was a failure. People without access to adequate land or income, regardless of their country of residence, remain ill fed.

- *Did the green revolution exacerbate hunger?* Answers to this more provocative and difficult question are hotly contested, with some arguing that it did and others that it didn't.[3] A more pertinent answer, however, is to note that the fundamental motivations for policies promoting the development of high-yielding cereal production were at best only tangentially aimed at alleviating hunger and poverty.

- *Was the green revolution an environmental disaster that left humankind dependent on fertilizers derived from fossil fuels, a narrowed genetic base in its major crop plants, and increased use of pesticides?* Some environmental criticisms of the green revolution seem to imply that the world would be better off if the underlying science and technology had not been developed.[4] Unfortunately, little evidence suggests that continuation of more traditional production practices would have ended hunger and left human beings more secure. Continuation of lower yields in the face of population growth would have created incentives for conversion of more land to agriculture and thus increased habitat destruction for wildlife. This is not to argue that the green revolution solved problems in a fully satisfactory way. If the high-yielding practices are to be rejected on environmental grounds, however, they must be rejected explicitly and with a discussion of the full effects of their withdrawal.

Looking Back at the Motivations for the Green Revolution

The argument presented in this book was that two major concerns—national security and population levels—prompted the U.S. government and major U.S. philanthropic foundations to support the modernization of agriculture in the third world. At this point it is appropriate to ask a very different kind of question: *Granted that concerns over national security and population may have spurred policies to promote the development of high-yielding agriculture in the third world and elsewhere, were the theories about these "correct"? Specifically, was the population–national security theory true?*

In other words, was it really true that U.S. national security was enhanced by the new agricultural practices? Similarly, was it really true that either population sizes or growth rates mandated an increase in food production in order to prevent starvation? Also, was it really true that overpopulation in the third world was the cause of susceptibility to communist insurrection, as claimed in the population–national security theory (see chapter 6)? We turn first to the questions on population and overpopulation.

Discussions about the effects of the size of the human population, and the population's growth rate, on the quality of human life almost always end in a morass of vitriolic disagreement. Unshakable and generally implicit presuppositions frequently form the framework of the analyst's position and thus affect the conclusions. Of most importance is the question of whether the analyst either harbors or rejects the Malthusian presupposition: human life tends toward misery because of the almost universal improvident exercise of human reproductive powers. Those who embrace Malthus see all human demographic data as supporting their pessimistic conclusions. Those who don't generally see the potential for the betterment of the human condition. Neither side can abide or accept the other side's point of view.

Despite the rude intrusion of gloom from Malthus, what is clear from a biological point of view is that *population levels and growth rates are of consequence.* All other things being equal, more people implies more human biomass, which implies an ability of the population to tap greater supplies of food, which implies an ability to harvest more of the earth's primary productivity. Similarly, rapidly growing populations change their perturbation of the surrounding ecosystem more rapidly than slow-growing or static populations, all other things being equal. In a word, population matters, even if we may have a hard time understanding the links among population size, growth rate, and the prospects for happiness.

This book does not attempt a reconstruction and critical reappraisal of the mass balance considerations between the human population of the 1940s and 1950s, on the one hand, and the abilities to tap photosynthetic processes at the time, on the other. Such an exercise could be done, but it is vastly beyond the scope of what is presented here. Without such a reconstruction, no definitive judgment can be made about whether the world's population of the 1940s and 1950s, or segments of the population in particular countries, was truly beyond the carrying capacity of the earth.

Several other statements can be made, however. First, influential leaders in the Rockefeller Foundation, the U.S. government, and elsewhere *firmly believed* that population levels and growth rates in certain third world countries were too high and thus insupportable. Second, these same leaders did not raise questions about the distribution of wealth and power within those countries. Nor did they raise questions about why some people were hungry, even when physical food supplies were adequate. Third, these same analysts did not raise questions about the historical legacy of imperialism on the agricultural productivity of the third world.

In other words, the population theories in the population–national security theory were too simplistic to provide insights into the reasons for food production levels in the third world or the distribution of the harvest already being achieved. Instead, based on the oversimplification of the population–national security theory, leaders from the United States embraced higher production of cereal grains as the key to solving hunger in the third world. With hunger solved, the threat they perceived from communist insurrection would diminish.

The popularity of the population–national security theory among policy strategists within the United States had its roots in the uncritical embrace of Malthus. So long as Malthusian despair was the only thing that could be seen, there was no other way of thinking about population and the accompanying requirements for agriculture.

It is in this sense, then, that the population components of the population–national security theory were false. Essentially the theory incorporated an uncritical acceptance of the Malthusian tragedy as a presupposition and used demographic data to reformulate the presupposition as a conclusion. Blinded to the immense importance of distributional factors, the theorists targeted aggregate physical yields as the major problem. High-yielding production systems solved that problem. Of course hunger was left untouched because all along hunger had been primarily a matter of distribution, not aggregate physical supplies.

In a similar vein, the population–national security theory posited a linear linkage between overpopulation, hunger, and a resultant political instability and susceptibility to communist subversion. Even if overpopulation were not the cause of hunger, is it possible that susceptibility to communist subversion was linked to hunger? This question, drawn from the links posited in the theory, is itself quite odd.

On the one hand, the answer might be yes. A poverty-stricken and hungry population, recognizing that its misery derived from the skewed distribution of wealth and power in a country, might embrace the promised redistributive effects of communist revolution. Thus it may be that the population–national security theory had a germ of truth in it. Maybe the susceptibility to communist influences was linked to hunger, even if not to overpopulation.

On the other hand, the answer might be no. The very form of the theory presumes that communist revolution was an evil foisted upon otherwise worthy and intelligent people by subversion, generally from unscrupulous infiltrators. With the possible exception of Eastern Europe, communist insurgencies that were successful in establishing revolutionary governments probably derived their successes not from the evils of infiltrators but from their abilities to marshal and focus domestic outrage at the abuses of wealth and power within the country.

In other words, the population–national security theory was also false in its presuppositions about the sources of revolutionary fervor. The theory could not envision that revolution might be seen as a rational, liberating act that garnered support from a significant portion of a population. Equally significantly, the theory could not accommodate the often burning desire of former colonies to expunge the stench of imperialism from the national psyche, which generally meant ending the privilege that a small portion of the population received from the imperial power in exchange for collaboration with that power.

At its core, therefore, the population–national security theory was blinded by an uncritical assumption of Malthusian misery and an inability to fathom the smoldering resentments born of unequal distributions of wealth and power in the third world. Put simply, the theory was not a sound and rational basis for the creation of U.S. policy. Perhaps in its avoidance of serious questions about the distribution of wealth and power, the theory may have satisfied its supporters, who saw no need or good stemming from raising questions about equity. For most Americans, however, avoidance of those important questions was harmful. It may be, for example, that variants of the theory lay behind the disaster of the Vietnam War and other armed incursions in the third world. Leaders who followed the theory built policies that shamed and directly harmed the vast majority of the American people.

Let us turn now to the other major question of whether the embrace of high-yielding agricultural practices was linked to the national security of nation-states. It is the argument of this book that the governments of the United States, Mexico, India, and the United Kingdom, each for their own different reasons, saw their respective national securities enhanced by higher agricultural yields. Were these perceptions really true? Put differently, would any or all of these countries have been materially less secure if they had not adopted high-yielding practices?

It is important to note here that this question is subtly but importantly different than the question about the population–national security theory. Here we are asking whether in fact a nation-state gained strength, and hence a better capacity to exercise its will on behalf of its own security, by having higher agricultural yields. Before, we were asking about the validity of a particular perception in the United States about the links between population and communist subversion in a third world country, which in turn was asserted to be linked to the vital interests of the United States.

For three reasons, the validity of the link between a nation's aggregate yields and its strength is secure. First, higher yields produce surplus stocks that can provide a national population with higher food reserves in times of adversity. Even if the nation is not entirely self-sufficient for calories and protein from its own produce, higher yields reduce dependency on a harvest somewhere else, outside the sovereignty of the nation.

Second, higher yields provide something to trade with other countries and thus serve as a potential earner of foreign exchange. With more income of foreign exchange, a country has more flexibility in how it deals with the rest of the world, and thus its options for maintaining national integrity are higher.

Third, higher yields from a given land base potentially provide the wherewithal to have a larger portion of the national population engaged in work outside of agricultural production. Such work may include production of other goods for trade, production of armaments, or direct service in the armed forces. Agricultural strength is one of the foundations upon which national leaders can protect and project their influence.

In each of our four countries, higher agricultural yields made the country a more formidable presence in the world. In the United States, higher yields from a steadily declining labor force produced the world's largest industrial establishment, the world's largest armaments industry, and one of the world's largest armed forces. Surplus agricultural goods were a keystone in the foreign exchange earnings of the United States, and large surpluses virtually guaranteed that no other nation could threaten the country with lack of food supplies. In turn, after 1945 the United States mounted an aggressive containment strategy against the socialist countries, a campaign that would have been far more difficult if the farm front at home had been less productive.

Mexico embraced the promise of high-yielding wheat production in order to soften a chronic shortage of foreign exchange earnings. With the decrease of wheat imports, Mexico's foreign exchange reserves could be used for other industrial expansion projects and for support of its armed forces. Although Mexico's military strength may appear small compared with that of the colossus on its northern border, the country exercises considerable influence within Latin America. A lower yielding agriculture would diminish Mexico's clout within the community of nations.

India faced an economy devastated by imperialism and shattered by partition in 1947. For complicated reasons (see chapter 8), India did not immediately embrace an effective program to increase its aggregate agricultural yields. As a result, it ended up putting itself at the sometimes untender mercies of the United States, from which it obtained major supplies of wheat needed for its survival. In addition to enduring the pressure born of hunger, India's low yields left its foreign exchange reserves exhausted, and it faced uncertainty in any project to draw labor out of agriculture into industrial production. War with Pakistan and two consecutive years of serious drought brought the matter to a head. India finally jumped for the higher yielding technologies, and as a result it eased its chronic dependency on foreign grain. Foreign exchange could go for other purposes, people could leave agriculture for industry, and the humiliation of pressure from Pakistan and the United States was eased.

Britain's economy faced problems of major reconstruction after 1945. Immense damage from the war, the loss of empire, and an acute foreign exchange crisis dramatized the weaknesses of the United Kingdom. Combined with the dreadful memories of being threatened with starvation in two wars with Germany, the postwar world was one where Britain had to either increase its yields or be exceptionally clever in other ways to earn the foreign exchange to rely on imported foods. Although the British economy never lost its heavy emphasis on foreign trade and extensive use of imported food, the changes embraced led its agriculture to substantially higher levels of production. Agricultural exports contributed to foreign exchange earnings, and the country conceivably could feed itself in time of crisis (although the diet might not be very popular).

Higher yielding agriculture was one of the major keys in each country's efforts to position itself during the cold war years. To be sure, cereal yields were not the only thing that mattered in determining a nation's strategic position against other nations, but just as surely they were among the most important contributors. Economic planners and top political leaders alike learned to recognize and draw upon the resources made available by high-yielding agriculture.

In a way, the situation was ironic for the United States. Although a major theory driving its foreign policy, the population–national security theory, was blind and fatally deficient as a guide to how the world worked, one dimension of the theory was true: higher yields had something to do with national security. What was ironic was that American strategists designed the theory to explain events in the third world, but in fact the one part of the theory that was true was true globally, including for the United States.

An Eye to the Future

The preceding argument emphasized that the population–national security theory was historically important but fatally flawed as a way of thinking about agricultural needs in the third world, even though higher yields themselves had considerable effect on a nation's security. The following question, however, is still at issue: *Do the links between agricultural research, agricultural yields, and national security have an effect on efforts to reform agriculture, in both its social and environmental consequences?* More specifically, what does the historical legacy of the linkages say about the four most prominent debates that surround contemporary agricultural affairs:

- Is agriculture sustainable for the indefinite future?
- Is the global human population at or beyond the carrying capacity or Malthusian limits of the earth based on sustainable agricultural practices?
- What is the appropriate economic and political balance between tillers of the soil who raise crops and consumers who must purchase them to survive?
- What is the appropriate balance between the industrialized nations ("the North") and the less industrialized nations ("the South")?

In order to address these questions, recall that the political ecology framework directs our attention to (1) human participation in ecosystem functions such as energy flows and biogeochemical cycles, (2) the use of agricultural technology to harvest the biosphere's primary productivity, (3) the creation of political economic structures to control the production and distribution of materials from the biosphere, and (4) the legacy of history that hangs over any effort to change either technology or political economic structures. Disciplined attention to these facets can shed light on the major debates about contemporary agriculture. Each is sketched briefly in the following.

On Sustainability

"Sustainability" has become an overworked and underinformative word since the report of the United Nations World Commission on Environment and Development (the Brundtland report) in 1987.[5] Despite its use in ways that hide rather than illuminate, the term clearly refers to something that has intuitive appeal. What is at issue centers in the standard definition: sustainable practices provide for the current generation without diminishing the ability of future generations to provide for themselves. In short, the term emphasizes the moral issue that people now living must consider the needs of people who are yet to come. An unsustainable practice by definition is one that creates intergenerational sin.

What is so often overlooked about sustainability, however, is an elaboration of details to focus on the major components at stake: What exactly is to be sustained? Over what area of the earth? Over what period of time? By whom? For whom?[6] Until and unless these questions are answered with precision, discussions of sustainability are empty.

Agriculture's many paradoxes and puzzles, noted at the start of chapter 1, make it particularly important to achieve precision on these questions in discussions of agricultural sustainability. At least four major potential candidates might answer the question of what exactly is to be sustained:

- The social community of farm families in an area
- The economic and political power of a particular group or nation
- The magnitude of yields of particular harvests
- The usability of particular natural resources, for example, soil or groundwater

In some cases, making one of these sustainable may make some or all of the other three also sustainable. Unfortunately, in other cases making one sustainable may mean at least one of the other three cannot be sustainable. Different interest groups may have radically different ideas about exactly what is to be sustained. Failure to specify explicitly the objectives of a particular candidate for a sustainable practice can lead to severe misunderstanding among groups of people.

A political ecological perspective helps the discussion on sustainability of agriculture in several ways. We are reminded to focus on the technology proposed for harvesting the biosphere's primary productivity. Such a focus instantly brings into relief questions about what resources are to be tapped, who owns them, what yields are to be expected, who will own and who will use the technology, and who will own the harvest. A focus on the technology also easily invites comparisons to other practices that might accomplish the same harvest by a different means.

In another, albeit indirect, way, political ecology also helps guide the discussion on sustainability by reminding us of the importance of history. From the narrative in this book, the most salient point in the history of the green revolution was the centrality of national security concerns in its genesis. This legacy still hangs over all efforts to reform agriculture in order to make it more sustainable. To put it bluntly, if sustainable means a diminution of aggregate yields, then questions may be raised about the threats of smaller yields to the integrity of nations. It may not matter that smaller yields might be more profitable for individual farmers and adequate for sound nutrition of the populace. If reduced surpluses are seen as a threat by economic and political leaders, then the sustainable practices responsible for the lowered yields may be viewed with extreme disfavor. It is this legacy that haunts the search for alternative modes of farm production.

On Carrying Capacity and Malthusian Limits

Assertions that the global human population is nearing, at, or beyond the carrying capacity of the earth are common in both the popular press and the scholarly literature. Grain harvest levels and stored grain reserves in 1995, for example, were lower than they had been in several previous years. Some analysts concluded from these statistics that the Malthusian day of reckoning was at hand.[7]

Unfortunately, it is impossible to refute an assertion that the judgment day is near. Maybe it is, for reasons we might or might not be able to understand. What is clear, however, is that no proof can show that the apocalypse is absent. For this reason, a prophesy of Malthusian doom may catch the public's imagination or interest but is nevertheless intellectually stale: if we cannot prove the impossibility of an apocalypse, then to predict one may be too easy in that the prophet of doom cannot be shown to be incorrect, except with the passage of time. After time, however, the prophet can launch a new warning, and most listeners have forgotten the old one.

Intellectual problems aside, however, both the public and political leaders rightly should get agitated over reasonable predictions that a serious problem exists. The questions, therefore, are whether predictions of Malthusian catastrophe are reasonable and whether we should take immediate steps to protect ourselves? Political ecology can usefully inform this debate. Two features of the framework are most important.

First, we are reminded that, as part of ecosystems, people participate in energy flows and biogeochemical cycles. Agriculture is, in fact, the set of practices or technologies by which we tap into those flows and cycles. Changes in agricultural technology can enhance or decrease our abilities to tap energy flows and biogeochemical cycles. Thus any prophecy of having reached the carrying capacity of the earth must be subjected to scrutiny on a specific point: Is it really true that no mechanism (i.e., no technology) exists to enhance the harvest? If a mechanism exists, can it be used indefinitely into the future with the same expected yields? If these questions are not adequately addressed, then a case exists that the prophecy of doom is not yet well justified.

Second, we are reminded that people create political economic systems to control both the production of material goods from the biosphere and their distribution and enjoyment. If the distribution and enjoyment of the agricultural harvest is not equitable, then some people may be hungry in the midst of physical supplies that greater than the amounts needed to provide an adequate diet to all. Thus any prophecy of limits must be examined for the question of equity. If this examination has not occurred, then the prophecy is not yet completely justified.

As population biologist Joel Cohen has recently argued, efforts to estimate the carrying capacity of the earth date from many years ago. It is extremely difficult to settle on a firm conclusion, however, because the number of people who can be supported depends on the agricultural technologies employed, the diets enjoyed, and the patterns of sharing the harvest.[8] Added to these complexities, of course, are the further difficulties in understanding sustainability.

Absence of the ability to predict a definitive carrying capacity does not mean that population is of no consequence. It does, however, impose on an analyst the obligation to specify assumptions explicitly. What was especially unfortunate was to have a conception such as the population–national security theory drive agricultural research and practice. The legacy of this theory in the United States suggests the need for caution in evaluating proposals for addressing population issues.

On the Balance Between Producers and Consumers

An increasingly small minority of the human population is needed for producing adequate food supplies for the entire population, provided mechanization is used and yields are high. Thus is born an inherent tension between two classes of people: those who till the soil and are directly responsible for the earth's agricultural harvest, and those who obtain a portion of this harvest in some other way. Each of these classes will want the terms of trade to be skewed in its favor.

Democracies based on majority rule, therefore, face an intriguing moral dilemma. Nontillers of the soil are or will soon be a substantial majority and thus can vote to create terms of trade in their own interest. Should an urban majority vote its approval of governments that bias policies against the interests of rural people who farm? A counterbalance to the powers of urban majorities stems from the fact that when agriculture becomes both mechanized and high-yielding, the few farmers left in business may become sufficiently powerful economically to exert a political influence disproportionate to their numbers. Should these "neoaristocrats" of farming exercise their muscle to gain privileges over the less economically powerful urban folks?

This is not the place to attempt a balancing formula between the two opposing classes. It is, however, pertinent to note several items of importance. For example, patterns of ownership and control of land are generally the most important political economic structure in any agrarian society and of great importance in industrial countries. High inequality in landholdings may have been derived from days long past. New technological practices such as the green revolution may exacerbate the inequitable distributions of the harvest and lead to a widening of socio-economic differences. Similarly, ancient cultural customs such as govern the relations between the sexes or between races may lead to very inequitable, even cruel, divisions of the harvest. One can agitate about the inequities and propose radical redistributions that would be more even. Unless, however, one recognizes the possibility of connections between the technology of agriculture and the division of its wealth, little hope attends the achievement of greater equality and social tranquility. At the same time, it is also necessary to recognize that continuation of traditional agriculture may entail equally poor prospects for equality and peace.

On the Balance between the Industrialized "North" and the Less Industrialized "South"

Just as individuals are divided by inequity within nations, nations are divided by inequity between those with intensive industrialization and those with less intensive industrial development. Under imperialism from the eighteenth to the twentieth century, it was assumed by the imperial powers that the home country with industry would manufacture and the colonies without would farm. Trade between the two got everybody what they needed, but imperial theorists generally ignored the fact that real wealth and power could lie only in the industrial heartland.

Since the end of formal imperialism in the second half of the twentieth century, the imbalance of power between the former colonies and former masters has often remained, even though "development aid" for agriculture and industry was ostensibly supposed to even the playing field. More recently, the global push to liberalize trade in all goods among all nations has been touted as the best way to ensure that all people will benefit maximally from modern technology.

It may be true that free trade will bring economic benefits, at least to some people, in every country. In fact, it probably will. Economic theory, after all, holds that in conditions of free trade, every country will produce those goods and services in which it has comparative advantage. Thus all things will be produced at lowest cost, which under free trade can be exchanged for those not produced in a country. In free-trade ideologies, therefore, over time countries will adjust their agriculture and industry to produce the products they produce most cheaply. For other things needed, they will trade. In theory, at least, everybody will be better off.

The problem with the free-trade mantra is not that it contains no reasonable ideas but that it directs our attention away from issues shown in this book to be of deep and abiding importance. For example, free-trade ideas neglect to inquire into the links among biological productivity (agricultural harvest), economic value, and the acquisition of political power. The physiological necessity for food means that any persons not in possession of food can be subjected to enormous coercion as they attempt to trade whatever they have produced for food. This is a relationship that is all but invisible in the ideas justifying free trade as a guide to agricultural policy.

It also happens that the current potential for the highest yields of grains lies in the heavily industrialized areas because it is these countries (Western Europe, the United States and Canada, and Japan) that are best able to marshal the coordinated use of improved varieties, fertilizer, pest control, and irrigation under highly mechanized conditions. With the exception of Canada, the major industrial areas are also the seat of old imperialisms, with their legacies of resentment still rummaging around the former colonies.

If free trade in agricultural goods ends up placing a less industrialized area of the world at the mercy of fierce competition from both the industrial and agricultural sectors of the industrialized world, the less industrialized country will be at a distinct disadvantage because of the heavy hand of history affecting its agricultural productivity. Should an industrialized country be able to use its power over food, the less advantaged country will indeed be in a bind and very likely will resent it immensely. Desire for food self-sufficiency helped create the embrace of high-yielding agriculture in countries such as India, Mexico, and the United Kingdom, and may continue to do so in the future, despite the ideological assertion that free trade is a better way to go.

Concluding Remarks

Humankind's endless quest for food is the motivating force behind the earth's premier industry, agriculture. From agricultural wealth, we have found the wherewithal to build many other fountains of economic production. By modifying the biosphere in order to tap the energy of photosynthesis, we have remade both the earth and ourselves. Our relentless obliteration of nonhuman ecosystems in favor of agricultural ecosystems is a major force determining the balance between humankind and the other species with whom we share the earth. An industry of this magnitude and importance demands every ounce of our intelligence and creativity to understand it. With such understanding, perhaps we will continue to share and enjoy the productivity of agricultural wealth in a way that is pleasing to the spirit. Without understanding, we are very likely to destroy both our own humanity and many of the other species with whom we need to live.

Notes

Abbreviations Used in Notes

FFA   Ford Foundation Archives, New York, New York
PRO   Public Records Office, Kew, London, United Kingdom
RFA   Rockefeller Foundation Archives, Rockefeller Archive Center, North Tarrytown, New York
USNA United States National Archives, Washington, D.C.

Chapter 1

1. I am indebted to a former colleague at Miami University, Nancy Nicholson, who first helped me see this relationship. The relationship between people and technology is the subject of a substantial literature in the history, sociology, philosophy, and economics of technology. For an important review that anticipated future developments, see Edwin T. Layton Jr., Technology as Knowledge, *Technology and Culture* 15, no. 1 (1974): 31–41. Most of the literature on the intellectual, political, and social context of technology, however, is not particularly concerned with questions of environmental quality and natural resource degradation. What Nicholson pointed me toward was a way to connect concerns about the environment with a rich discussion of technological development.

2. Paul R. Ehrlich and Anne H. Ehrlich, *The Population Explosion* (New York: Simon and Schuster, 1990), 320 pp.

3. Barry Commoner, *The Closing Circle* (New York: Knopf, 1971), 326 pp.

4. One of the most influential statements that developed the theme of overconsumption was by the editors of *The Ecologist, Blue Print for Survival* (Boston: Houghton Mifflin, 1972), 189 pp. This book intertwined overconsumption with overpopulation.

269

5. Julian Simon, *The Ultimate Resource* (Princeton, N.J.: Princeton University Press, 1981), 415 pp.

6. William Leiss, *Domination of Nature* (New York: Braziller, 1972), 242 pp.

7. Carolyn Merchant, *The Death of Nature* (San Francisco: Harper and Row, 1980), 348 pp.

8. Richard White, *Land Use, Environment, and Social Change* (Seattle: University of Washington Press, 1980), 234 pp.

9. Carolyn Merchant, *Ecological Revolutions* (Chapel Hill: University of North Carolina Press, 1989), 379 pp.

10. David Ehrenfeld, *Beginning Again: People and Nature in the New Millennium* (New York: Oxford University Press, 1993), 216 pp.; Luc Ferry, *Le Nouvel Ordre écologique: L'Arbre, l'animal et l'homme* (Paris: Bernard Grasset, 1992), 275 pp.

11. Norman Myers, *Ultimate Security: The Environmental Basis of Political Security* (New York: Norton, 1993), 308 pp.

12. Jonathon Porritt, *Seeing Green: The Politics of Ecology Explained* (Oxford: Basil Blackwell, 1984), 249 pp.

13. Georg Borgstrom, *Harvesting the Earth* (New York: Abelard-Schuman, 1973), 237 pp.

14. Colin Clark, *Starvation or Plenty?* (New York: Taplinger, 1970), 180 pp.

15. Otto T. Solbrig and Dorothy J. Solbrig, *So Shall You Reap: Farming and Crops in Human Affairs* (Washington, D.C.: Island Press, 1994), 284 pp. The Solbrigs' study provides a broad overview of agricultural changes on a global basis, while the current study focuses on how wheat yields increased and how the embrace of higher yields was tied to strategic political events. I came across the Solbrigs' excellent book very late in the stages of preparing this manuscript, so it was not influential in how I set up the analytical framework of political ecology. Nevertheless, it is clear that the Solbrigs thought along similar lines, and their work clearly set a precedent for mine.

16. Kristin S. Shrader-Frechette and E. D. McCoy, *Method in Ecology* (Cambridge: Cambridge University Press, 1993), 328 pp.

17. Robert Heilbroner and William Milberg, *The Crisis of Vision in Modern Economic Thought* (New York: Cambridge University Press, 1995), pp. 20–22; E. Ray Canterbery, *The Making of Economics*, 3rd ed. (Belmont, Calif.: Wadsworth, 1987), pp. 34–84.

18. Heilbroner and Milberg, *Crisis of Vision*, and Canterbery, *Making of Economics*, pp. 185–212.

19. Willard W. Cochrane, *The Development of American Agriculture* (Minneapolis: University of Minnesota Press, 1979), pp. 387–395.

20. A general summary of the social constructionist research agenda is in Wiebe E. Bijker, *Of Bicycles, Bakelites, and Bulbs: Towards a Theory of Sociotechnical Change* (Cambridge, Mass.: MIT Press, 1995), pp. 1–17. See also Wiebe E. Bijker, Sociohistorical Technology Studies, in *Handbook of Science and Technology Studies*, ed. Shiela Jasanoff, Gerald E. Markle, James C. Petersen, and Trevor Pinch (Thousand Oaks, Calif.: Sage, 1995), pp. 229–256. An important critical review of the social constructionist agenda argued that social constructionism had a potential to be morally vacuous because the mode of inquiry failed to pay attention to the consequences of innovation, especially to individuals who were not directly involved in the creation of new technologies (Langdon Winner, Upon Opening the Black Box and Finding It Empty: Social Constructivism and the Philosophy of Technology, *Science, Technology and Human Values* 18, no. 3 (summer 1993): 362–378. Political ecology embraces the notion that technology is best understood as a social artifact, and a major goal is to understand the environmental consequences of technical innovation.

21. The links between capital accumulation and innovation through plant breeding are derived from Heilbroner and Milberg, *Crisis of Vision*, pp. 106–107.

22. Joseph A. Schumpeter, *Capitalism, Socialism and Democracy*, 3d ed.; (New York: Harper and Row, 1950; reprint 1976), pp. 81–86.

23. E. J. Wellhausen, Rockefeller Foundation Collaboration in Agricultural Research in Mexico, *Agronomy Journal* 42, no. 4 (April 1950): 167–175; Norman E. Borlaug, The Impact of Agricultural Research on Mexican Wheat Production, *Transactions of the New York Academy of Sciences* 20, no. 3 (1958): 278–295; E. C. Stakman, Richard Bradfield, and Paul C. Mangelsdorf, *Campaigns against Hunger* (Cambridge, Mass.: Harvard University Press, 1967), 328 pp.; Sterling Wortman and Ralph W. Cummings Jr., *To Feed This World* (Baltimore: Johns Hopkins University Press, 1978), 440 pp.; Haldore Hanson, Norman E. Borlaug, and R. Glenn Anderson, *Wheat in the Third World* (Boulder, Colo.: Westview Press, 1982), 174 pp.; Robert D. Stevens and Cathy L. Jabara, *Agricultural Development Principles: Economic Theory and Empirical Evidence* (Baltimore: Johns Hopkins University Press, 1988), 478 pp.

24. Chapter 6 contains a more detailed discussion of population's relationship to plant breeding. Norman E. Borlaug, one of the most prominent plant breeders, passionately believed that plant breeding was worth little in the long run unless the "population monster" were curbed. See Challenges for Global Food and Fiber Production, *K. Skogs-och Lantbruksakademien Tidskrift, Suppl.* 21 (1988): 15–55.

25. See, for example: Francine R. Frankel, *India's Green Revolution* (Princeton, N.J.: Princeton University Press, 1971), 232 pp.; Frankel, *India's Political Economy, 1947–1977* (Princeton, N.J.: Princeton University Press, 1978), 600 pp.; Kathryn G. Dewey, Nutritional Consequences of the Transformation from Subsistence to Commercial Agriculture in Tabasco, Mexico, *Human Ecology* 9, no. 2 (1981): 151–187; Billie R. DeWalt, Halfway There: Social Science in Agricultural Development and the Social Science of Agricultural Development, *Human Organization* 47, no. 4 (1988): 343–353; Jack Ralph Kloppenburg Jr., *First the Seed: The Political Economy of Plant Biotechnology, 1492–2000* (Cambridge: Cambridge University Press, 1988), 349 pp.; Michael Lipton, *New Seeds and Poor People* (Baltimore: Johns Hopkins University Press, 1989), 473 pp.; and Vandana Shiva, *The Violence of the Green Revolution* (London: Zed Books, 1991), 264 pp. Contrary views dispute the pessimism and criticism on social grounds, for example: R. K. Sharma, *Technical Change, Income Distribution and Rural Poverty* (Delhi: Shipra, 1992), 187 pp.; and Rita Sharma and Thomas T. Poleman, *The New Economics of India's Green Revolution* (Ithaca, N.Y.: Cornell University Press, 1993), 272 pp.

26. Efforts to promote sustainable, alternative, and organic agriculture are all based on the premise that the high-input practices of the green revolution are unsustainable and environmentally destructive. See, for example, Wes Jackson, *Altars of Unhewn Stone: Science and the Earth* (San Francisco: North Point Press, 1987), 158 pp. Jackson's concerns have been largely with American agriculture. Dahlberg studied the green revolution from frameworks of policy, economics, and evolution and concluded that in the long run we would be better off with smaller, more ecologically sound agricultural technologies. See Kenneth A. Dahlberg, *Beyond the Green Revolution* (New York: Plenum Press, 1979), pp. 47–90. Bruce Jennings provided an excellent critique of the Mexican Agricultural Program of the Rockefeller Foundation in *Foundations of International Agricultural Research* (Boulder, Colo.: Westview Press, 1988), 196 pp. His concern was that investigations in agricultural natural science frequently assumed changes in the social realm as well. For a critique of British intensive agriculture and its environmental and social changes, see, for example, Richard Cottrell, *The Sacred Cow: The Folly of Europe's Food Mountains* (London: Grafton Books, 1987), 192 pp. For an environmental and social critique from India, see, for example, Shiva, *Violence of the Green Revolution*. Conway and Pretty have assembled a large amount of data about environmental damage from agriculture on a global basis in Gordon R. Conway and

Jules N. Pretty, *Unwelcome Harvest: Agriculture and Pollution* (London: Earthscan Publications, 1991), 645 pp. Contrary views can also be found—for example, Stephen B. Brush, Mauricio Bellon Corrales, and Ella Schmidt, Agricultural Development and Maize Diversity in Mexico, *Human Ecology* 16, no. 3 (1988): 307–328.

Chapter 2

1. Descriptive material on wheat is drawn largely from E. J. M Kirby and Margaret Appleyard, *Cereal Development Guide* (Stonleigh, England: National Agricultural Centre, 1987), 96 pp.

2. John Percival, *The Wheat Plant* (New York: Dutton, 1921), p. 147.

3. Ibid., pp. 147–153.

4. T. E. Miller, Systematics and Evolution, in *Wheat Breeding: Its Scientific Basis*, ed. F. G. H. Lupton (London: Chapman and Hall, 1987), pp. 1–30.

5. "Ploidy" refers to the concept that an organism has a basic number of chromosomes, which is the set of chromosomes that has at least one copy of each gene for the organism. In wheat the basic set of chromosomes is 7. Diploids have $2 \times 7 = 14$, tetraploids have $4 \times 7 = 28$, and hexaploids have $6 \times 7 = 42$ chromosomes.

6. Miller, Systematics and Evolution.

7. Before the firm acceptance of ploidy as the basis for the botanical classifications of wheat, many of the botanical classifications were derived from agronomic or use characteristics. See Percival, *The Wheat Plant*, pp. 147–163.

8. W. T. Yamazaki, M. Ford, K. W. Kingswood, and C. T. Greenwood, Soft Wheat Production, in *Soft Wheat: Production, Breeding, Milling, and Uses*, ed. W. T. Yamazaki and C. T. Greenwood (St. Paul, Minn.: American Association of Cereal Chemists, 1981), pp. 1–32.

9. N. L. Kent, *Technology of Cereals* (Oxford: Pergamon Press, 1983), p. 133.

10. ICI, *Growing Cereals* (ICI, no date), p. 28.

11. Yamazaki et al., *Soft Wheat Production*.

12. Percival, *The Wheat Plant*, pp. 17–18.

13. A. J. Worland, M. D. Gale, and C. N. Law, Wheat Genetics, in *Wheat Breeding*, ed. F. G. H. Lupton (London: Chapman and Hall, 1987), pp.129–171.

14. H. J. Loving and L. J. Brenneis, Soft Wheat Uses in the United States, in *Soft Wheat: Production, Breeding, Milling, and Uses*, ed. W. T. Yamazaki and C. T. Greenwood (St. Paul: American Association of Cereal Chemists, 1981), pp. 169–207.

15. Percival, *The Wheat Plant*, p. 335.

16. Miller argues that *Triticum dicoccoides* is a wild allotetraploid that arose by amphiploidy between *Triticum urartu* and a species of the genus *Aegilops*, section *sitopsis* (Miller, Systematics and Evolution, p. 21).

17. Daniel Zohary and Marcia Hopf, *Domestication of Plants in the Old World* (Oxford: Clarendon Press, 1988), pp. 13–14.

18. G. D. H. Bell, The History of Wheat Cultivation, in *Wheat Breeding: Its Scientific Basis*, ed. F. G. H. Lupton (London: Chapman and Hall, 1987), pp. 31–49.

19. Miller, Systematics and Evolution, pp. 1–30.

20. Zohary and Hopf, *Domestication of Plants in the Old World*, p. 44.

21. Ibid.

22. Ibid., pp. 15–16.

23. Ibid., pp. 28–37.

24. Julian Thomas, Neolithic Explanations Revisited: The Mesolithic-Neolithic Transition in Britain and South Scandinavia, *Proceedings of the Prehistory Society* 54 (1988): 59–66.

25. Graeme Barker, *Prehistoric Farming in Europe* (Cambridge: Cambridge University Press, 1985), p. 198; Miller, Systematics and Evolution.

26. Stanley Wolpert, *A New History of India* (New York: Oxford University Press, 1982), pp. 4–9.

27. Ibid., pp. 14–20.

28. Timothy Champion, Clive Gamble, Stephen Shennan, and Alasdair Whittle, *Prehistoric Europe* (London: Academic Press, 1984), p. 27, places *Homo* in Europe after 350,000 years B.P. Harold E. Driver, *Indians of North America*, 2d ed. (Chicago: University of Chicago Press, 1969), pp. 3–4, places people in North America via land bridges from Siberia about 40,000 B.C. and again between 25,000 and 10,000 B.C. See Wolpert *New History of India*, pp. 4–9, for reference to human habitation of India between 200,000 and 400,000 years ago.

29. Barbara Pickersgill and Charles B. Heiser Jr., Origins and Distribution of Plants Domesticated in the New World Tropics, in *Origins of Agriculture*, ed. Charles A. Reed (The Hague: Mouton, 1977), pp. 803–835.

30. Irvin Milburn Atkins, *A History of Small Grain Crops in Texas: Wheat, Oats, Barley, Rye, 1582–1976* (College Station: Texas A & M University Press, 1980), pp. 1–6.

31. Henry Adolf Knopf, Changes in Wheat Production in the United States, 1607–1960, (Ph.D. diss., Cornell University, 1967), p. 30.

32. Mark N. Cohen, Population Pressure and the Origins of Agriculture: An Archeological Example from the Coast of Peru, in *Origins of Agriculture*, ed. Charles A. Reed (The Hague: Mouton, 1977), pp. 135–177.

33. Richard W. Redding, A General Explanation of Subsistence Change: From Hunting and Gathering to Food Production, *Journal of Anthropological Archeology* 7 (1988): 56–97.

34. For arguments that agriculture was a technological response to higher population levels, see Ester Boserup, *The Conditions of Agricultural Growth* (Chicago: Aldine, 1965), 124 pp., and Cohen, Population pressure and the Origins of Agriculture, pp. 135–177. For arguments that "population pressure" can be only one of a complex of factors that induces people to take up agriculture, see Bennet Bronson, The Earliest Farming: Demography as Cause and Consequence, in *Origins of Agriculture* ed. Charles A. Reed (The Hague: Mouton, 1977), pp. 23–48.

35. An important pair of assumptions lies behind this conclusion that farming is necessary: first, the argument assumes that it is important to preserve those now living; second, it assumes that everyone has a right to reproduce. It is beyond the scope of this study to justify these two assumptions, but I believe they are both appropriate. It is not clear to me that other scholars, particularly those who believe that the size of the human population must shrink, would necessarily grant the worthiness of these assumptions.

36. Zohary and Hopf, *Domestication of Plants in the Old World*, pp. 49–50.

37. Ibid., p. 52.

38. Mark Overton, Estimating Crop Yields from Probate Inventories: An Example from East Anglia, 1585–1735, *Journal of Economic History* 39 (1979): 363–378.

39. Barker, *Prehistoric Farming in Europe*, pp. 50–54.

40. Lynn White Jr., *Medieval Technology and Social Change* (Oxford: Oxford University Press, 1962), pp. 69–75.

41. Stanhill estimates wheat yields in England at 0.5 tons per hectare in 1200. See G. Stanhill, Trends and Deviations in the Yield of the English Wheat Crop During the Last 750 Years, *Agro-ecosystems* 3, no. 1 (1976): 1–10. Overton provides a review of wheat yield estimates beginning in 1270, and uses probate inventories to make estimates of yields in Norfolk and Suffolk from 1585 to 1735. He also summarizes and evaluates some of the yield

figures in the late eighteenth century. See Overton, Estimating crop yields from probate inventories, 363–378). Biffen and Engledow quote a figure of 10 bushels per acre for manorial wheat yields in England. R. H. Biffen and F. L. Engledow, *Wheat Breeding Investigations at the Plant Breeding Institute, Cambridge* (London: His Majesty's Stationary Office, 1926), 114 pp.

Conversion of bushels per acre to kilograms per hectare: assume 1 bushel of wheat = 60 pounds, 1 kilogram = 2.2046 pounds, 1 hectare = 2.471 acre.

42. E. Anthony Wrigley, Urban Growth and Agricultural Change: England and the Continent in the Early Modern Period, *Journal of Interdisciplinary History* 15, no. 4 (spring 1985): 683–728.

43. Overton, Estimating Crop Yields from Probate Inventories.

44. Stanhill, Trends and Deviations, 1–10.

45. R. V. Jackson, Growth and Deceleration in English Agriculture, 1660–1790, *Economic History Review*, 2d Ser., 38, no. 3 (August 1985): 333–351.

46. Stanhill, Trends and Deviations.

47. Jackson, Growth and Deceleration in English Agriculture.

48. Wrigley, Urban Growth and Agricultural Change.

49. H. H. Lamb, *Climate, History and the Modern World* (London: Methuen, 1982), pp. 202, 219–220.

50. Carolyn Merchant, *The Death of Nature* (San Francisco: Harper and Row, 1980), 348 pp.

51. F. G. H. Lupton, History of Wheat Breeding, in *Wheat Breeding, Its Scientific Basis*, ed. F. G. H. Lupton (London: Chapman and Hall, 1987), pp. 51–70; see also discussion of Knight in Herbert Fuller Roberts, *Plant Hybridization before Mendel* (Princeton, N.J.: Princeton University Press, 1929; reprint New York: Hafner, 1965), pp. 85–87, 114.

52. Lupton, History of Wheat Breeding, pp. 51–70.

53. Patrick Shirreff, *Improvement of the Cereals and an Essay on the Wheat-fly* (Edinburgh: William Blackwood, 1873), 112 pp; Lupton, History of Wheat Breeding, pp. 51–70.

54. Frederic F. Hallett, On "Pedigree" in Wheat as a Means of Increasing the Crop, *Journal of the Royal Agricultural Society of England* 22 (1861): 371–381.

55. Lupton, History of Wheat Breeding, pp. 51–70; Roberts, *Plant Hybridization before Mendel*.

Chapter 3

1. For example, in the United States two closely related societies are probably home to most plant breeders: the American Society of Agronomy (ASA) and the Crop Science Society of America (CSSA). In 1994 the former reported 12,433 total members and the latter 4,971. Membership in the CSSA automatically includes membership in the ASA, so some members may be counted twice. Members indicating a first-priority interest in crop breeding were 2,242 from the ASA and 2,146 from the CSSA. Thus it is reasonable to estimate that approximately 2,000 scientists in the United States are currently focused on plant breeding. See American Society of Agronomy, Crop Science Society of America, and Soil Science Society of America, *1994 Membership Report* (Madison, Wis.: The Societies, 1994), pp. 3, 41; copy supplied by Cleo C. Tindall, membership and certification registrar of the Societies.

2. The history of evolution and variation is complex. For an interesting treatment that reviews much of this work, especially the contributions of Lamark and Darwin, see Ernst Mayr, *The Growth of Biological Thought: Diversity, Evolution, and Inheritance* (Cambridge, Mass.: Harvard University Press, 1982), esp. pp. 343–362, 394–417.

3. Charles Darwin, *The Variation of Animals and Plants under Domestication*, vol. 2. (New York: D. Appleton, 1896; reprint New York: AMS, 1972), p. 370.

4. Agricultural prices were high in Britain from the 1790s to the end of the Napoleonic Wars. In order to prevent disastrous reductions in prices, the landed interests of England were able to pass in Parliament the Corn Laws of 1815, which excluded the import of wheat until prices in Britain reached a sufficiently high level. A new Corn Law passed in 1828 provided a sliding scale of import duties in which duties were high when domestic prices were low but lowered as home wheat prices went up. See David Thomson, *England in the Nineteenth Century*, rev. ed. (Harmondsworth, England: Penguin, 1978), pp. 35–37.

5. E. J. Hobsbawm, *Industry and Empire* (Harmondsworth, England: Penguin, 1968), pp. 99–100.

6. Thomson, *England in the Nineteenth Century*, pp. 61–62.

7. P. J. Perry, ed., *British Agriculture, 1875–1914* (London: Methuen, 1973), pp. xiv–xix.

8. C. S. Orwin, *A History of English Farming* (London: Thomas Nelson and Sons, 1949), pp. 74–77.

9. David Grigg, *English Agriculture: An Historical Perspective* (Oxford: Basil Blackwell, 1989), pp. 5–9.

10. E. J. T. Collins, Dietary Change and Cereal Consumption in Britain in the nineteenth century, *Agricultural History Review* 23, pt. 2 (1975): 97–115.

11. Perry, *British Agriculture*, p. xxx.

12. Colin McEvedy and Richard Jones, *Atlas of World Population History* (Harmondsworth, England: Penguin, 1978), pp. 19–39; Carlo M. Cipolla, *The Economic History of World Population* (Harmondsworth, England: Penguin, 1974), pp. 115–116.

13. Alfred W. Crosby, *Ecological Imperialism: The Biological Expansion of Europe, 900–1900* (Cambridge: Cambridge University Press, 1986), pp. 2–7.

14. Wilfred Malenbaum, *The World Wheat Economy, 1885–1939* (Cambridge, Mass.: Harvard University Press, 1953), pp. 172–175.

15. Ibid., pp. 154–155.

16. Ibid., pp. 154–161.

17. Paul W. Gates, *The Farmer's Age: Agriculture, 1815–1860* (New York: Holt, Rinehart and Winston, 1960; reprint White Plains, N.Y.: M. E. Sharpe, 1976), pp. 285–286; Willard W. Cochrane, *The Development of American Agriculture* (Minneapolis: University of Minnesota Press, 1979), p. 68.

18. Clarence H. Danhof, *Changes in Agriculture: The Northern United States, 1820–1870* (Cambridge, Mass.: Harvard University Press, 1969), pp. 228–229.

19. I am indebted to Pat Labine for bringing this insight to my attention.

20. Danhof, *Changes in Agriculture*, pp. 228–249.

21. Comparative studies of American and Japanese agriculture in the nineteenth century suggest that people have a tendency to invent things that allow them to more efficiently use those resources that are in shortest supply. Japanese farmers were short of land but had plenty of labor. They were most adept at inventing new farming practices that economized on the use of land. American farmers, by contrast, saw their situation as being one of plentiful land and little labor. They tended to find ways of stretching the limited labor supplies. See Yujiro Hayami and Vernon W. Ruttan, *Agricultural Development: An International Perspective* (Baltimore: Johns Hopkins University Press, 1985), 506 pp.

22. Laborsaving machinery also became important in other countries during the nineteenth century. For example, British wheat producers experimented with threshing and harvesting machinery throughout the period, but the impetus to innovate lay more in labor shortages caused by war or by loss of labor to higher paying factory jobs. Britain had no excess or new land to speak of to bring into cultivation, especially at production costs comparable

to those for North American lands. See, for example, E. J. T. Collins, The Rationality of "Surplus" Agricultural Labour: Mechanization in English Agriculture in the Nineteenth Century, *Agricultural History Review* 35, no. 1 (1987): 36–46; and Stuart Macdonald, The Progress of the Early Threshing Machine, *Agricultural History Review* 23, no. 1 (1975): 63–77.

23. Malenbaum, *World Wheat Economy*, p. 153.

24. Ibid., pp. 156–157.

25. Ibid., p. 104.

26. Andrew Denny Rodgers III, *Liberty Hyde Bailey: A Story of American Plant Sciences* (New York: Hafner, 1965), pp. 1–18, 52–86.

27. Ibid., pp. 86–88, 118–122.

28. L. H. Bailey, *Plant-Breeding, Being Six Lectures upon the Amelioration of Domestic Plants*, 4th ed. (New York: Macmillan, 1906), pp. 394–396; also see Conway Zirkle, The Role of Liberty Hyde Bailey and Hugo de Vries in the Rediscovery of Mendelism, *Journal of the History of Biology* 1 (1968): 205–218.

29. L. H. Bailey, *Plant-Breeding, Being Five Lectures upon the Amelioration of Domestic Plants* (New York: Macmillan, 1895), p. vi.

30. Ibid., pp. 1–2.

31. Ibid., p. 32.

32. From Dr. G. N. Lauman, quoted in Rodgers, *Liberty Hyde Bailey*, p. 242.

33. Bailey, *Plant-Breeding, Five Lectures*, p. 13.

34. L. H. Bailey, *Plant-Breeding* (New York: Macmillan, 1915), pp. vii–viii.

35. Bailey, *Plant-Breeding, Six Lectures*, pp. viii–x, 155–156.

36. L. C. Dunn, Mendel, His Work and His Place in History, *Proceedings of the American Philosophical Society* 109, no. 4 (1965): 189–198. Mendel's paper was Versuche über Pflanzenhybriden, in *Verhandlungen NaturForschender Verein, Brünn* 4 (1866): 3–47.

37. Zirkle, Role of Liberty Hyde Bailey and Hugo de Vries, 205–218.

38. Alfred H. Sturtevant, The Early Mendelians, *Proceedings of the American Philosophical Society* 109 (1963): 199–204; Herbert Fuller Roberts, *Plant Hybridization before Mendel* (Princeton, N.J.: Princeton University Press, 1929; reprint New York: Hafner, 1965), 374 pp.

39. Sturtevant, The Early Mendelians, 199–204.

40. Nils Roll-Hansen, Svalöf and the Origins of Classical Genetics, in *Svalöf 1886–1986: Research and Results in Plant Breeding*, ed. Gösta Olsson (Stockholm: LTs Förlag, 1986), pp. 35–43.

41. Augustine Brannigan, The Reification of Mendel, *Social Studies of Science* 9, no. 4 (1979): 423–454.

42. William B. Provine, *The Origins of Theoretical Population Genetics* (Chicago: University of Chicago Press, 1971), pp. 1–89. See also Daniel J. Kevles, Genetics in the United States and Great Britain, 1890–1930: A Review with Speculations, *Isis* 71 (1980): 441–455, who reinforces skepticism about Provine's account of the debate between biometricians and Mendelians as being just a matter of personality clashes.

43. Provine, *Origins of Theoretical Population Genetics*, p. 66.

44. Diane B. Paul and Barbara A. Kimmelman, Mendel in America: Theory and Practice, 1900–1919, in *The American Development of Biology*, ed. Ronald Rainger, Keith R. Benson, and Jane Maienschein (Philadelphia: University of Pennsylvania Press, 1988), pp. 281–310.

45. Hugo de Vries, *Plant-Breeding: Comments on the Experiments of Nilsson and Burbank* (Chicago: Open Court, 1907), 360 pp.

46. E. John Russell, *A History of Agricultural Science in Great Britain, 1620–1954* (London: George Allen and Unwin, 1966), pp. 208–211.

47. Loren R. Graham, *Science and Philosophy in the Soviet Union* (New York: Knopf, 1972), pp. 211–218. Graham notes that Vavilov lost the struggle to defeat Lysenko, was arrested in 1940, and died sometime after that in prison camp.

48. Edgar Anderson *Plants, Man and Life* (Berkeley: University of California Press, 1969), p. 77.

49. R. H. Biffen, Mendel's Laws of Inheritance and Wheat Breeding, *Journal of Agricultural Science* 1 (1905): 4–48.

50. Russell, *History of Agricultural Science in Great Britain*, p. 209.

51. W. Bateson, *Mendel's Principles of Heredity* (Cambridge: Cambridge University Press, 1913), p. 1.

52. Biffen, Mendel's Laws of Inheritance and Wheat Breeding, 4–48.

53. Russell, *History of Agricultural Science in Great Britain*, p. 213; G. D. H. Bell, *The Plant Breeding Institute, Cambridge* (United Kingdom: Massey-Ferguson, 1976), p. 10

54. Rodgers, *Liberty Hyde Bailey*, pp. 79–80.

55. Alfred D. Chandler Jr., *The Visible Hand: The Managerial Revolution in American Business* (Cambridge, Mass.: Harvard University Press, 1977), 608 pp.

56. Fred A. Shannon, *The Farmers's Last Frontier: Agriculture, 1860–1897* (New York: Holt, Rinehart and Winston, 1945; reprint White Plains, N.Y.: M. E. Sharpe, 1973), pp. 291–309. John Opie reviews the development of railroads and the settlement of the wheat-growing areas of the Great Plains in *The Law of the Land: Two Hundred Years of American Farmland Policy* (Lincoln: University of Nebraska Press, 1987), pp. 70–111.

57. Shannon, *The Farmer's Last Frontier*, pp. 309–311, 329–332.

58. Lawrence Goodwyn, *The Populist Revolt: A Short History of the Agrarian Revolt in America* (Oxford: Oxford University Press, 1978), 349 pp.

59. Alan Marcus, *Agricultural Science and the Quest for Legitimacy* (Ames: Iowa State University Press, 1985), pp. 25–26.

60. Other members of the commission included Kenyon L. Butterfield, president of the Massachusetts State College of Agriculture; Gifford Pinchot, chief forester in the USDA; Walter Hines Page, editor of *World's Work*; Henry Wallace, a farmer and editor of *Wallace's Farmer*; Charles S. Barrett, president of the Farmers' Cooperative and Educational Union of America; and William A. Beard, editor of *Great West Magazine* and chairman of both the Sacramento Valley Improvement Association and the National Irrigation Society. See William L. Bowers, *The Country Life Movement in America, 1900–1920* (Port Washington, N.Y.: Kennikat Press, 1974), pp. 24–27, 45–61.

61. L. H. Bailey, *The Country-Life Movement in the United States* (New York: Macmillan, 1911), pp. 6–7.

62. David B. Danboom reviews the many complexities of the rural reform efforts and the country life movement in his book *The Resisted Revolution, Urban America and the Industrialization of Agriculture, 1900–1930* (Ames: Iowa State University Press, 1979), pp. 23–50. I have been heavily influenced by his argument that the country life movement had a strong tendency to be the work of urban reformers who wanted the countryside to be remade in their interests. Well-being for farmers, assumed the reformers, would come, but the reform was fundamentally imposed by the urban powers on rural people, some of whom saw no benefit in the reforms. In this light, it is important to note that the Commission on Country Life included only one farmer, Henry Wallace, but Wallace was perhaps more known as a publisher and in any case was hardly a typical farmer. Other members of the commission were academics, publishers, and proponents of federal and state support for scientific and technical agriculture.

63. Bailey, *The Country-Life Movement in the United States*, pp. 9–13.

64. Bowers, *The Country Life Movement in America*, pp. 27, 139 n. 31.

65. Bailey, *The Country-Life Movement in the United States*, pp. 17–21.

66. Ibid., pp. 55–60.

67. Ibid., pp. 14–15.

68. Ibid., pp. 61–84.

69. Liberty Hyde Bailey, Country Living in the Next Generation, *Independent* 85 (1916): 336–338.

70. In 1906 Bailey published survey results of Cornell students, from both rural and urban backgrounds, about their interests in agriculture. Reasons for entering agriculture included liking the work and a disregard of obtaining low financial returns. Those leaving emphasized the hard work, low pay, and lack of social amenities. See L. H. Bailey, Why Do Boys Leave the Farm? *Century* 72 (1906): 410–416; and Bailey Why Some Boys Take to Farming, *Century* 72(1906): 612–617.

71. Rodgers, *Liberty Hyde Bailey*, pp. 405–410.

72. As examples see Oliver Mayo, *The Theory of Plant Breeding* (Oxford: Oxford University Press, 1987), pp. 8–12; and Fred N. Briggs and P. F. Knowles, *Introduction to Plant Breeding* (New York: Reinhold, 1967), pp. 70–71.

73. Bailey, *Plant-Breeding, Five Lectures*, pp. 1–38.

74. Mayr, *Growth of Biological Thought*, pp. 251–297, esp. pp. 251–265.

75. Bailey, *Plant-Breeding, Five Lectures*, pp. 31–34. In his glossary Bailey defined "species": "An indefinite term applied to all individuals of a certain kind which come or are supposed to come from a common parentage" (p. 285).

76. Ibid., pp. 106, 122–127.

77. Ibid., pp. 1–38.

78. See, for example, de Vries's discussion in *Plant-Breeding*, pp. 104–105: "A second discovery made at Svalof, and equally valuable for practice and for science, was that of the almost astonishing richness in elementary species among our agricultural crops. Every cultivated species seems to embrace something like a hundred of them, and the cereals were found to include even several hundreds in each of the older species."

East made the ontological reality of de Vries's elementary species even more explicit: "Some of Linnaeus' species were relatively uniform, others contained many groups of plants which differed from each other by one or more constant characters. . . . De Vries has called these groups of plants, 'elementary species.' . . . De Vries' work shows that in the working of the principle of natural selection, it is more likely a contest in which the fittest elementary species survives." See Edward M. East, *The Relation of Certain Biological Principles to Plant Breeding*, bulletin no. 158 (New Haven: Connecticut Agricultural Experiment Station, 1907), pp. 31–33.

79. Provine, *Origins of Theoretical Population Genetics*, p. 68.

80. De Vries, *Plant-Breeding*, pp. 353, 357.

81. Bailey, *Plant-Breeding, Six Lectures*, pp. 145–155, 179–181, 189–202; East, *Relation of Certain Biological Principles to Plant Breeding*, pp. 30–35, 45–72; Mark Alfred Carleton, *The Small Grains* (New York: Macmillan, 1924), pp. 210–211.

82. Liberty Hyde Bailey's series of textbooks and articles published between 1895 and 1920 provide one useful measure for the transition in the development of plant-breeding theory. The 1895 text included extensive discussions of variation, Darwin, and the nature of species. At that time he had no awareness of Mendel's work. By 1906 Bailey still began his discussion with Darwin, but he also included an extensive discussion of Mendelism and de Vries's ideas of evolution. He also had de Vries write sections on hybridization for the third and fourth editions. By the fifth edition in 1915, Bailey still began with Darwin, but Mendelism was firmly entrenched as the major way to study heredity. In fact, plants and animals were composed of unit characters, which were the smallest heritable part or

attribute of an organism. Mendel's work allowed the behavior of unit characters in crosses to be understood.

Shortly after the last edition of Bailey's book, Herbert Kendall Hayes and Ralph John Garber put out an American textbook, *Breeding Crop Plants* (New York: McGraw-Hill, 1921). By the second edition in 1927, Darwin got a short mention after a longer discussion of the understanding of sexuality in plants. Other than noting that artificial selection took the place of natural selection in plant breeding, the book simply acknowledged evolution as an important idea. The concept of "species" was not important enough even to merit an index entry, and the book spends no time discussing its philosophy of species. Hayes and Garber move immediately to a thorough discussion of Mendelian heredity and the study of factors of inheritance.

In 1942 Hayes again produced a major textbook, *Methods of Plant Breeding* (New York: McGraw-Hill, 1942), with Forrest Rhinehart Immer. In this treatment evolutionary theory does not even appear for a brief acknowledgment, Darwin is mentioned only in the context of his work on self-fertilization, and species does not emerge as a troublesome concept. Students instead are taken directly into a study of cytology and the chromosome theory of inheritance, the sexual physiology of plants, and then a thorough discussion of the many complexities of Mendelian genetics.

83. Bailey, *Plant-Breeding, Five Lectures*, pp. vi–vii.

84. Rodgers, *Liberty Hyde Bailey*, pp. 165–170

85. East, *Relation of Certain Biological Principles to Plant Breeding*, p. 44.

86. Darwin, *Variation of Animals and Plants under Domestication*, vol. 1, p. 473.

87. Ibid., vol. 2, p. 242.

88. Ibid., vol. 2, p. 348.

89. Ernest Brown Babcock and Roy Elwood Clausen, *Genetics in Relation to Agriculture* (New York: McGraw-Hill, 1918), p. 15.

90. Provine, *Origins of Theoretical Population Genetics*, pp. 92–96.

91. W. Johannsen, Heredity in Populations and Pure Lines, reprinted and translated in *Classic Papers in Genetics*, ed. James A. Peters (Englewood Cliffs, N.J.: Prentice-Hall, 1959), pp. 20–26; Babcock and Clausen, *Genetics in Relation to Agriculture*, pp. 250–255.

92. Edmund W. Sinnott and L. C. Dunn, *Principles of Genetics: A Textbook, with Problems*, 2d ed. (New York: McGraw-Hill, 1932), p. 24.

93. Provine, *Origins of Theoretical Population Genetics*, pp. 115–118; Roll-Hansen, Svalöf and the Origins of Classical Genetics, pp. 35–43.

94. Provine, *Origins of Theoretical Population Genetics*, pp. 118–121.

95. Babcock and Clausen, *Genetics in Relation to Agriculture*, pp. 32–56, 419–436.

96. Provine, *Origins of Theoretical Population Genetics*, pp. 98–99.

97. Theodor Boveri, *Dictionary of Scientific Biography*, vol. 2 (New York: Scribner's, 1970), pp. 361–365; and Walter Stanborough Sutton, *Dictionary of Scientific Biography*, vol. 13 (New York: Scribner's, 1976), pp. 156–158.

98. Garland E. Allen, *Thomas Hunt Morgan: The Man and His Science* (Princeton, N.J.: Princeton University Press, 1978), p. 130.

99. Ibid., pp. 116–144.

100. Ibid., pp. 144–164.

101. Evelyn Fox Keller, *A Feeling for the Organism: The Life and Work of Barbara McClintock* (New York: Freeman, 1983), pp. 55–59; Harriet B. Creighton and Barbara McClintock, A Correlation of Cytological and Genetical Crossing-over in *Zea mays*, in *Classic Papers in Genetics*, ed. James A. Peters (Englewood Cliffs, N.J.: Prentice-Hall, 1959), pp. 155–160.

102. L. C. Dunn, *A Short History of Genetics* (New York: McGraw-Hill, 1965), p. 110.

103. Hitoshi Kihara, *Wheat Studies, Retrospect and Prospect* (Amsterdam: Elsevier, 1982), 308 pp.

104. M. D. Gale and T. E. Miller, The Introduction of Alien Genetic Variation into Wheat, in *Wheat Breeding: Its Scientific Basis*, ed. F. G. H. Lupton (London: Chapman and Hall, 1987), pp. 188–204.

105. Herbert Kendall Hayes and Forrest Rhinehart Immer, *Methods of Plant Breeding* (New York: McGraw-Hill, 1942), p. 11.

106. Carleton, *The Small Grains*, pp. 160–230.

107. De Vries, *Plant-Breeding*, pp. 48–90.

108. Bailey, *Plant-Breeding, Five Lectures*, pp. 95–128; Hayes and Immer, *Methods of Plant Breeding*, 432 pp.

109. Bailey, *Plant-Breeding, Five Lectures*, p. 54. Darwin was not sure exactly how wheat was fertilized (*The Variation of Animals and Plants under Domestication*, vol. 1, p. 334).

110. Bailey, *Plant-Breeding, Lectures*, pp. 282–283.

Chapter 4

1. E. John Russell, *A History of Agricultural Science in Great Britain, 1620–1954* (London: George Allen and Unwin, 1966), pp. 20–26, 55–65.

2. Lucile H. Brockway, *Science and Colonial Expansion* (New York: Academic Press, 1979), pp. 6–10, 78–81.

3. Maryanna S. Smith, and Dennis M. Roth, *Chronological Landmarks in American Agriculture*. AIB-425 (Washington, D.C.: U.S. Department of Agriculture, 1990), pp. 2–3.

4. M. S. Randhawa, *A History of Agriculture in India*, vol. 3 (New Delhi: Indian Council of Agricultural Research, 1983), pp. 33–47.

5. N. P. W. Goddard, The Royal Agricultural Society of England and Agricultural Progress, 1838–1880 (Ph.D. diss., University of Kent, 1981), pp. 7–12, 74–98.

6. Russell, *History of Agricultural Science in Great Britain*, pp. 110–112.

7. A. D. Hall, *The Book of the Rothamsted Experiments* (New York: Dutton, 1905), pp. 14, 31–62; Russell, *History of Agricultural Science in Great Britain*, pp. 88–107.

8. Susan Foreman, *Loaves and Fishes: An Illustrated History of the Ministry of Agriculture, Fisheries and Food, 1889–1989* (London: Her Majesty's Stationary Office, 1989), pp. 1–3; H. E. Dale, Agriculture and the Civil Service, in *Agriculture in the Twentieth Century* (Oxford: Clarendon Press, 1939), pp. 1–20.

9. Foreman, *Loaves and Fishes*, pp. 4–9.

10. Dale, Agriculture and the Civil Service, pp. 1–20.

11. Willard W. Cochrane, *The Development of American Agriculture: A Historical Analysis*, 2d ed. (Minneapolis: University of Minnesota Press, 1993), pp. 241–242, 478.

12. Willard W. Cochrane, *The Development of American Agriculture* (Minneapolis: University of Minnesota Press, 1979), pp. 96–97, 104–105.

13. Zoya Hasan, Power and Mobilization: Patterns of Resilience and Change in Uttar Pradesh Politics, in *Dominance and State Power in Modern India*, ed. Francine R. Frankel and M. S. A. Rao (Delhi: Oxford University Press, 1989), pp. 134, 140–143. For a more general discussion arguing that British imperial policy was designed to extract revenue in order to finance the empire and benefit certain Englishmen, see K. S. Shelvankar, *The Problem of India* (Harmondsworth, England: Penguin, 1940; reprint 1943), pp. 49–80; and Jadunath Sarkar, *Economics of British India* (Calcutta: M. C. Sarkar and Sons, 1917), pp. 362–365. More specialized arguments of this kind are also in Thomas R. Metcalf, From Raja to Landlord: The Oudh Talukdars, 1850–1870; and Metcalf, Social Effects of British

Land Policy in Oudh; both in *Land Control and Social Structure in Indian History*, Ed. Robert Eric Frykenburg (New Delhi: Manohar, 1979), pp. 123–141, 143–162.

14. Randhawa, *History of Agriculture in India*, vol. 3, pp. 172–186. See also Shelvankar, *The Problem of India*, pp. 49–80.

15. The first provincial department of agriculture was established in what was then called the Northwestern Province (now in Uttar Pradesh) in 1875. Other provincial departments were primarily inspired by the report in 1880 of the Famine Enquiry Commission. See N. Srinivasan and S. P. Singh, The Growth of the State Departments of Agriculture, in *Agricultural Administration in India*, ed. N. Srinivasan (New Delhi: Indian Institute of Public Administration, 1969), pp. 46–66.

16. Randhawa, *History of Agriculture in India*, vol. 3, pp. 232–238.

17. Ibid., pp. 270–271.

18. For a complete review of the famines of 1896–97 and 1899–1900, see Premansukumar Bandyopadhyay, *Indian Famine and Agrarian Problems: A Policy Study on the Administration of Lord George Hamilton, Secretary of State for India, 1895–1903* (Calcutta: Star, 1987), 259 pp. This study lays the majority of blame for the tragedies of these famines on Lord George Hamilton. See also Stanley Wolpert, *A New History of India*, 2d ed. (New York: Oxford University Press, 1982), p. 267.

19. David Dilks, *Curzon in India*, vol. 1 (New York: Taplinger, 1969), pp. 221–248.

20. Randhawa, *History of Agriculture in India*, vol. 3, pp. 262–271.

21. Ibid., pp. 271–273.

22. Louise E. Howard, *Sir Albert Howard in India* (London: Faber and Faber, 1953), pp. 266–267.

23. Ibid., pp. 21–22.

24. Albert Howard and Gabrielle L. C. Howard, *Wheat in India: Its Production, Varieties and Improvement* (Calcutta: Imperial Department of Agriculture in India, 1909), 288 pp.

25. Randhawa, *History of Agriculture in India*, vol. 3, pp. 342–344.

26. Howard and Howard, *Wheat in India*, pp. 133–141.

27. Albert Howard is perhaps the better known name than Gabrielle Howard, but this is due largely to his work after retirement from IARI in 1924 to accept a position at the newly formed, private Indore Institute of Plant Industry. There he developed his comprehensive theories of soil fertility and farming systems. In 1931, after Gabrielle's death in 1930, he returned to England, married Gabrielle's younger sister, Louise, and launched a sixteen-year crusade to reform English agriculture. He created the mode of farming known now as "organic agriculture," which was based on the use of composted manures and was heavily influenced by his experiences in India. He also urged reform of the agricultural experiment station, which he felt had become too regimented into departments and was out of touch with growers. For her accomplishments, Gabrielle Howard was given a salaried post as personal assistant to Albert in 1910 and in 1913 was named second imperial economic botanist to the Government of India. These were remarkable achievements for the male-dominated and conservative tenor of the Government of India. These aspects of the Howards' life are dealt with in detail in Louise Howard's biography, *Sir Albert Howard in India* (London: Faber and Faber, 1953), 272 pp. His ideas were dismissed as "neo-vitalism" by E. John Russell, director of the Rothamsted Experiment Station. Today, Howard is remembered by many primarily for his ideas on organic agriculture, but his earlier scientific reputation was based on his pioneering work in the botany and genetics of Indian wheats. See Philip Conford, All Things Organic, *Times Higher Education Supplement*, 17 July 1987, p. 16; Russel, *History of Agricultural Science in Great Britain*, pp. 467–468.

28. Albert Howard and Gabrielle L. C. Howard, *The Development of Indian Agriculture* (Bombay: Oxford University Press, 1929), p. 23.

29. T. O. Lloyd, *Empire to Welfare State: English History 1906–1985*, 3d ed. (Oxford: Oxford University Press, 1986), pp. 1–10.

30. Howard Newby, *Green and Pleasant Land?* (London: Wildwood House, 1979), pp. 31–38.

31. Jonathan Brown, *Agriculture in England: A Survey of Farming, 1870–1947* (Manchester: Manchester University Press, 1987), pp. 1–13.

32. Alan Armstrong, *Farmworkers: A Social and Economic History, 1770–1980* (London: B. T. Batsford, 1988), pp. 246–247.

33. Lloyd, *Empire to Welfare State*, pp. 17–23.

34. Brown, *Agriculture in England*, pp. 85–86.

35. Lloyd, *Empire to Welfare State*, pp. 17–18.

36. Thomas C. H. Jones, *Lloyd George* (London: Oxford University Press, 1951), p. 37.

37. Russel, *History of Agricultural Science in Great Britain*, pp. 268–272.

38. G. W. Cooke, ed., *Agricultural Research, 1931–1981* (London: Agricultural Research Council, 1981), p. 8.

39. School of Agriculture, 1899–1949, *University of Cambridge School of Agriculture Memoirs*, no. 21, 1 October 1948–30 September 1949, pp. 5–7.

40. G. D. H. Bell, Plant Breeding Institute: Historical Review, 1912–1948, *University of Cambridge School of Agriculture Memoirs*, no. 20, 1 October 1947–30 September 1948, pp. 5–10; F. G. H. Lupton, Historical Survey, in *The Plant Breeding Institute: 75 Years, 1912–1987* (Cambridge: Plant Breeding Institute, 1987), pp. 7–19; Cooke, *Agricultural Research*, p. 10.

41. Russel, *History of Agricultural Science in Great Britain*, pp. 271–272.

42. John Fryer, The Organisation for Agricultural Research, *Agricultural Progress* 22 (1947): 11 pp., in "ARC Constitution Charter," Library of the Agricultural Research Council, London; Russel, *History of Agricultural Science in Great Britain*, pp. 269–272.

43. Quotation taken from W. Watkin Davies, *Lloyd George, 1863–1914* (London: Constable, 1939), p. 314.

44. A. Whitney Griswold, *Farming and Democracy* (New Haven, Conn.: Yale University Press, 1948), pp. 47–85.

45. Ferdinand Mount, The New Song of the Land, *Spectator*, 10 January 1987, pp. 9–13.

46. Samuel P. Hays has developed the most thorough statement of how conservation was an ideology to promote efficiency. He emphasizes that it was not, in the eyes of its creators, meant to promote democracy. Instead, it was intended to bring professional scientists and experts into a position of leadership and decision-making power. See Hays, *Conservation and the Gospel of Efficiency: The Progressive Conservation Movement 1890–1920* (Cambridge, Mass.: Harvard University Press, 1959), 297 pp.

47. Charles E. Rosenberg, Science, Technology, and Economic Growth: The Case of the Agricultural Experiment Station Scientist, 1875–1914, *Agricultural History* 45 (January 1971): 1–20.

48. Murray R. Benedict, *Farm Policies of the United States, 1790–1950* (New York: Twentieth Century Fund, 1953), pp. 112–137.

49. John A. Stevenson, Plants, Problems, and Personalities: The Genesis of the Bureau of Plant Industry, *Agricultural History* 28, no. 4 (October 1954): 155–162. See also Clayton Coppin, James Wilson, and Harvey Wiley: The Dilemma of Bureaucratic Entrepreneurship, *Agricultural History* 64, no. 2 (spring 1990): 167–181.

50. Alfred Charles True, *A History of Agricultural Experimentation and Research in the United States, 1607–1925*, Miscellaneous Publication no. 251 (Washington, D.C.: Gov-

ernment Printing Office, 1937), p. 171. Alan I. Marcus argues that passage of the Hatch Act worked to create a professional cadre of agricultural scientists and that farmers thereby became firmly identified as businessmen who were dependent on scientists. This trend became even stronger with passage of the Adams Act in 1906. See Alan I. Marcus, *Agricultural Science and the Quest for Legitimacy* (Ames: Iowa State University Press, 1985), 269 pp.

51. True, *History of Agricultural Experimentation and Research in the United States*, p. 212.

52. L. Margaret Barnett, *British Food Policy during the First World War* (Boston: George Allen and Unwin, 1985), pp. 1–19.

53. Ibid., pp. 102–105.

54. Benedict, *Farm Policies of the United States*, pp. 158–159.

55. Lloyd, *Empire to Welfare State*, 75–77.

56. Foreman, *Loaves and Fishes*, pp. 11–14.

57. Russell, *History of Agricultural Science in Great Britain*, pp. 202–206.

58. Thomas Hudson Middleton, *Food Production in War* (Oxford: Clarendon, 1923), pp. 10–12.

59. Barnett, *British Food Policy during the First World War*, pp. 64–65.

60. Middleton, *Food Production in War*, pp. 136–138.

61. Barnett, *British Food Policy during the First World War*, pp. 195–198; Pamela Horn, *Rural Life in England in the First World War* (New York: St. Martin's, 1984), pp. 51–52.

62. Middleton, *Food Production in War*, pp. 312–316.

63. Barnett, *British Food Policy during the First World War*, pp. 205–206.

64. Ibid., p. 195; Horn, *Rural Life in England in the First World War*, p. 51.

65. Barnett, *British Food Policy during the First World War*, p. 14.

66. Middleton, *Food Production in War*, pp. 9, 85–86, 261.

67. Frank M. Surface and Raymond L. Bland, *American Food in the World War and Reconstruction Period* (Stanford, Calif.: Stanford University Press, 1931), pp. 189–194.

68. Vincent A. Smith, *The Oxford History of India*, 4th ed. (Delhi: Oxford University Press, 1981), p. 779.

69. Imran Ali, *The Punjab under Imperialism, 1885–1947* (Delhi: Oxford University Press, 1988), pp. 237–244.

70. Dharma Kumar, ed., *The Cambridge Economic History of India*. Vol. 2, C. 1757–c. 1970 (Hyderabad: Orient Longman, 1982), pp. 135–137.

71. Bipan Chandra, *India's Struggle for Independence, 1857–1947* (New Delhi: Penguin, 1989), pp. 146–169.

72. Smith, *Oxford History of India*, pp. 779–780.

73. Benedict, *Farm Policies of the United States*, pp. 160–161; Benjamin H. Hibbard, *Effects of the Great War upon Agriculture in the United States and Great Britain* (New York: Oxford University Press, 1919; reprint Wilmington, Del.: Scholarly Resources, 1974), pp. 68–90, 100–105; Surface and Bland, *American Food in the World War and Reconstruction Period*, p. 5.

74. Hibbard, *Effects of the Great War upon Agriculture*, pp. 70–72.

75. Ibid., pp. 23–24.

76. A summary of the acreage and total production figures for the period 1914–1920 is in Benedict, *Farm Policies of the United States*, pp. 166–167 n. 84. These figures differ slightly from those compiled during and immediately after the war in Hibbard, *Effects of the Great War upon Agriculture in the United States and Great Britain*, pp. 24–25, and in O. C. Stine, The World's Supply of Wheat, in *Yearbook of the United States Department of Agriculture 1917* (Washington, D.C.: Government Printing Office, 1918), pp. 461–480. In this section,

Benedict's figures are used, but the conclusions reached are not sensitive to the variations in the different sources.

77. Average yields calculated from figures in Benedict, *Farm Policies of the United States*, pp. 166–167 n. 84.

78. Ibid., p. 159. Prices quoted are for No. 1 Northern Spring wheat.

79. Ibid., pp. 166–167 n. 84.

80. Hibbard, *Effects of the Great War upon Agriculture in the United States and Great Britain*, pp. 24–25.

81. Benedict, *Farm Policies of the United States*, pp. 166–167 n. 84. Over 60 percent of the winter wheat in Kansas was killed in the winter of 1916–17; see E. G. Heyne, The Development of Wheat in Kansas, in *The Rise of the Wheat State: A History of Kansas Agriculture, 1861–1986*, ed. George E. Ham and Robin Higham (Manhattan, Kans.: Sunflower University Press, 1987), p. 43.

82. Hibbard, *Effects of the Great War upon Agriculture in the United States and Great Britain*, pp. 31–33; Benedict, *Farm Policies of the United States*, p. 165. Although wheat consumption fell in the United States during the war, this fall is not reported to have led to starvation or malnutrition beyond what was considered "normal" at the time. The food security of the United States thus stood in stark contrast to the severe famine conditions that existed in Europe, particularly in central and eastern Europe.

83. Benedict, *Farm Policies of the United States*, pp. 161–164.

84. Elvin Charles Stakman, A Study in Cereal Rusts: Physiological Races (Ph.D. diss., University of Minnesota, 1913), 56 pp. Stakman found evidence that physiological races were reasonably stable in most cases, even though race might change over time as the rust grew on a particular variety of wheat. He cited Biffen's work, and so was clearly aware of early Mendelian papers in plant breeding. See also C. M. Christensen, *E. C. Stakman, Statesman of Science* (St. Paul: American Phytopathological Society, 1984), pp. 3–44.

85. Christensen, *E. C. Stakman*, pp. 48–50.

86. Ibid., pp. 50–51.

87. K. F. Kellerman to R. W. Thatcher, 21 February 1918; W. A. O[rtman] to E. C. Stakman and G. H. Coons, 15 February 1918; E. M. Freeman to Karl F. Kellerman, 8 March 1918; and E. C. Stakman to William Taylor, 14 April 1918; all in Bureau Chief's Correspondence, 1908–39 (7800–7895), Bureau of Plant Industry, Soils, and Agricultural Engineering, Record Group 54, USNA.

88. Christensen, *E. C. Stakman*, pp. 24, 51–55, 139.

89. Ibid., pp. 23–24.

90. Ohio, Michigan, Indiana, Illinois, Wisconsin, Iowa, Minnesota, Nebraska, South Dakota, North Dakota, Colorado, Wyoming, and Montana.

91. Christensen, *E. C. Stakman*, pp. 52–61.

92. Carol E. Windels and Carl J. Eide, eds., *Aurora Sporealis, 75th Anniversary Edition*, vol. 54 (St. Paul: Department of Plant Pathology, University of Minnesota, 1983), pp. 14–15.

93. Christensen, *E. C. Stakman*, pp. 55–56.

94. F. M. Crosby to Walter H. Newton, 10 July 1929, Bureau Chief's Correspondence, 1908–39, file 6072 (1908–39), box 585, Record Group 54, USNA.

95. Carleton R. Ball to W. A. Taylor, 8 April 1929, Bureau Chief's Correspondence, 1908–39, file 7992 (1910–34), box 612, Record Group 54, USNA.

96. A. P. Roelfs, personal communication, 4 November 1991.

97. Christensen, *E. C. Stakman*, pp. 37–44, 62–64.

98. Brown, *Agriculture in England*, pp. 76–79.

99. A. D. Hall to E. J. Russell, F. B. Wood, Rol. Biffen, B. T. E. Barker, C. S. Orwin, C. Crowther, W. Bateson, 25–1–18, "Reconstruction of Agricultural Research," MAF 33/I2672/1918, file I, PRO.

100. Alfred Daniel Hall, *Agriculture after the War* (London: John Murray, 1916), pp. 127–131; Hall, *Reconstruction of the Land* (London: Macmillan, 1941), pp. 257–260; Russell, *History of Agricultural Science in Great Britain*, p. 278.

101. R. H. Biffen to A. D. Hall, 5 May 1918, "Reconstruction of Agricultural Research," MAF 33/I2672/1918, file I, PRO.

102. "Two Million Pounds for Agricultural Education and Research for the Five Years Beginning 1rst October, 1919," MAF 33/59/TE/942/1922, PRO.

103. "Reconstruction of Agricultural Research," MAF 33/TG/1490/1920, Public Records Office, Kew, London.

104. "Agricultural Research Council—Constitution, 1919. Reconstitution, 1931," MAF 33/749, PRO.

105. Andrew Fenton Cooper, *British Agricultural Policy, 1912–1936: A Study in Conservative Politics* (Manchester: Manchester University Press, 1989), pp. 64–89.

106. Lloyd, *Empire to Welfare State*, p. 156.

107. These are the priorities articulated by Dr. Christopher Addison, who became minister of Agriculture and Fisheries in June 1930. See Kenneth Owen Morgen and Jane Morgan, *Portrait of a Progressive: The Political Career of Christopher, Viscount Addison* (Oxford: Clarendon Press, 1980), pp. 192–194.

108. Clearing all of the bureaucratic details was more involved than indicated here. MacDonald's first appointment was of a Subcommittee of the Committee of Civil Research. The Committee of Civil Research was absorbed into the Economic Advisory Council on 27 January 1930, and the subcommittee became the Committee on Agricultural Research Organisation. Discussions of the committee's report in the cabinet occurred during May and June and revealed some strenuous objections to the report from the minister of agriculture and fisheries, Noel Buxton. Buxton's major problems were that the report recommended the new council have funds of its own to administer, independent of the ministry, and that the council report to the Privy Council, not the ministry. In May, Buxton was replaced by Christopher Addison, who worked out a compromise, which left the new council reporting to the Privy Council but obliged to consult with the minister and the Department of Agriculture for Scotland on all of its activities. In addition, the primary duty of the council was to advise the Ministry and the Department, and it had only a small amount of funds to spend under its own authority. See "Formation of Agricultural Research Council," CAB/123/275, Public Records Office, Kew, London, for copies of all reports, relevant memoranda, and cabinet minutes.

109. G. W. Cooke, ed., *Agricultural Research, 1931–1981* (London: Agricultural Research Council, 1981), pp. 21–25.

110. Ibid., pp. 35–37.

111. "Agricultural Research Council—Constitution, 1919. Reconstitution, 1931," ARC minutes, 9 December 1930, Minutes no. 20, and Conference of Directors of Agricultural Research Institutes, Minutes no. 21, 8 December 1931, MAF 33/749, PRO.

112. Cooke, *Agricultural Research*, pp. 18–19. Medical research in the Medical Research Committee came under state sponsorship as a result of the Liberals' National Insurance Act (1911). The committee became the Medical Research Council in 1920, simultaneous with the creation of the Committee of the Privy Council for Medical Research. A Committee of the Privy Council for Scientific and Industrial Research was organized in 1915, and in 1916 it became the Department of Scientific and Industrial Research. Agricultural research was

not considered important enough to merit comparable organization until 28 July 1930, when MacDonald's government organized the Committee of the Privy Council for the Organisation and Development of Agricultural Research, and the Privy Council intended the organizational scheme for agricultural research to be parallel to those for medicine and industry.

113. From 1910 to 1931, about £6.6 million were paid into the Development Fund by Parliament; of this sum, £4.5 million (68 percent) went to the Ministry of Agriculture and Fisheries for agricultural education and agricultural research. The following table shows the expenditures for these activities from 1910–11 to 1932. All information from "Agricultural Research Council, Relations with Ministry," MAF 33/753, PRO.

Expenditures (£) for Agricultural Education and Research

Year	Education	Research
1910–11	—	—
1911–12	—	9,712
1912–13	—	12,853
1913–14	4,458	34,269
1914–15	23,949	42,343
1915–16	31,583	42,779
1916–17	14,137	33,366
1917–18	7,558	34,543
1918–19	11,031	32,797
1919–20	4,093	74,727
1920–21	—	177,742
1921–22	—	123,896
1922–23	30,831	174,037
1923–24	—	180,129
1924–25	20,780	237,460
1925–26	88,355	261,882
1926	86,291	245,502
1927	4,961	250,163
1928	8,555	237,865
1929	9,079	235,903
1930	4,203	265,001
1931 est.	5,039	260,645
1932 est.	3,375	249,655

Data from "Agricultural Research Council, Relations with Ministry." MAF 33/753, Public Records Office, Kew, London.

114. "Formation of Agricultural Research Council," notes of meeting, 26 May 1930, CAB/123/275, PRO.

115. The argument presented here is parallel to the argument advanced by Vannevar Bush on why the National Science Foundation in postwar America should be a quasi-autonomous body, controlled by a National Science Board made up of scientists. Only the director and the staff of the foundation became government employees, responsible to the president, not a regular department. Scientists receiving grants were thus not government employees, not subject to bureaucratic control, and free to work on subjects agreed to by their peer scientists. See Vannevar Bush, *Science: The Endless Frontier* (Washington, D.C.: Government Printing Office, 1945), pp. 25–34.

116. Hall, *Agriculture after the War*, 137 pp.

117. Morgen and Morgan, *Portrait of a Progressive*, pp. 102, 143, 154, 157, 163–166, 171–173.

118. Brown, *Agriculture in England*, pp. 116–117.

119. Cooper, *British Agricultural Policy, 1912–1936*, pp. 127–139, 160–180.

120. Chandra, *India's Struggle for Independence*, pp. 177–183; Wolpert, *New History of India*, pp. 297–300.

121. Chandra, *India's Struggle for Independence*, pp. 184–187; Wolpert, *New History of India*, pp. 301–303.

122. Wolpert, *New History of India*, p. 307.

123. Chandra, *India's Struggle for Independence*, pp. 195–196.

124. Wolpert, *New History of India*, pp. 307–308.

125. Ibid., p. 311.

126. Randhawa, *History of Agriculture in India*, vol. 3, pp. 376–377.

127. Royal Commission on Agriculture in India, *Report* (London: Her Majesty's Stationary Office, 1928, Cmd. 3132), pp. [ii], 678–696.

128. Ibid., p. 5.

129. Royal Commission on Agriculture in India, *Minutes of Evidence Taken before the Royal Commission on Agriculture, 1927–1928*, vol. 10 (London: Her Majesty's Stationary Office, 1927), p. 88.

130. Joseph Stancliffe Davis, *Wheat and the AAA* (Washington, D.C.: Brookings Institution, 1935), pp. 7–27.

131. Cochrane, *Development of American Agriculture*, pp. 126–149, 316–319, 341–343.

Chapter 5

1. Randall E. Stross, *The Stubborn Earth: American Agriculturalists on Chinese Soil, 1898–1937* (Berkeley: University of California Press, 1986), pp. 143–160.

2. James C. Thomson Jr., *While China Faced West: American Reformers in Nationalist China, 1928–1937* (Cambridge, Mass.: Harvard University Press, 1969), pp. 122–150.

3. Stross, *Stubborn Earth*, p 13. Stross reports that Lossing Buck, based at the University of Nanking in agricultural economics in the 1920s, was quite hostile to the Communists and their insistence on land reform. See also Thomson, *While China Faced West*, p. 150.

4. My understanding of the Mexican Agricultural Program (MAP) was greatly enhanced by the work of several others. William C. Cobb, a former staff member of the Rockefeller Foundation's Office of Publications, wrote "The Historical Background of the Mexican Agricultural Program," March 1956, RG1.2, series 323, box 10, RFA. Although I describe the emergence of the program in ways that are quite different from Cobb's approach, I am indebted to his work and insights on a number of points. Cynthia Hewitt de Alcantara wrote an excellent analysis of the changes in Mexican agriculture and their roles in fostering the development of Mexico as an industrial economy. Her work was especially important in helping me see the relationships between agricultural change and changes in the broader

political economy. See Cynthia Hewitt de Alcantara, *Modernizing Mexican Agriculture: Socioeconomic Implications of Technological Change, 1940–1970*, Report no. 76.5 (Geneva: U.N. Research Institute for Social Development, 1976), 350 pp. Bruce Jennings has provided an excellent analysis of the entire MAP (not just the efforts devoted to wheat breeding) in the context of the growth of international agricultural research. He has also provided a provocative analysis of the impacts of MAP on Mexico. See Bruce H. Jennings, *Foundations of International Agricultural Research* (Boulder, Colo.: Westview Press, 1988), 196 pp. Angus Wright's *Death of Ramón González: The Modern Agricultural Dilemma* (Austin: University of Texas Press, 1990), 337 pp., provides the most sweeping historical analysis to place changes in the twentieth century into a broad perspective. To each of these four major authors I am very deeply indebted.

5. Trustees of the Rockefeller Foundation, *Plans for the Future* (New York: The Rockefeller Foundation, 1963), 9 pp.

6. Ibid., p. 5.

7. James E. Austin and Gustavo Esteva, *Food Policy in Mexico: The Search for Self-Sufficiency* (Ithaca, N.Y.: Cornell University Press, 1987), p. 25.

8. Ibid. See also Charles Curtis Cumberland, *Mexican Revolution: Genesis under Madero* (Austin: University of Texas Press, 1952), pp. 22–23.

9. Steven E. Sanderson, *Agrarian Populism and the Mexican State: The Struggle for Land in Sonora* (Berkeley: University of California Press, 1981), pp. 40–41.

10. Charles C. Cumberland, *Mexican Revolution: The Constitutionalist Years* (Austin: University of Texas Press, 1972), pp. 209–211; Ronald Atkin, *Revolution! Mexico 1910–1920* (New York: John Day, 1970), pp. xiii–xiv.

11. Roger D. Hansen estimates that 17.9 million hectares were distributed to 811,000 people (*The Politics of Mexican Development* [Baltimore: Johns Hopkins University Press, 1971], pp. 33–34). Venezian and Gamble estimate that 20 million hectares were given to 750,000 people during the Cárdenas administration (Eduardo L. Venezian and William K. Gamble, *The Agricultural Development of Mexico: Its Structure and Growth Since 1950* [New York: Praeger, 1969], pp. 54–62).

12. Leonel Duran, ed., *Lázaro Cárdenas, Ideario Politico* (Mexico: Ediciones Era, 1972), p. 9.

13. Government of Mexico, *The True Facts about the Expropriation of the Oil Companies' Properties in Mexico* (Mexico: Government of Mexico, 1940), 271 pp.

14. Purport files, M 973, roll 371, RG 59, National Archives, Washington, D.C.

15. Hull to Daniels, 12 April, 1938, 812.6363/344OA, RG 59, National Archives, Washington, D.C.

16. Albert L. Michaels, *The Mexican Election of 1940* (Buffalo: Council on International Studies, State University of New York, 1971), 52 pages, mimeo.

17. Many dispatches in 812.00, RG 59, National Archives, between 1936 and 1939 speak of threats to United States' interests from the Left and the Right in Mexico. John Ferrell of the Rockefeller Foundation staff noted that Vice President Henry Wallace, in urging the Rockefeller Foundation to get involved in Mexico, "emphasized the importance of our southern neighbor from the standpoint of our national defense" (Memorandum of JAF of conference, 3 February, 1941).

18. Two incidents of pressure on the State Department to adopt a more aggressive stance toward Mexico came in September 1938. John Thompson, foreign editor of the *San Francisco News*, had an interview with former Mexican president, Plutarco Elías Calles, in San Diego. Thompson then wrote a personal letter to Secretary of State Cordell Hull and reported that Calles felt Mexico was headed for chaos. Calles apparently was thinking of

trying to get back into power and displacing Cárdenas. Calles favored vast irrigation schemes and agricultural development as a way to raise the living standard of the Mexican people (812.0013625, John Thompson to Mr. Secretary, 29 September, 1938, RG 59, National Archives, Washington, D.C.).

Senator William G. McAdoo introduced Frank McLaughlin, former state director of the Works Progress Administration for California and then assistant to the president of Golden Gate International Exposition, to Secretary Hull. During McLaughlin's interview at the State Department, he urged that the first effort should be to reach a settlement on the seized properties but that if negotiations failed the United States should force the Mexicans to comply (812.00/30626 and 812.00/30627, 19 September 1938 and 21 September 1938, RG 59, National Archives, Washington, D.C.).

19. In the summer of 1934, President Cárdenas visited in the Yaqui Valley with general and former president Plutarco Elías Calles. Support for water development in this valley was strong, and surveying began in 1935 (812.6113/94, 8(?) July 1934; 812.6113/96, 4 September 1934; and 812.6113/100, 30 January 1935; all in RG 59, USNA.)

20. Michaels, The Mexican Election of 1940.

21. Ibid. A thorough, early review of industrialization begun in 1940 is Sanford A. Mosk, Industrial Revolution in Mexico (Berkeley: University of California Press, 1950), esp. pp. vii, 307–308.

22. Hull to Daniels, 12 November 1940, 812.001 Camacho Manuel A 122A, RG 59, USNA.

23. Daniels to State, 26 November 1940, 812.00/31568, RG 59, USNA.

24. J. A. Ferrell to Mr. Fosdick, 16 October 1936, Rockefeller Foundation Archives, RG 1.2, series 323, box 10, folder 63, RFA. William C. Cobb, "The Historical Background of the Mexican Agricultural Program" (annotated edition), March 1956, RG 1.2, series 323, box 10, folder 62, RFA.

25. Memorandum of JAF of conference: Vice President Wallace, RBF and JAF, regarding Mexico—its problems and remedies, 3 February 1941, RG1.1, series 323, box 1, folder 2, and William C. Cobb, "The historical background of the Mexican Agricultural Program," March, 1956, RG1.2, series 323, box 10, folder 62, RFA. John Ensor Harr and Peter J. Johnson, The Rockefeller Century (New York: Scribner's, 1988), pp. 440–441.

26. Raymond B. Fosdick, The Story of the Rockefeller Foundation (London: Odhams Press, 1952), pp. 184–185; Cobb, "The Historical Background of the Mexican Agricultural Program."

27. Warren Weaver, Scene of Change (New York: Scribner's, 1970), pp. 94–95.

28. E. C. Stakman, Richard Bradfield, and Paul C. Mangelsdorf, Campaigns against Hunger (Cambridge, Mass.: Harvard University Press, 1967), pp. 22–23; The Reminiscences of Elvin C. Stakman, pp. 941–943, RG 13, RFA.

29. "Agricultural Conditions and Problems in Mexico," Report of the Survey Commission of the Rockefeller Foundation, 1941, RG 1.1, series 323, box 5, folder 37, RFA.

30. Weaver, Scene of Change, pp. 96–97; Oral History, Rockefeller Foundation Program in Agriculture, Interview with J. George Harrar, series 923/Oral History/vol. 13, RFA. Documents in the Rockefeller Foundation Archives suggest that the foundation did not consult with Mexican officials about the hiring of Harrar. See Raymond B. Fosdick to Marte Gomez R., 17 March 1943, RG1.2, series 323, box 10, Folder 63, RFA.

31. Report on a trip to Mexico, P. C. Mangelsdorf, [November?], 1943, RG1.2, series 323, box 10, Folder 61, and Report of Mexican Trip with Confidential Supplement Regarding Mexican Agricultural Improvement Project and Personnel, 20 September 1945, E. C. Stakman, RG1.2, series 323, box 10, Folder 60, RFA.

32. Norman E. Borlaug, personal interview, 21–22 June 1989.

33. H. M. Miller Jr. to F. B. Hanson, 7 February 1943, RG1.2, series 323, box 10, folder 63, and Report on Agricultural Activities in Mexico, 3 February to 20 May 1943, E. C. Stakman, RG1.2, series 323, box 10, folder 60, RFA.

34. Report on Agricultural Activities in Mexico, 3 February to 20 May 1943.

Stakman's wording in the report revealed the surprise at the inclusion of wheat rust and the importance of Ávila Camacho's ideas:

Rather unexpectedly, the Secretary of Agriculture and others . . . [believed] the control of wheat rust as the most important single problem. . . . Although it is doubtful whether this actually is the most important single problem, it probably is the most important one which has received virtually no attention. . . . [T]he President of the Republic is very desirous of increasing the acreage; therefore this is the policy of the Department of Agriculture also. Obviously then, the control of wheat rust should be the beginning of a general project for wheat improvement and expansion in acreage.

35. Proposals for a Memorandum of Understanding between the Secretaria de Agricultura of Mexico and the Rockefeller Foundation, n.d., RG 1.2, series 323, box 10, folder 63, RFA. Stakman drew up the draft of the Memorandum of Agreement for the Foundation (H. M. Miller Jr. to F. B. Hanson, 7 February 1943, RG 1.2, series 323, box 10, folder 63, RFA.)

36. Henry M. Miller Jr. to Frank Blair Hanson, 10 February 1943, Rg 1.2, series 323, box 10, folder 63, RFA.

37. Raymond B. Fosdick to Marte Gomez R., 17 March 1943, RG 1.2, series 323, box 10, folder 63, RFA.

38. OSS is often referred to as "OEE," the acronym for its Spanish name, Oficina de Estudios Especiales.

39. CIMMYT is the Spanish acronym for Centro International de Mejoramiento de Maíz y Trigo. See CIMMYT, *CIMMYT and Mexico* ([Mexico City]: CIMMYT, 1985), 14 pp.

40. Extensive correspondence in the Rockefeller Foundation Archives demonstrates two points about the OSS. First, the memorandum of understanding between the foundation and Mexico was revised several times between 1948 and 1954. The foundation always resisted taking on extension responsibilities as a part of the formal agreement. In 1954 the foundation felt the Mexican government wanted to have control of foundation activities, which the foundation did not accept. Second, the OSS was clearly controlled by the New York office of the foundation. Although the New York office had a high level of trust in Harrar's leadership, approval of major decisions was always made in New York. In this correspondence Harrar specifically rejected a proposal that the OSS be made part of the Institute of Agricultural Research, directed by Edmundo Taboada. See series 323, boxes 1-2, folders 6, 7, 8, 11–15, and box 11, folder 64, RFA.)

41. Joseph Cotter, The Origins of the Green Revolution in Mexico: Continuity or Change? in *Latin America in the 1940s: War and Postwar Transitions*, ed. David Rock (Berkeley: University of California Press, 1994), pp. 224–247.

42. Venezian and Gamble, *Agricultural Development of Mexico*, pp. 54–62; Sanderson, *Agrarian Populism and the Mexican State*, pp. 129–135, 158–159.

43. Survey Commission, *Agricultural Conditions and Problems in Mexico*, 1941, RG 1.1, series 323, box 5, folder 37, RFA.

44. Clifton B. Kroeber, *Man, Land, and Water: Mexico's Farmlands Irrigation Policies, 1885–1911* (Berkeley: University of California Press, 1983), pp. 219–220.

45. Venezian and Gamble, *Agricultural Development of Mexico*, pp. 54–62.

46. Michaels, *The Mexican Election of 1940*.

47. Venezian and Gamble, *Agricultural Development of Mexico*, pp. 54–62.

48. Dispatches 812.6113/94, 8(?) July 1934; 812.6113/96, 28 July 1934; 812.6113/101, 22 May 1935; and 812.6113/106LH, in RG 59, USNA.

49. Dispatch 812.6113/98, 1(?) October 1934, RG 59, USNA. Cárdenas sent Calles into exile in California in 1936, where the former president stayed until 1941.

50. Evelyn Hu-Dehart, *Yaqui Resistance and Survival: The Struggle for Land and Autonomy, 1821–1910* (Madison: University of Wisconsin Press, 1984), pp. 211–219.

51. Norman E. Borlaug, personal interview, 21 June 1989; Norman E. Borlaug, Oral History Interview, RFA, p. 169. See also Samuel N. Dicken, Corn and Wheat in Mexico's Economy, *Journal of Geography* 38 (March 1939): 99–109.

52. Michaels, *The Mexican Election of 1940*.

53. Albert L. Michaels, The Crisis of Cardenismo, *Journal of Latin American Studies* 2, no. 1 (1970): 51–79.

54. Dicken, Corn and Wheat in Mexico's Changing Economy, 99–109.

55. Michaels, The Crisis of Cardenismo, 51–79; Michaels, The Mexican Election of 1940; Norman E. Borlaug, The Impact of Agricultural Research on Mexican Wheat Production, *Transactions of The New York Academy of Sciences* 20, no. 3 (1958): 278–295.

56. Deborah Fitzgerald, Exporting American Agriculture: The Rockefeller Foundation in Mexico, 1943–1953, *Social Studies of Science* 16, no. 3 (1986): 457–483.

57. Rockefeller Foundation, *Strategy for the Conquest of Hunger: Proceedings of Symposium*, 1–2 April 1968, Rockefeller University (New York: Rockefeller Foundation, [1968]), p. 8.

58. Venezian and Gamble, *Agricultural Development of Mexico*, p. 62.

59. W. M. Miner, Mexico, *Agriculture Abroad* 14 (December 1959): 13–17; see also Borlaug, Impact of Agricultural Research on Mexican Wheat Production, 278–295.

60. Cynthia Hewitt de Alcantara, *La modernización de la agricultura mexicana, 1940–1970*, 5th ed. (Mexico: Siglo Veintiuno Editores, 1985), 319 pp.

61. Lourdes Arizpe and Carlota Botey, Mexican Agricultural Development Policy and Its Impact on Rural Women, in *Rural Women and State Policy: Feminist Perspectives on Latin American Agricultural Development*, ed. Carmen Diana Deere and Magdalena Leon (Boulder, Colo.: Westview Press, 1987), pp. 67–83.

62. R. B. Fosdick, *The Story of the Rockefeller Foundation* (New York: Harper, 1952), p. 185.

63. Some Promising Fields for Foundation Activity, compiled by the officers of the Rockefeller Foundation, 21 October 1938, Rockefeller Foundation Archives, RG 3, series 906, box 1, folder 8, Rockefeller Archive Center.

64. Natural Science Program, General Summary, by Warren Weaver, 3 December 1948, RG 3, series 915, box 2, folder 14, RFA.

65. Fosdick, *Story of the Rockefeller Foundation*, p. 185.

66. The early interest in an expansion of agricultural assistance came in correspondence between foundation staff regarding E. C. Stakman. Stakman was the first choice of the foundation to be head of the MAP. Between January and July 1942, he was solicited and gave the matter serious attention. One of his concerns was whether the foundation intended to expand beyond Mexico. President Raymond B. Fosdick told him that the trustees had made no commitments beyond Mexico but that John D. Rockefeller III was interested. Stakman eventually decided to remain at his research at the University of Minnesota, but he recommended J. G. Harrar to be the head of the MAP and agreed to serve as chair of the advisory committee to the MAP. See Frank Blair Hanson to George C. Payne, 29 January 1942, folder 2; FBH, Diary, 1 April 1942, folder 3; Raymond B. Fosdick to E.C. Stakman, 8 April 1942,

folder 3; and E. C. Stakman to Raymond B. Fosdick, 3 July 1942, folder 3; all in RG 1.1, series 323, box 1, RFA.

67. Weaver, *Scene of Change*, p. 99.

Chapter 6

1. Thomas Robert Malthus, *An Essay on the Principle of Population*, ed. Philip Appleman (New York: Norton, 1976), 260 pp.

2. The distinction between "Malthusianism" and "neo-Malthusianism" is this: the former refers to Malthus's sense that human life is likely to remain miserable and poor for most people because of the tendency for population growth rates to exceed growth rates of food production. No hint of political instability or ecological collapse from environmental degradation attends Malthus's original formulation of the population problem. Malthus simply believed that living conditions could not be improved for most people. Neo-Malthusianism, in contrast, posits a consequence of rapid population growth: political and/or ecological instability and collapse. In neo-Malthusianism not only will most people tend to be miserable, hungry, and poor due to population growth rates exceeding productivity growth rates but the very social and environmental systems upon which they depend are likely to collapse to the detriment of everyone, both poor and rich. Keynes initiated a tradition of neo-Malthusianism concerned about *political* instability. Later authors added concerns about environmental or *ecological* instability.

3. The literature that can be considered "neo-Malthusian" is vast. For a few representative samples, see the edited anthology Malthus, *Essay on the Principle of Population*, pp. 230–256.

4. Joseph J. Spengler, The World's Hunger: Malthus 1948, *Academy of Political Science, Proceedings* 23 (1949): 149–167.

5. John Maynard Keynes, *Economic Consequences of the Peace* (New York: Harcourt, Brace, and Howe, 1920), Preface.

6. Ibid., pp. 12–26, 252–298.

7. See preface by J. M. Keynes in Harold Wright, *Population* (New York: Harcourt, Brace, 1923), pp. v–vii.

8. Raymond Pearl, *The Nation's Food: A Statistical Study of a Physiological and Social Problem* (Philadelphia: Saunders, 1920), 274 pp.; Edward M. East, *Mankind at the Crossroads* (New York: Scribner's, 1923) 360 pp.; J. S. Davis, The Specter of Dearth of Food: History's Answer to Sir William Crookes, in *Facts and Factors in Economic History*, ed. Arthur H. Cole, Edwin F. Gay, Arthur Louis Dunham, and Norman Scott Brien Gras, (Cambridge, Mass.: Harvard University Press, 1932), pp. 733–754; and E. Parmalee Prentice, *Hunger and History* (New York: Harper, 1939), 269 pp.

9. A. M. Carr-Saunders, *Population* (London: Oxford University Press, 1925), 112 pp.; Enid Charles, *The Twilight of Parenthood* (New York: Norton, 1934), 226 pp.; Carr-Saunders, *World Population: Past Growth and Present Trends* (Oxford: Clarendon, 1936), 336 pp.; and David Victor Glass, *The Struggle for Population* (Oxford: Clarendon, 1936), 148 pp.

10. Cedric Dover, Population Biology in Bengal, *Population* 2, No. 1 (November 1935): 90–96; J. N. L. Baker, Some Problems of Population in India, *Scottish Geography Magazine* 52 (July 1936): 231–240; D. G. Karve, *Poverty and Population in India* (Oxford: Oxford University Press, 1936), 127 pp.; R. T. Young, Problems of population in India, *Commercial Intelligence Journal* 54 (25 January 1936): 155–157; Birendranath Ganguli, *Trends of Agriculture and Population in the Ganges Valley: A Study in Agricultural Economics* (London: Methuen, 1938), 315 pp.; and Gyan Chand, *India's Teeming Millions: A Contribution to the Study of the Indian Population Problem* (London: George Allen and Unwin, 1939), 374 pp.

11. Raymond Pearl, *The Biology of Population Growth* (New York: Knopf, 1925), 260 pp.; Robert R. Kuczynski, *The Measurement of Population Growth* (London: Sidgwick and Jackson, 1935), 255 pp.; S. Vere Pearson, *The Growth and Distribution of Population* (New York: Wiley, Inc., 1935), 448 pp.

12. Frank Lorimer and Frederick Osborn, *Dynamics of Population: Social and Biological Significance of Changing Birth Rates in the United States* (New York: Macmillan, 1934), 461 pp.; Anonymous, Eugenics, Socialism, and Capitalism, *Eugenics Review* 27 (July 1935): 109–119; A. M. Carr-Saunders, Eugenics in the Light of Population Trends, *Eugenics Review* 27 (April 1935): 11–20.

13. Carr-Saunders, *World Population*, pp. 328–329.

14. It is important to note that the high yields of hybrid maize are a result of the phenomenon called *heterosis*, or the vigor often associated with offspring from genetically different parents. The offspring do not breed true; instead, their progeny segregate into classes that are parental types or recombinations of traits from the two parents. No satisfactory explanation of how heterosis works has ever been provided, so the term is simply a descriptive label without explanatory power.

High-yielding varieties of wheat, the primary focus of this study, have been the result of incorporating new genes into varieties. Because of wheat's strong tendency to self-fertilize within its floral structures, high-yielding wheats breed true (i.e., the offspring of a high-yielding plant will have the same characteristics as its parent). Many of the genes that confer high-yielding capacities to wheat tend to either confer disease resistance to the plant or modify plant architecture such that a higher proportion of the plant is grain rather than stem and leaves. Exactly how the genes function physiologically and biochemically is usually not yet clear, but in contrast to maize the genes responsible for high yields are clearly identifiable by simple Mendelian crosses. Various efforts to create high-yielding wheat with heterosis have been attempted but to date have shown little success compared with the genes conferring disease resistance and modified plant architecture.

15. Biographical sketches of East are found in Deborah Fitzgerald, *The Business of Breeding: Hybrid Corn in Illinois, 1890–1940* (Ithaca, N.Y.: Cornell University Press, 1990), pp. 17–22, 30–41; and Jack Ralph Kloppenburg Jr., *First the Seed: The Political Economy of Plant Biotechnology, 1492–2000* (Cambridge: Cambridge University Press, 1988), pp. 98–100.

16. East, *Mankind at the Crossroads*, p. 69. East put the maximum supportable population size at 5.2 billion. Hybrid maize became a practical reality at about the time East wrote *Crossroads*. Early successful varieties, however, had limited ranges of adaptability, and it was not until after World War II that all of the American maize production areas had suitable varieties of hybrid maize that could outperform the traditional varieties. In addition, heavy rates of nitrogen fertilization were not economically feasible until after the war, and much of the yield increases with hybrid maize are the result of varieties that successfully respond to heavy doses of nitrogen fertilizer. Thus, at the time East wrote, hybrid maize was still more of a potential for yield increases, not a reality.

17. Ibid., pp. 346–351.

18. East argued that the *quality* of population was decreasing because the "best families" were not reproducing themselves. East was fully a part of the racist eugenic thought that so permeated the early twentieth century. See ibid., pp. 110–145.

19. Kenneth M. Ludmerer, *Genetics and American Society: A Historical Appraisal* (Baltimore: Johns Hopkins University Press, 1972), 222 pp.; Daniel J. Kevles, *In the Name of Eugenics: Genetics and the Uses of Human Heredity* (New York: Knopf, 1985), 426 pp.

20. Warren S. Thompson, Population: A Study in Malthusianism. (New York: Columbia University Press, 1915), 216 pp. Biographical details from *American Men of Science*, (New York: R. R. Bowker Company, 1956), Vol. 3, *Social and Behavioral Sciences*, 9th ed., p. 678;

James R. Peacock, The History of the Scripps Foundation (unpublished paper from the Scripps Foundation, May 1989).

21. Warren S. Thompson, *Danger Spots in World Population* (New York: Knopf, 1929), pp. 327–328.

22. Peacock, History of the Scripps Foundation; Frederick Osborn, American Foundations and Population Problems, in *U.S. Philanthropic Foundations: Their History, Structure, Management, and Record,* ed. Warren Weaver (New York: Harper and Row, 1967), pp. 365–374; for biographical details on Whelpton, see P. K. Whelpton, 71, Researcher, Dies, *New York Times,* 7 April 1964, p. 35.

23. Thompson, *Danger Spots in World Population,* pp. 10–15.

24. Warren S. Thompson, *Population and Peace in the Pacific* (Chicago: University of Chicago Press, 1946; reprint Freeport, N.Y.: Books for Libraries Press, 1972), 397 pp.

25. Biographical details on Notestein are from *American Men and Women of Science, Social and Behavioral Sciences,* 13th ed. (New York: Bowker, 1978), p. 894; and *Who Was Who in America,* (Chicago: Marquis, 1985), Vol. 8, p. 303.

26. Regine K. Stix and Frank W. Notestein, Effectiveness of Birth Control: A Study of Contraceptive Practice in a Selected Group of New York Women, *Milbank Memorial Fund Quarterly* 12, no. 1 (January 1934): 57–68.

27. Warren Thompson was well supported by the Rockefeller Foundation, receiving $117,500 between 1922 and 1948. Similarly, Notestein's Office of Population Research received $305,000 between 1944 and 1955 from the foundation. See Population Studies Supported by the RF and Related Rockefeller Boards, November 17, 1948, prepared for CIB by FMR. In RG 3.2, series 900, box 57, folder 310, RFA.

28. Documents 812.5018/27, 21 June 1943; 812.5018/30, 22 July 1943; and 812.5018/35, 4 September 1943, all in RG 59, USNA.

29. Document 812.61/108, 16 July 1940, RG 59, USNA.

30. James H. Kempton and Harry T. Edwards were the Bureau of Plant Industry personnel assigned to do the survey. Their report in September had many of the same recommendations that were contained in the survey report of the Rockefeller Foundation, which was done at the same time. Kempton and Edwards emphasized improvement of technical staff, more agricultural education, expanded plantings of complementary crops, better production methods of domestic crops, better storage of grain, relief from taxes for farmers, and the use of new crops. The conduct of the survey was apparently approved by President Ávila Camacho. See documents 812.61/119, 11 April 1941; 812.61, 2 May 1941; and 812.61/30.5, September 1941, RG 59, USNA.

31. A major problem was the production in Mexico of crops that competed with American crops. For example, in 1942 Mexico proposed a program to increase production of sugar, alcohol, oranges, mangoes, pears, pineapple, flowers, rubber/guayule, and cattle, all for export to the United States. The USDA indicated that only pineapple, flowers, and rubber/guayule were of interest. The other crops would compete with American farmers, would be subject to strict quarantine for Mediterranean fruit fly, or would require machinery that the War Production Board would not approve. Document 812.61/141, 6 May 1942, RG 59, USNA.

32. Documents 812.61/145.5 and 812.61/147, RG 59, USNA.

33. The American embassy in Mexico estimated that sesame increased from 33,000 tons in 1941 to an estimated 59,000 tons in 1943. Peanuts went from 11,000 tons to 58,000 tons, and linseed from 4,000 tons to 9,000 tons during the same time period, Document 812.5018/59, 20 October 1943, RG 59, USNA. The total increase was thus in the neighborhood of 78,000 tons, compared with the 15,000 to 65,000 tons of maize that Mexico wanted from the United States.

34. Document 812.5018/49, 23 September 1943.

35. Documents 812.5018/49, 13 October 1943; 812.5018/57, /57A, /65, all on 15 October 1943; and 812.5018/51, 16 October 1943; RG 59, USNA.

36. Stettinius to FDR, 26 October 1943, 812.5018/60A; Stettinius to Messersmith, 25 October 1943, 812.5018/72A; and document 812.5018/69, 9 November 1943; all in RG 59, USNA.

37. Document 812.5018/150, 6 May 1944, RG 59, USNA. A complete review of the impact on Mexico of cooperation with American war efforts is Stephen R. Niblo, *The Impact of War: Mexico and World War II*, Occasional Paper no. 10 (Melbourne: La Trobe University, Institute of Latin American Studies, 1988), 39 pp.

38. Murray R. Benedict, *Farm Policies of the United States, 1790–1950* (New York: Twentieth Century Fund, 1953), pp. 440–445.

39. Ibid., pp. 447–449.

40. The World Larder, *Economist* 148 (26 May 1945): 688–689.

41. Zones of Occupation, *Economist* 148 (16 June 1945): 798–800.

42. London Bread Shortage, *London Times*, 4 June 1945, p. 2.

43. T. O. Lloyd, *Empire to Welfare State: English History 1906–1985* (Oxford: Oxford University Press, 1986), pp. 266–269.

44. Labour for the Harvest, and Next Year's Crops, *London Times*, 21 July 1945, pp. 2, 5.

45. J. C. R. Dow, *The Management of the British Economy, 1945–60* (Cambridge: Cambridge University Press, 1964), pp. 9, 17–18; Labour Party, National Executive Committee, "Crisis Quiz," 24 September 1947 (Sussex: Harvester Press, 1975), microfiche.

46. CAB 128/5, Conclusion 1(46), 1 January 1946, PRO.

47. CAB 128/5, Conclusion 2(46), 3 January 1946, PRO.

48. CAB 128/5, Conclusion 9(46), 28 January 1946, PRO.

49. CAB 129/6, Paper (46)30, 30 January 1946, PRO.

50. CAB 129/6, Paper (46)30, "The Indian Food Situation," 30 January 1946; CAB 129/6, Paper (46)31, "Home Production of Wheat and Annual Feeding Stuffs Rations in 1946," 30 January 1946; CAB 129/6, Paper CP(46)33, "World Rice Situation," 30 January 1946; and CAB 128/5, Conclusion 14(46), 11 February 1946; all in PRO.

51. CAB 128/5, Conclusion 10(46), 31 January 1946, and CAB 128/5, Conclusion 14(46)4, 11 February 1946, PRO.

52. CAB 129/6, Paper (46)26, "The Effect upon the United Kingdom Wheat/Flour Position of the Agreed Reduction in Shipments to This Country during the Six Months Ending 30th June, 1946," 28 January 1946, PRO.

53. CAB 128/5, Conclusion 19(46)8, 28 February 1946; CAB 128/5, Conclusion 20(46)4, 3 March 1946; CAB 129/7, Paper CP(46)90, "World Food Supplies—Proposed Political Discussions with United States and Canadian Governments," 1 March 1946; and CAB 128/5, Conclusion 21(46), 5 March 1946; all in PRO.

54. The World Food Situation, *London Times*, 22 March 1946, p. 4.

55. CAB 129/8, CP(46)127, "Draft White Paper on the World Food Shortage," 29 March 1946, PRO.

56. CAB 128/5, Conclusion 32(46), 10 April 1946, PRO.

57. CAB 129/9, CP(46)159, "World Wheat Supplies and Their Repercussions upon the United Kingdom Position," 16 April 1946, PRO.

58. CAB 128/5, Conclusion 32(46), 10 April 1946; CAB 128/5, Conclusion 34(46), 12 April 1946; both in PRO.

59. Maurice Hutton to Ben Smith, 23 April 1946, in CAB 129/9, Paper CP(46)183, "United Kingdom Wheat and Flour Position," 1 May 1946, PRO.

60. America's Solemn Obligation in World Famine Crisis, Address by the President, *Department of State Bulletin* 14 (28 April 1946): 716.

61. Report of the Hoover Mission, May 13, 1946, in "May 13, 1946," RG 16 Office of the Secretary of Agriculture, Famine Emergency Program, 1946, General Correspondence, box 6, USNA. The same document was published in *Department of State Bulletin* 14 (May 26, 1946): 897–900.

62. Food crisis: 1946, in "Food Crisis-1946," RG 16, Office of the Secretary of Agriculture, Famine Emergency Program 1946, General Correspondence, box 6, USNA.

63. James W. Young to Famine Emergency Committee, June 1, 1946, in "PEC & DEFP Data," RG 16, Office of the Secretary of Agriculture, Famine Emergency Program, 1946, General Correspondence, box 6, USNA.

64. Amy L. Bentley, Uneasy Sacrifice: The Politics of United States Famine Relief, 1945–1948, *Agriculture and Human Values* 11 no. 4 (fall 1994): 4–18.

65. Charles Wolf Jr., *Foreign Aid: Theory and Practice in Southern Asia* (Princeton, N.J.: Princeton University Press, 1960), pp. 23–26.

66. Virgil W. Dean, Farm Policy and Truman's 1948 Campaign, *Historian* 55, no. 3 (spring 1993): 501–516.

67. William Reitzel, Morton A. Kaplan, and Constance G. Coblenz, *United States Foreign Policy, 1945–1955* (Washington, D.C.: Brookings Institution, 1956), pp. 58–64, 104–107; Wilson D. Miscamble, *George F. Kennan and the Making of American Foreign Policy, 1947–1950* (Princeton, N.J.: Princeton University Press, 1992), pp. 31–32.

68. This formulation of the political strategies of scientific experts is based on the assumption that they seek the power, prestige, and reward that come with assuming a leadership role in political economic life. Experts lose much of their claim to special privilege unless they can accurately predict and solve problems in their field long before nonexperts even know a problem exists. The argument essentially is, "Why bother with experts if their expertise can't keep you out of trouble?" My assumptions about the motivations of scientific experts is based largely on my understanding of professionalism, derived from Magali Sarfatti Larson, *The Rise of Professionalism: A Sociological Analysis* (Berkeley: University of California Press, 1977), 309 pp.

69. Thompson, *Danger Spots in World Population*; Peter J. Donaldson, On the origins of the United States Government's International Population Policy, *Population Studies* 44 (1990): 385–399.

70. Dudley Kirk, Population Changes and the Postwar World, *American Sociological Review* 9 (1944): 28–35.

71. Donaldson, On the Origins of the United States Government's International Population Policy, 385–399. At the time of writing this article, Donaldson was based in Bangkok as the senior associate and representative for south and east Asia of the Population Council.

72. Ibid., 385–399.

73. Population Studies Supported by the Rockefeller Foundation and Related Rockefeller Boards, 17 November 1948, prepared for CIB by FMR, RG 3.2, series 900, box 57, folder 310, RFA; Daniel J. Kevles, *In the Name of Eugenics* (Harmondsworth, England: Penguin, 1985), pp. 200, 202, 208–209, 210. The interest of John D. Rockefeller III in population is detailed in John Ensor Harr and Peter J. Johnson, *The Rockefeller Century* (New York: Scribner's, 1988), pp. 452–467.

74. Raymond B. Fosdick to Thomas Perran, 3 October 1946, RG 3.2, series 900, box 57, folder 310, RFA.

75. Raymond B. Fosdick to Thomas Perran, 3 October 1946, RG 3.2, series 900, box 57, folder 310, RFA.

76. GKS to RBF, 25 November 1946, RG 3.2, series 900, box 57, folder 310, RFA.

77. Harr and Johnson, *The Rockefeller Century*, p. 462.

78. GKS Diary, 2–3 December 1947, RG 3.2, series 900, box 57, folder 310 RFA; Rockefeller Foundation Trustees Meeting, 2–3 December 1947, RG 3.2, series 900, box 57, folder 310, RFA.

79. Minutes, Scientific Directorate, IHD, 14 June 1948, RG 3.2, series 900, box 57, folder 310, RFA. Harr and Johnson, *The Rockefeller Century*, pp. 462–463.

80. Excerpt from Dr. Strode's Diary, 26 October 1948, RG 3.2, series 900, box 57, folder 310, RFA. Marston Bates, *The Prevalence of People* (New York: Scribner's, 1962), p. 1.

81. Special Report to the Board of Scientific Directors of the International Health Division of the Rockefeller Foundation, 4 November 1949, RG 3.1, series 915, box 3, folder 23, Pro-Agr-3, RFA.

82. Special Report to the Board of Scientific Directors of the International Health Division of the Rockefeller Foundation, 4 November 1949.

83. WW to GKS, 7 November 1949, RG 3.2, series 900, box 57, folder 310, RFA.

84. Henry Allen Moe to Chester I. Barnard, 30 November 1949, RG 3.2, series 900, box 57, folder 310, RFA.

85. Chester I. Barnard to Henry Allen Moe, 2 December 1949, RG 3.2, series 900, box 57, folder 310, RFA.

86. Harr and Johnson, *The Rockefeller Century*, pp. 462–463.

87. Marshall C. Balfour, Roger F. Evans, Frank W. Notestein, and Irene B. Taeuber, Public Health and Demography in the Far East, pp. 40–49, in Record Group l.l, series 600, box 2, folder 10, RFA.

88. See various documents, especially WW to CIB, 28 November 1949, RG 3.2, series 900, box 57, folder 310, RFA.

89. Report of the Rockefeller Foundation Commission on Review of the International Health Division, November 1951, RG 3, series 908, P&P, Pro-CR-9, IHD, box 14, folder 147, RFA.

90. For a complete summary of all Rockefeller Foundation grants on population through June 1962, see Population Research Supported by the Rockefeller Foundation to June 1962, RG 3.2, series 900, box 57, folder 312, RFA. For the purposes of this book, the most important were a series of grants to Harvard University, directed by Dr. John E. Gordon. Critics believe these grants were designed to limit population growth in India, although the announced objectives stated that they were intended (1) to study variables affecting fertility, (2) to test a particular mode of fertility control, and (3) to train physicians and health aides. See Note on grants to Dr. John E. Gordon for Research on Population in India, P 56006, EC1/27/56, RG 1.2, series 200, Harvard University, box 45, folder 369, RFA. For the criticism of Gordon's study, see Mahmood Mamdani, *The Myth of Population Control* (New York: Monthly Review Press, 1972), 173 pp.

91. Fairfield Osborn, *Our Plundered Planet* (Boston: Little, Brown, 1948), pp. 49, 67–86.

92. Details of Osborn's life are taken from Fairfield Osborn, the Zoo's No. 1 Showman, Dies, *New York Times*, 17 September 1969; The Talk of the Town, *New Yorker*, 9 March 1957, pp. 23–24; and *Current Biography 1949* (New York: H. W. Wilson, 1950), pp. 463–465.

93. Nelson Rockefeller proposed the office to Franklin Roosevelt and then became the coordinator.

94. Details about Vogt's life are taken from *Current Biography 1953* (New York: H. W. Wilson, 1954), pp. 638–640; and William Vogt, Former Director of Planned Parenthood, Is Dead, *New York Times*, 12 July 1968.

95. William Vogt, *Road to Survival* (New York: William Sloane, 1948), pp. 16, 284–288.

96. CIB to WW, 31 August 1948, RG 3.2, series 900, box 57, folder 310, RFA.

97. Hayes's affection for East was reflected in his dedication to East of his book *Breeding Crop Plants*, 2d ed. (New York: Mc-Graw Hill, 1927), p. vii. In contrast to East, however, Hayes's major interests lay entirely in applied plant breeding, especially in the identification of wheat varieties resistant to rust. See H. K. Hayes, Green Pastures for the Plant Breeder, *Journal of the American Society of Agronomy* 27 (1935): 957–962.

98. The World Food Problem, Agriculture, and the Rockefeller Foundation, by Advisory Committee for Agricultural Activities, 21 June 1951, RG 3, series 915, box 3, folder 23, RFA. The committee consisted of Warren Weaver, Elvin Stakman (chair), Richard Bradfield, and Paul Mangelsdorf, that is, the same people who had originally planned the Mexican Agricultural Program. Also involved with the committee's work were J. George Harrar and Harry Miller.

99. The World Food Problem, Agriculture, and the Rockefeller Foundation, by Advisory Committee for Agricultural Activities, pp. 3–7; Warren Weaver to Richard Bradfield, Paul C. Mangelsdorf, E. C. Stakman, J. G. Harrar, Harry M. Miller Jr., 14 June 1951, RG 3, series 915, box 3, folder 20, RFA.

100. Karl T. Compton to Chester I. Barnard, 25 June 1951, RG 3, series 915, box 3, folder 20, RFA.

Chapter 7

1. Frederick D. Richey, Why Plant Research? *Journal of the American Society of Agronomy* 29 (1937): 969–977.

2. W. M. Jardine, The Agronomist of the Future, *Journal of the American Society of Agronomy* 9 (1917): 385–390.

3. See, for example, the presidential addresses of Charles E. Thorne, The Work of the American Agronomist, *Journal of the American Society of Agronomy* 7 (1915): 257–265, and of F. S. Harris, The Agronomist's Part in the World's Food Supply, *Journal of the American Society of Agronomy* 12 (1920): 217–225.

4. Richard Bradfield, Our Job Ahead, *Journal of the American Society of Agronomy* 34 (1942): 1065–1075.

5. Food and Agriculture Organization of the United Nations, *FAO: The First 40 Years* (Rome: Food and Agriculture Organization of the United Nations, 1985), p. 5.

6. Murray R. Benedict, *Farm Policies of the United States, 1790–1950* (New York: Twentieth Century Fund, 1953), pp. 397–398.

7. Food and Agriculture Organization of the United Nations, *FAO*, pp. 6–11; John Abbott, *Politics and Poverty: A Critique of the Food and Agriculture Organization of the United Nations* (London: Routledge, 1992), pp. 1–6.

8. Harry S. Truman, inaugural address, reprinted in the *New York Times*, 21 January 1949.

9. Ibid.

10. Ibid.

11. It is interesting to note that former vice president Henry A. Wallace, who ran against Truman in 1948, was reported to have reacted quite negatively to the militaristic and war-like tone he saw in Truman's address (*New York Times*, 21 January 1949). Wallace, it will be remembered, was of prime importance in helping the Rockefeller Foundation begin the Mexican Agricultural Program (see chapter 5).

12. See, for example, Roger C. Riddell, *Foreign Aid Reconsidered* (London: James Currey, 1987), 309 pp.; Brian H. Smith, *More Than Altruism: The Politics of Private Foreign Aid* (Princeton, N.J.: Princeton University Press, 1970), 352 pp.; Robert E. Wood, *From Marshall Plan to Debt Crisis: Foreign Aid and Development Choices in the World Economy* (Berkeley: University of California Press, 1986), 400 pp.

13. Wood, *From Marshall Plan to Debt Crisis*, pp. 29–67.

14. Shivaji Ganguly, *U.S. Policy toward South Asia* (Boulder, Colo.: Westview Press, 1990), pp. 20–23.

15. "Hoover's Statement on India's Food Situation," New Delhi, 27 April 1946, 3 pp., press release, 845.6131/4–2746, Record Group 59, USNA.

16. W. L. Clayton to the Agent General for India, Sir Girja, 2 May 1946, 845.61311/5–246; WLC to Mr. Stillwell, 2 May 1946, 845.61311/5–246; E. G. Gale to Mr. Clayton, 14 October 1946, 845.6131/10–1446; all in Record Group 59, USNA.

17. George Blyn, *Agricultural Trends in India, 1891–1947: Output, Availability, and Productivity* (Philadelphia: University of Pennsylvania Press, 1966), pp. 240–243, Appendix Table 5c, p. 334.

18. V. K. R. V. Rao, to J. A. Stilwell, 27 January 1947, 845.61311/1–2347, Record Group 59, USNA.

19. Howard Donovan to the Secretary of State, 17 June 1947, Dispatch no. 1153, New Delhi, 845.6131/6–1747, Record Group 59, USNA.

20. Macdonald to Secretary of State, 17 June 1947, 845.61311/6–1747; Howard Donovan to the Secretary of State, 17 July 1947, 845.61311/7–17447; Grady to the Secretary of State, 14 October 1947, 845.61311/10–1447; John H. Macdonald to the Secretary of State, 28 October 1947, 845.61/10–2847; all in Record Group 59, USNA.

21. George R. Merrell to the Secretary of State, 11 January 1946, Dispatch no. 436, 845.61A/1–1146, Record Group 59, USNA.

22. George R. Merrell to the Secretary of State, 19 September 1946, Dispatch no. 766, 845.61A/9–1946, Record Group 59, USNA.

23. Howard Donovan to the Secretary of State, 16 December 1947, Dispatch no. 420, 845.61A/12–1647; Charge d'Affaires to Secretary of State, 11 June 1948, 845.61A/6–1148; Marshall to AMEMBASSY 19 July 1948, 845.61A/6–1148; Secretary of State to Ambassador of India, 1 November 1948, 845.61A/11–1648; Ambassador of India to the Secretary of State, 3 March 1949, 845.61A/3–349; all in Record Group 59, USNA.

24. Howard Donovan to The Secretary of State, 26 April 1949, Dispatch No. 341, 845.61/4–2649, Record Group 59, USNA. The conversations with Raja were on 20 April and 22 April 1949.

25. Waynick to Direct Point 4 Program, *New York Times*, 18 May 1950.

26. Howard Donovan to the Secretary of State, 26 April 1949, Dispatch no. 344, 845.61A/4–2649, Record Group 59, USNA.

27. Loy W. Herderson to the Secretary of State, 6 May 1949, Dispatch no. 371, 845.61A/5–649, Record Group 59, USNA.

28. Howard Donovan to the Secretary of State, 13 June 1949, Dispatch no. 502, 845.61/6–1349, Record Group 59, USNA.

29. Howard Donovan to the Secretary of State, 17 June 1949, Dispatch no. 515, 845.61A/6–1749, Record Group 59, USNA.

30. Texts of Truman orders to Implement Point 4 Plan, *New York Times*, 9 September 1950.

31. Joe Alex Morris, *Nelson Rockefeller, A Biography* (New York: Harper and Brothers, 1960), pp. 111–133.

32. Ibid., pp. 184–188, 227–233.

33. Ibid., pp. 141–142; John Ensor Harr and Peter J. Johnson, *The Rockefeller Century* (New York: Scribner's, 1988), pp. 440–441.

34. Martha Dalrymple, *The AIA Story: Two Decades of International Cooperation* (New York: American International Association for Economic and Social Development, 1968), pp. 10–11.

35. Morris, *Nelson Rockefeller*, pp. 271–272; Harr and Johnson, *The Rockefeller Century*, pp. 443–444.

36. Morris, *Nelson Rockefeller*, pp. 270–278.

37. Weighed as Head Adviser to the Point 4 Program, *New York Times*, 3 November 1950.

38. H. G. Bennett Heads the Point 4 Program, *New York Times*, 15 November 1950.

39. Nelson A. Rockefeller Is Named by Truman to Head Advisory Board on Point Four Plan, *New York Times*, 25 November 1950.

40. International Development Advisory Board, *Partners in Progress* (New York: Simon and Schuster, [1951]), 120 pp.

41. Ibid., p. 25.

42. Ibid., pp. 16–27.

43. Ibid., pp. 29–40.

44. Ibid., p. 39.

45. *Partners in Progress* was clear that underdeveloped areas should receive economic assistance at about $500 million per year, exclusive of Greece, Turkey, and Korea. In addition, famine relief for India was excluded from this figure, as were programs for economic recovery in Europe, Japan, and funds for NATO (International Development Advisory Board, *Partners in Progress* pp. 69–70). In May 1951 the administration sent Congress a proposed budget for mutual security programs costing $8.5 billion. Of this, an estimated $512 million was for the Technical Cooperation Administration that handled the Point Four program. Most of the money requested, $6.3 billion, was for military assistance. See Hoffman Outlines 4-Front Program to Win "Cold War," *New York Times*, 18 July 1951.

46. Point 4 transfer to E.C.A. Favored, *New York Times*, 13 June 1951.

47. Point 4 Faces Loss of Identity by Agency Shift to E.C.A., *New York Times*, 11 June 1951; Point 4 May Be Lost in an Expanded E.C.A., *New York Times*, 24 June 1951; Hoffman Outlines 4-Front Program to Win "Cold War," *New York Times*, 18 July 1951.

48. Rockefeller Quits Point Four Board, *New York Times*, 6 November 1951.

49. Point 4 Policy to Be Kept, Will Continue along Lines Set by Bennett before Death, *New York Times*, 27 December 1951.

50. Text of Truman's state of the union message, *New York Times*, 10 January 1952.

51. Ford Fund Offers Its Own "Point 4," *New York Times*, 2 August 1951.

52. Douglas Ensminger, Oral history transcript, A.1., Introduction, FFA.

53. Charles F. Brannan to Everett McKinley Dirksen, April 26, 1951, Record Group 16, Office of the Secretary of Agriculture, Correspondence 1951, Foreign Relations 2 Foreign Assistance, box 2031, USNA.

54. New Delhi to Department of State, 7 June 1951, 891.20/6–751, Record Group 59, USNA.

55. Douglas Ensminger, Oral history transcript, A.1. Introduction, FFA.

56. JHW to RBF, 22 May 1947; RBF to JHW, 27 May 1947; AG to RBW, 16 January 1948; GKS Diary, 25 March 1948; GKS Diary, 1 June 1948; GKS Diary, 23 February 1949; Staff Conference—Minutes, 16 May 1949; Karl T. Compton to WW, 3 November 1949; WW to Karl Compton, 9 November 1949; all in Arch 2, 460 Program and Policy, unprocessed material, RFA.

57. John H. Perkins, The Rockefeller Foundation and the Green Revolution, 1941–1956, *Agriculture and Human Values* 7 (summer–Fall 1990): 6–18.

58. Excerpt from W. Weaver to A. T. Mosher, 26 December 1951, Arch 2, 460 Program and Policy, unprocessed material, RFA.

59. J. G. Harrar, Paul C. Mangelsdorf, and Warren Weaver, Notes on Indian Agriculture ([New York: Rockefeller Foundation], 1952), 29 pp. Copy in "I.A.P., Admin, Org & Policy, II, 1959–1960," box 10c18a, R.G. 6.7, RFA.

60. Perkins, The Rockefeller Foundation and the Green Revolution, 6–18.

61. Ibid., 6–18.

62. Details of these visits may be found in RG 1.2, series 464, box R962, folder: Agriculture 1953–1955, RFA.

63. Memorandum of conversation between Drs. Uppal and Pal and Drs. Weaver & Harrar: Delhi, Saturday, 10 October 1953, RG 1.2, series 464, box R962, file: Agriculture 1953–1955, RFA.

64. Items included: construct tubewells, improve the supply of fertilizer, provide iron and steel for agricultural implements, conduct locust and malaria control, construct large-scale dams, perform soil surveys and forestry research, train village level workers, and develop fisheries (New Delhi to Department of State, 21 January 1953, RG 59, 891.20/1–2153, USNA).

65. C. B. Mamoria, *Agricultural Problems of India*, 9th ed. (Allahabad-Delhi: Kitab Mahal, 1979), pp. 833–837; M. S. Randhawa, *A History of Agriculture in India*, vol. 4 (New Delhi: Indian Council of Agricultural Research, 1986), pp. 61–68.

66. Notes on Discussion, *India Conference*, Including Pakistan and Ceylon, 26 March 1952, pp. 11–12, Arch 2, 460, Program and Policy, unprocessed material, RFA.

67. E. C. Stakman, Richard Bradfield, and Paul C. Mangelsdorf, *Campaigns against Hunger* (Cambridge, Mass.: Harvard University Press, 1967), p. 34.

68. P. N. Thapar to Warren Weaver, 5 October 1955, RG 1.2, series 464, box R962, file: Agriculture 1953–1955, RFA.

69. J. G. Harrar to Richard Bradfield, 1 November 1955, RG 1.2, series 464, box R962, file: Agriculture 1953–1955; J. G. Harrar to P. N. Thapar, 5 January 1956, RG 1.2, series 464, box R962, file: Agriculture 1956, RFA.

70. FMR to JGH, RFC, KW, AHM, JA, 5 April 1956, RG 1.2, series 464, box R962, file: Agriculture 1956, RFA.

71. Survey Shows How Point Four Program Tackles Basic Problems in Under-developed East, *New York Times*, 12 January 1953.

72. Virgil W. Dean, Farm Policy and Truman's 1948 Campaign, *Historian* 55, no. 3 (spring 1993): 501–516.

73. David E. Wright, Alcohol Wrecks a Marriage: The Farm Chemurgic Movement and the USDA in The Alcohol Fuels Campaign in the Spring of 1933, *Agricultural History* 67 no. 1 (1993): 36–66.

74. Edward C. Banfield, Planning under the Research and Marketing Act of 1946: A Study in the Sociology of Knowledge, *Journal of Farm Economics* 31, pt. 1 (1949): 48–75.

75. Charles F. Brannan, Point Four Depends on Research, Secretary Brannan Tells Conference, *Chemurgic Digest* 9 (May 1950): 7–10.

76. Vernon Ruttan, The Politics of U.S. Food Aid Policy: A Historical Review, in *Why Food Aid?*, ed. Vernon W. Ruttan (Baltimore: Johns Hopkins University Press, 1993), pp. 2–36. Ruttan notes that the State Department at first resisted the use of food for strategic purposes because it favored policies of free trade. In time, however, the use of food to get other countries to fall in line with U.S. policy came to be seen as useful.

Chapter 8

1. George Blyn, *Agricultural Trends in India, 1891–1947: Output, Availability, and Productivity* (Philadelphia: University of Pennsylvania Press, 1966), Table 5.1, p. 96, Table 5.4, p. 104.

2. Ibid., Table 5.1, p. 96.

3. British India was a net exporter of food grains until the crop year 1919–20. Only in 1896–97, 1897–98, 1900–1901, and 1908–9 were food grain imports greater than exports.

After 1919–20, British India was a net exporter in only four crop years: 1922–23, 1923–24, 1924–25, and 1942–43. It is interesting to note that the last year, 1942–43, included the period of the Bengal famine of 1943 (i.e., exports were larger than imports at a time of horrible starvation in Bengal). The turning point in Indian food policy resulting from the Bengal famine is discussed in more detail later in the chapter. Figures are from Blyn, *Agricultural Trends in India*, Appendix Table 5C, p. 334.

4. M. Zarkovic, *Issues in Indian Agricultural Development* (Boulder, Colo.: Westview Press, 1987), pp. 8–9.

5. R. N. Chopra, *Food Policy in India: A Survey* (New Delhi: Intellectual Publishing House, 1988), pp. 25–26. Daniel and Alice Thorner concur with the analysis that commercial crops, especially groundnut, cotton, sugarcane, and tea, nearly doubled in the period 1893–94 to 1945–46: "What is striking is that this increase in commercial crops was achieved largely at the expense of food crop production." See Daniel Thorner and Alice Thorner, *Land and Labour in India* (Bombay: Asia Publishing House, 1962), pp. 82–112, quote on p. 103.

6. Figures calculated from Blyn, *Agricultural Trends in India*, Appendix Table 3A, pp. 253, 258, and Appendix Table 4D, p. 326. The population figure used for 1896 was an average of the population estimates for 1891 and 1901: 216.79 million. The estimated population for 1941 was 289.21 million. Rice and wheat production figures used were the average annual production for 1891–1900 and for 1937–46.

7. This calculation of energy supplies is based on long-grained raw rice at 12 percent water having 670 kilocalories per 185 grams. U.S. Department of Agriculture, *Nutritive Value of Foods*, Home and Garden Bulletin no. 72 (Washington, D.C.: Government Printing Office, 1981, Table 2, item no. 483, p. 22.

8. Jean Mayer, Management of Famine Relief, *Science* 188 (9 May 1975): 571–577, esp. Table 1, p. 576.

9. India, Famine Inquiry Commission, *Report on Bengal* (Delhi, 1945), pp. 103–107; Chopra, *Food Policy in India*, pp. 41–43.

10. The Great Bengal Famine has been reviewed in many places, for example, Chopra, *Food Policy in India*, pp. 28–60; Amartya Sen, *Poverty and Famines: An Essay on Entitlement and Deprivation* (Delhi: Oxford University Press, 1981), pp. 52–83; and Henry Knight, *Food Administration in India, 1939–1947* (Stanford, Calif.: Stanford University Press, 1954), pp. 45–105.

11. Knight, *Food Administration in India*, p. 15.

12. Ibid., pp. 7–10.

13. Sen, *Poverty and Famines*, p. 56; Knight, *Food Administration in India*, p. 77.

14. Jawaharlal Nehru, *The Discovery of India* (New Delhi: Indraprastha Press, 1981), p. 16.

15. Chopra, *Food Policy in India*, pp. 28–32.

16. Ibid., pp. 69, 72.

17. Ibid., pp. 55–58.

18. Shriram Maheshwari, *Rural Development in India: A Public Policy Approach* (New Delhi: Sage, 1985), pp. 29–30.

19. N. Srinivasan and S. P. Singh, The Growth of the State Departments of Agriculture, in *Agricultural Administration in India*, ed. N. Srinivasan (New Delhi: Indian Institute of Public Administration, 1969), p. 47.

20. Imran Ali, *The Punjab under Imperialism, 1885–1947* (Delhi: Oxford University Press, 1989), pp. vii–5.

21. J. J. De Blij and Peter O. Muller, *Geography, Regions and Concepts*, 6th ed. (New York: Wiley, 1991), pp. 451–459.

22. Stanley Wolpert, *A New History of India*, 2d ed. (New York: Oxford University Press, 1982), pp. 2–14

23. Ali, *The Punjab under Imperialism*, pp. 3–5, 109–157.

24. V. Norman Brown, *The United States and India and Pakistan* (Cambridge, Mass.: Harvard University Press, 1963), pp. 167–169.

25. Quote and information on division of irrigation works is from S. Thirumalai, *Post-War Agricultural Problems and Policies in India* (New York: Institute of Pacific Relations, 1954), pp. 48–55. Other authors concur with Thirumalai's conclusions and note that the Indian section of Punjab had insufficient food due to poor irrigation works, poor livestock, low literacy, and low urban population compared with the Pakistani section of Punjab. See Gurdev Singh Gosal and B. S. Ojha, *Agricultural Land-Use in Punjab: A Spatial Analysis* (New Delhi: Indian Institute of Public Administration, 1967), pp. 1–2. For another concurrence, see also Bhupendra B. Hooja, Planning Priorities and Development Policies with Reference to Indian Agriculture, in *Agriculture Administration in India*, ed. Hoshiar Singh (Jaipur: Printwell Publishers, 1986), pp. 22–60.

26. K. L. Dua, ed., *Review of Agricultural Research in the Punjab, from 1947 to 1963.* vol. 6, pt. 9, *Plant Breeding and Agronomy* (Ludhiana: Punjab Agricultural University, 1967), pp. i–ii.

27. T. M. P. Mahadevan, Social, Ethical, and Spiritual Values in Indian Philosophy, in *The Indian Mind: Essentials of Indian Philosophy and Culture*, ed. Charles. A. Moore (Honolulu: University of Hawaii Press, 1967), pp. 152–172.

28. William Harrison Moreland, *Agrarian System of Moslem India: A Historical Essay with Appendices*, 2d ed. (Delhi: Oriental Books Reprint, 1968), pp. 3, 207–208.

29. K. S. Shelvankar, *The Problem of India* (Harmondsworth, England: Penguin, 1940), pp. 49–62.

30. N. C. Mehta, State Leadership in Agrarian Economy, in *Developing Village India: Studies in Village Problems* ed. M. S. Randhawa (Bombay: Orient Longmans, 1951), pp. 23–26.

31. Quotation is from Singh's Foreword to Randhawa, *Developing Village India*, p. vii. In a biographical sketch, Singh was noted to own one of the largest farms in the Punjab.

32. India, National Commission on Agriculture, *Report, Part II, Policy and Strategy* (New Delhi: Ministry of Agriculture and Irrigation, 1976), p. 31; Francine R. Frankel, *India's Political Economy, 1947–1977: The Gradual Revolution* (Princeton, N.J.: Princeton University Press, 1978), pp. 113–155.

33. Frankel, *India's Political Economy*, pp. 71–76.

34. Chopra, *Food Policy in India*, pp. 44–53; Joshwantrai Jayantilal Anjaria, D. T. Lakdawala, and D. R. Samant, *Price Control in India with Special Reference to Food Supply* (Bombay: Popular Book Depot, 1946), p. 43; S. C. Chaudri, Price Support: Only the First Step, in *Foundations of Indian Agriculture*, ed. Vadilal Dagli (Bombay: Vora, 1968), pp. 255–260.

35. Chopra, *Food Policy in India*, pp. 53–60. Chopra notes that over 3 million tons of food grains were needed, compared with the over 2 million imported.

Legalization of the Congress Party, agitation over the illegitimacy of British judgment of the Indian National Army (that had sided with Japan and Germany), a rebellion in the Royal Indian Navy, and eventual formation of an interim government by Nehru in September 1946 were some of the major highlights of this intense period in Indian politics. (See Bipan Chandra, Mridula Mukherjee, Aditya Mukherjee, K. N. Panikkar, and Sucheta Mahajan, *India's Struggle for Independence* (New Delhi: Penguin, 1989), pp. 473–493; Stanley Wolpert, *A New History of India* (New York: Oxford University Press, 1982), pp. 339–345.

36. India, Fiscal Commission, *Report of the Fiscal Commission 1949–1950*, vol. 1 (Delhi: Manager of Publications, 1950), pp. 22–23. After partition in 1947, Pakistan held about 40 percent and 80 percent of cotton and jute production, respectively, compared with pre-partition British India (see ibid., pp. 13–14).

37. J. C. R. Dow, *The Management of the British Economy, 1945–1960* (Cambridge: Cambridge University Press, 1964), pp. 42–44; Text of Cripps' Speech Announcing Devaluation of the Pound, *New York Times*, 29 September 1949.

38. India, Fiscal Commission, *Report of the Fiscal Commission* vol. 1, pp. 43–44; V. Norman Brown, *The United States and India and Pakistan* (Cambridge, Mass.: Harvard University Press, 1963), pp. 169–170

39. Office of the Economic Adviser to the Government of India *Monthly Survey of Business Conditions in India*, 19, no. 2 (1951): 49–52.

40. Brown, *The United States and India and Pakistan*, pp. 180–181; Judith M. Brown, *Modern India: The Origins of an Asian Democracy* (Delhi: Oxford University Press, 1985), pp. 366–369; Wolpert, *New History of India*, pp. 352–354, 374–376, 387–390.

41. Chopra, *Food Policy in India*, pp. 62–64; Virendra Kumar, *Committees and Commissions in India, 1947–73*, vol. 1 (Delhi: K. K. Publishing House, 1975), pp. 24–33.

42. Some uncertainty surrounds the question of how large was the deficit in domestic production. In its *Interim Report*, the Foodgrains Policy Committee of 1947 stated that the system of controlled purchases and distribution of food was likely to perpetuate a sense of food crisis "in an artificial manner" (p. 11). The committee believed the statistics of production were too small and the estimates of deficit too large. Furthermore, the prices at which the central government was procuring grains were too low, which exacerbated the tendency of producers of grains not to report and deliver their surplus production. Government announcements of over 140 million persons being under rationing also vastly exaggerated the sense of crisis because the actual government procurement of around 4 million tons of grain per year was in no way adequate to feed such a multitude. In fact, most people said to be under rationing were obtaining much of their food in the uncontrolled market. Building domestic production, reducing imports, saving foreign exchange, and gradual disengagement from running the food system were the committee's preferences for reform of India's food situation. See Foodgrains Policy Committee, *Interim Report* (Delhi: Manager of Publications, 1948), pp. 3–13.

43. Chopra, *Food Policy in India*, pp. 64–67, 71.

44. Frankel, *India's Political Economy*, pp. 94–100.

45. Approximately half of the owned, rural land in India was held by about 8 percent of the rural households. About 60 percent of rural households each held 2.5 acres or less and about 6 percent of the owned area. Twenty-two percent of households each owned less than 0.005 acres. Thus landownership in India at independence was highly skewed and unequal. Comprehensive statistics on landownership did not become available until 1955. See India, The National Sample Survey, *First Report on Land Holdings, Rural Sector* (Delhi: Government of India, [1955]), eighth round: July 1954–March 1955, no. 10, pp. iv, 14.

46. Maheshwari, *Rural Development in India*, pp. 147–149.

47. S. Thirumalai, *Post-War Agricultural Problems and Policies in India* (Bombay: Indian Society of Agricultural Economics, and Institute of Pacific Relations, 1954), 280 pp.; Daniel Thorner and Alice Thorner, Emergence of an Indian Economy, 1760–1960, in *Land and Labour in India* (Bombay: Asia Publishing House, 1962, pp. 51–69).

48. Lloyd I. Rudolph and Susanne Hoeber Rudolph, *In Pursuit of Lakshmi: The Political Economy of the Indian State* (Chicago: University of Chicago Press, 1987), pp. 314–315.

49. Ibid., p. 315.

50. Department of State, Office of Intelligence Research, *Land Reform in India*, OIR Report no. 5390, 10 November 1950, Record Group 59, USNA.

51. Quoted from Maheshwari, *Rural Development in India*, p. 144.

52. Although the Grow More Food campaign was not successful in ending imports of food grains by India, several commentators noted that the steady increase in the Indian population plus some improvement in living standards both tended to increase demand for food. Thus the GMF program was always aiming at a target—"needed domestic production"—that itself was continually increasing.

53. Chopra, *Food Policy in India*, pp. 68–74, 395.

54. Records of the Indo-U.S. discussions in 1949 are in 845.61311, starting with 7–2249, Record Group 59, U.S. National Archives. See also a summary report, Department of State, *India: Problems and Prospects*, OIR Report no. 5052, 4 October 1949, 53 pp.

55. W. Burns, *Technological Possibilities of Agricultural Development in India* (Lahore: Government Printing, 1944), 127 pp.

56. B. P. Pal, Organizations, Management and Progress of Agricultural Research in India, *Indian Journal of Public Administration* 15, no. 3 (July–September 1969): 374–384; Indian Council of Agricultural Research, *50 Years of Agricultural Research and Education* (New Delhi: Indian Council of Agricultural Research, 1979), pp. 163–169.

57. Advisory Board, Imperial Council of Agricultural Research, *Memorandum on the Development of Agriculture and Animal Husbandry in India* (Delhi: Government of India Press, 1944), 75 pp.

58. Indian Council of Agricultural Research, *50 Years of Agricultural Research and Education*, Appendix 4, pp. 166–167. Research expenditures for 1941–42 through 1944–45 were not available in the preparation of *50 Years*. A figure of 1.4 million rupees was spent by ICAR institutes in 1940–41, and to this was added an estimated 0.4 million rupees spent by the ICAR on ad hoc schemes (latter figure calculated from Appendix 1, pp. 163–164); thus the total for 1940–41 was estimated to be 1.8 million rupees.

59. Figures are calculated from Indian Council of Agricultural Research, *50 Years of Agricultural Research and Education*, Appendices 2 and 4, pp. 163–167.

60. Ibid., Appendix 2, p. 165.

61. Figures calculated from ibid.

62. B. P. Pal, Organisation and Growth, in Indian Council of Agricultural Research, *50 Years of Agricultural Research and Education*, pp. 5–14; Pal, Organization, Management and Progress of Agricultural Research in India, 374–384. As developed later, Pal served as the first director general of a vastly reorganized and strengthened ICAR starting in 1965.

63. Srinivasan and Singh, The growth of the state departments of agriculture, pp. 46–66.

64. Rates of increase are calculated from ibid.

65. Dua, *Review of Agricultural Research in the Punjab*, pp. 2–6; J.P. Tandon and M.V. Rao, Organisation of Wheat Research in India and Its Impact, in *Twenty-five Years of Co-ordinated Wheat Research, 1961–1986*, ed. J. P. Tandon and A. P. Sethi (New Delhi: Indian Agricultural Research Institute, [1986]), p. 1.

66. E. W. Sprague, N. L. Dhawan, and L. R. House, Increased Food Production with Hybrid Maize, *Current Science* 29, no. 8 (1960): 295–297.

67. Indian Agricultural Research Institute, *Report of the Achievement Audit Committee on Indian Agricultural Research Institute, 1966–1971* (New Delhi: Indian Agricultural Research Institute, n.d.), pt. 1, p. 3; India, National Commission on Agriculture, *Report, Part XI, Research, Education and Extension* (New Delhi: Ministry of Agriculture and Irrigation, 1976), pp. 195–205; O. P. Gautam, Agricultural Education: Development of Agricultural

universities, in Indian Council of Agricultural Research, *50 Years of Agricultural Research and Education*, pp. 135–141.

68. M. K. Gandhi, *Food Shortage and Agriculture* (Ahmedabad: Navajivan Publishing House, 1949), pp. iii–v. In *Harijan* of 22 September 1946, Gandhi addressed the need for improved production methods and decreased production of commercial crops (article reprinted in *Food Shortage and Agriculture*). See also C. D. Deshmukh, *Economic Developments in India, 1946–1956* (Bombay: Asia Publishing House, 1957), p. 23.

69. J. S. Mathur and A. S. Mathur, *Economic Thought of Mahatma Gandhi* (Allahabad: Chaitanya Publishing House, 1962), pp. 230–231; Shanti S. Gupta, *Economic Philosophy of Mahatma Gandhi* (Delhi: Ashok Publishing House, n.d.), 224 pp.

70. New Delhi to Department of State, 16 January 1950, 891.20/1–1650, Record Group 59, USNA. This note from the American embassy reported a speech by K. L. Panjabi on problems of Indian agricultural production.

71. Sahay's statement is attached to Howard Donovan to the Secretary of State, 17 June 1949, 845.6131/6–1747, Record Group 59, USNA. That the amount of grain claimed to be needed by India was viewed skeptically by the Americans is suggested by conversations and correspondence in late 1946 and early 1947 between Indian and American government officials in Washington (Memo of conversation, 17 December 1946, 845.6131/12–1746; V. K. R. V. Rao to J. A. Stilwell, 23 January 1947, 845.6131/1–2347; both in Record Group 59, USNA.)

72. Macdonald to the Secretary of State, 17 June 1947, 845.6131/6–1747; Howard Donovan to the Secretary of State, 17 July 1947, 845.6131/7–1747; Grady to the Secretary of State, 14 October 1947, 845.6131/10–1447; Macdonald to the Secretary of State, 28 October 1947, 845.61/10–2847; all in Record Group 59, USNA.

73. Memo of conversation, 22 July 1949, 845.6131/7–2249, Record Group 59, USNA.

74. Howard Donovan to the Secretary of State, 17 August 1949, 845.61311/8–1749, Record Group 59, USNA.

75. Howard Donovan to the Secretary of State, September 7, 1949, 845.61311/9–749, Record Group 59, USNA.

76. For example, in his response to the 3 September meeting, Taylor reported that wheat at Allahabad's open market price was 29 rupees per maund but procurement price was 13 rupees 8 paise per maund (1 maund was 82.28 pounds).

77. Howard Donovan to the Secretary of State, 17 August 1949, 845.61311/8–1749, Record Group 59, USNA National Archives, Washington, D.C.

78. Memo of conversation, 18 October 1949, Subj: Barter Deal with India—Wheat for Strategic Materials, 845.61311/10–1849; Acheson to AMEMBASSY, New Delhi, 20 October 1949, 845.61311/10–2049; Donovan to the Secretary of State, 27 October 1949, 845.61311/10–2749 (this dispatch specifically mentioned that American grain shipments would have a significant effect politically in India and would allow the government of India to answer its socialist and communist critics that it was doing something); Memo of conversation, 7 November 1949, Discussion of the Wheat-Manganese Barter Deal with India, 845.61311/11–749; Webb (Acting) to AMEMBASSY, New Delhi, 8 November 1949, 845.61311/10–2749; Donovan to the Secretary of State, 13 November 1949, 845.61311/11–1349; Acheson to AMEMBASSY, 17 November 1949, 845.61311/11–1749; Memo of conversation, 18 November 1949, 845.61311/11–1849; all in Record Group 59, USNA.

79. B. K. Nehru to Donald D. Kennedy, 28 December 1949, 845.61311/12–2949, Record Group 59, USNA.

80. New Delhi to Department of State, 6 February 1950, 891.20/2–650; Madras to Department of State, May 2, 1950, 891.20/5–250; Madras to Department of State, 18 May 1950, 891.20/5–1850; all in Record Group 59, USNA.

81. Calcutta to Department of State, 21 November 1950, 891.20/11–2150, Record Group 59, USNA.

82. New Delhi to Department of State, 30 December 1950, 891.20/12–3050, Record Group 59, USNA.

83. New Delhi to Department of State, 26 September 1950, 891.20/9–2650; and Memo of conversation, 4 October 1950, 891.20/10–450; both in Record Group 59, USNA.

84. Chopra, *Food Policy in India*, p. 124.

85. Mr. McGhee (NEA) through Mr. Matthews (G) to the Secretary, 27 December 1950, 891.20/12–1950, Record Group 59, USNA.

86. Ibid.

87. Chopra, *Food Policy in India*, pp. 76–94, 103–124, 384.

88. Frankel, *India's Political Economy*, pp. 113–132.

89. S. K. Dey, *Nilokheri* (Bombay: Asia Publishing House, 1962), 128 pp.

90. K. A. P. Menon, *Indian Agriculture: Administrative and Organisational Constraints* (New Delhi: Sreedeep Publications, 1987), p. 39; J. G. Harrar, Paul C. Mangelsdorf, and Warren Weaver, "Notes on Indian Agriculture," 11 April 1952, "I.A.P., Admin, Org & Policy, II, 1959–60," Box 10c18a, Record Group 6.7, RFA. Holmes was not entirely happy with the conditions of his work in Etawah, especially his payment in rupees rather than dollars, according to his reports to officials at the American embassy in New Delhi. See Howard Donovan to the Secretary of State, 26 April 1949, 845.61A/4–2649, Record Group 59, USNA. The conclusion that the Etawah project formed the model for the subsequent efforts in community development is given by Maheshwari, *Rural Development in India*, pp. 179–180. For Nehru's interest, see New Delhi to Department of State, 7 June 1951, 891.20/6–751, Record Group 59, USNA. Daniel and Alice Thorner noted that although yields increased in Etawah District, one of the most innovative and cooperative farmers (Anand Madho Shukla) did not extend his innovation to social reform. He owned 90 acres, 40 of which were sharecropped, with him taking 60 to 70 percent of the crop. See Daniel Thorner and Alice Thorner, The Agrarian Problems in India Today, in *Land and Labour in India* (Bombay: Asia Publishing House, 1962), pp. 3–13.

91. Douglas Ensminger, Oral history transcript, A.1., Introduction, FFA.

92. V. T. Krishnamachari, as quoted in Randhawa, *History of Agriculture in India*, vol. 4, pp. 62–63.

93. Clifford C. Taylor to Department of State, 30 December 1950, 891,20/12–3050, Record Group 59, USNA; "I.A.R.I.—Creation of the post of Assistant Director—Appointment of Dr. B. P. Pal," Department of Agriculture, Crops Section, File F.21–58/46–crops, National Archives of India, New Delhi; Green Revolution Scientist Dead, *The Times of India*, 15 September 1989; B. P. Pal, personal interview, 1–2 April 1987.

94. Douglas Ensminger, *Rural India in Transition* (Delhi: All India Panchayat Parishad, 1972), 115 pp.

95. Maheshwari, *Rural Development in India*, pp. 36–37.

96. Menon, *Indian Agriculture*, p. 56.

97. India, National Commission on Agriculture, *Report, Part XI, Research, Education and Extension*, pp. 380–383.

98. Menon, *Indian Agriculture*, p. 59.

99. Appointment of the First Indo-American Team brought to fruition a recommendation made by the University Education Commission in 1949 to create agricultural universities. This idea languished for a number of years, perhaps in deference to Gandhian ideas about simple village self-sufficiency. The team was appointed in 1954 and included five Indians and three Americans, who made visits to each other's countries in 1955. Their report in 1955 endorsed the notion that each Indian state should establish an agricultural uni-

versity. See B. P. Pal, Organization, management and progress of agricultural research in India, *Indian Journal of Public Administration* 15 (1969): 374–384; Randhawa *History of Agriculture in India*, vol. 4, pp. 183–184; and Hadley Read, *Partners with India: Building Agricultural Universities* (Urbana-Champaign: University of Illinois College of Agriculture, 1974), pp. 25–26.

100. John H. Perkins, The Rockefeller Foundation and the Green Revolution, 1941–1956, *Agriculture and Human Values* 7, nos. 3 and 4 (1990): 6–18.

101. Frankel, *India's Political Economy*, pp. 122.

102. Ibid., pp. 113–155.

103. Ibid., pp. 142–143, 153.

104. Ibid., pp. 157–159.

105. Douglas Ensminger, Oral history transcript, A.8., Relationships with Nehru, FFA.

106. The Agricultural Production Team, *Report on India's Food Crisis and Steps to Meet It* (New Delhi: Ministry of Food and Agriculture and Ministry of Community Development and Cooperation, 1959), p. 11. Report can also be found in Report no. 000520, FFA.

107. Douglas Ensminger, Oral history transcript, A.8. Relationships with Nehru, FFA.

108. Douglas Ensminger, Oral history transcript, B.3., The Foundation's Persistent Concern and Role in Assisting India Achieve a Status of Food Enough for Its People, 5 January 1972, pp. 27–35, FFA.

109. Ibid., pp. 29–54.

110. Frankel, *India's Political Economy*, pp. 223–245.

111. Ibid., pp. 241–247.

112. Ibid., pp. 250–251.

113. Ibid., pp. 250, 256–257.

114. India, Department of Agriculture, Ministry of Food and Agriculture, *Report of the Agricultural Prices Commission of Price Policy for Kharif Cereal for 1965–66 Season, May–July, 1965* ([Delhi]: Ministry of Food and Agriculture, 1965), 48 pp. See also M. L. Dantwala, Price Policy and Farm Incomes, in *Agricultural Administration in India*, ed. N. Srinivasin (Delhi: Indian Institute of Public Administration, 1969), pp. 329–336; and S. C. Chaudri, Price Support: Only the First Step, in *Foundations of Indian Agriculture*, ed. Vadilal Dagli (Bombay: Vora, 1968), pp. 255–260.

115. Chopra, *Food Policy in India*, pp. 135–139.

116. Bhanu Pratap Singh, *Betrayal of Rural India* (Delhi: B. R. Publishing, 1988), p. 19; Bhanu Pratap Singh, personal interview, 13 September 1989. Singh was union minister of state for agriculture during the rule of the opposition Janata Party, 1977–80, and was a longtime critic of Congress policies. A different sort of criticism was that the price support system increased incomes of the wealthier farmers and landowners at the expense of the poor, landless Indians who paid higher prices for their staples. See Alain de Janvry and K. Subbarao, *Agricultural Price Policy and Income Distribution in India* (Delhi: Oxford University Press, 1986), 113 pp.

117. Frankel, *India's Political Economy*, pp. 261–274.

118. Ibid., pp. 274–277.

119. Ibid., p. 277.

120. Ibid., p. 285.

121. Chopra, *Food Policy in India*, pp. 138, 140, 162.

122. James D. Gavan and John A. Dixon, India: A Perspective on the Food Situation, *Science*, 9 May 1975, pp. 541–549; Chopra, *Food Policy in India*, pp. 140–141.

123. Frankel, *India's Political Economy*, pp. 298–299.

Chapter 9

1. Jonathan Brown, *Agriculture in England: A Survey of Farming, 1870–1947* (Manchester: Manchester University Press, 1987), p. 109, provides the acreage for England and Wales in 1931; the figure for 1939, drawn from Central Statistical Office, *Statistical Digest of the War* (London: Her Majesty's Stationary Office, 1951), Table 55, p. 57, is for all of Great Britain, i.e., England, Scotland, and Wales. The figures are roughly comparable because very little wheat was grown in Scotland.

2. Central Statistical Office, *Statistical Digest of the War*, Table 55, p. 57.

3. E. J. Hobsbawm, *Industry and Empire* (Harmondsworth, England: Penguin, 1968), p. 204

4. *Agriculture in England*, p. 116.

5. The Ministry of Agriculture had opposed the formation of the animal disease research center under the council, but the cabinet was persuaded to let the council have £40,000 for the initiation of the project. "Agricultural Research Council, Corres. 1931–1937," CAB 123/277, PRO.

6. "Cambridge University, School of Agriculture, Plant Breeding Institute—General Developments," MAF 33/177, PRO. Indications that PBI was looking to ARC for guidance are contained in a memorandum from V. E. W[ilkins] of MAF, 11 May 1936, following the site visit of ARC and MAF officials to PBI.

7. For biographical details on Engledow, I am deeply indebted to G. D. H. Bell, Frank Leonard Engledow, 1890–1985, *Biographical Memoirs of Fellows of the Royal Society* 32 (1986): 189–219. I am also deeply indebted to Engledow's daughter for personal impressions: Ruth [Engledow] Stekete, personal communication, 6 June 1987.

8. Bell's biographical sketch of Engledow mentions this service but notes that Engledow never talked about the details of his activities other than to note observations on agriculture, the crops, and their botanical characteristics. Engledow is not mentioned in Arnold T. Wilson's *Mesopotamia, 1917–1920*, a thorough documentation by the acting civil commissioner of the British Civil Administration in Iraq. His absence from Wilson's record suggests that Engledow's work in Mesopotamia occurred while the Agriculture Department was still in the military's domain, not in the Civil Administration. Transfer of the Agriculture Department to civilian hands occurred on 1 March 1919. (See Arnold T. Wilson, *Mesopotamia, 1917–1920* (London: Oxford University Press, 1931), pp. 345–346.

By all accounts, the Mesopotamian campaign, designed to wrest control of the area now know as Iraq from the Ottoman Empire (a German ally) during the First World War, was a politically complex and militarily bloody affair. Engledow may well have found the experience in Mesopotamia both uninteresting and unpromising for an agricultural scientist. At the time, he was young (twenty-eight), without an advanced postgraduate degree or research expertise, and inexperienced in political maneuvering. Thus Engledow's return to Cambridge in 1919 was probably a welcome relief from a difficult situation.

9. The four varieties and the dates of their uses were Rampton Rivet (1939–57), Squareheads Master 13/4 (1940–60), Holdfast (1936–58), and Steadfast (1941–53). See Bell, Frank Leonard Engledow, 200.

10. In the 1920s Engledow published twenty-eight research papers; in the 1930s, this rate dropped to twelve, which probably reflected his increased responsibilities in teaching and government advising. See publication list in Bell, Frank Leonard Engledow, 189–219.

11. E. John Russell, *A History of Agricultural Science in Great Britain, 1620–1954* (London: George Allen and Unwin, 1966), pp. 198–206.

12. B. P. Pal, personal communication, 1–2 April 1987; "Biographical sketch of Dr. B. P. Pal," mimeo, no date, copy provided by B. P. Pal.

13. Ruth [Engledow] Stekete, personal communication, 6 June 1987. Ruth Mildred Engledow, born in 1928, was the third daughter of the Engledow family.

14. G. W. Cooke, ed., *Agricultural Research 1931–1981: A History of the Agricultural Research Council and a Review of Developments in Agricultural Science during the Last Fifty Years* (London: Agricultural Research Council, 1981), p. 342.

15. "Joint Advisory Committee on Higher Agricultural Education: (Loveday Committee Reconstituted) and Joint Advisory Committee on Agricultural Education: (Junior Loveday Committee), Constitution of Committees," MAF 33/404, PRO.

16. Bell, Frank Leonard Engledow, pp. 189–219.

17. Ronald W. Clark, *A Biography of the Nuffield Foundation* (London: Longman, 1972), pp. 3–15.

18. Papers of F. L. Engledow, box 3, "USA and Canada 1943," St. John's College, University of Cambridge.

19. H. T. Williams, ed., *Principles for British Agricultural Policy: A Study Sponsored by the Nuffield Foundation* (London: Oxford University Press, 1960), pp. v–vii, 303–304.

20. Brown, *Agriculture in England*, pp. 125–129.

21. Central Statistical Office, *Statistical Digest of the War*, Table 57, p. 59.

22. *Nutrition in War*, Fabian Society Tract Series no. 251, April 1940, 14 pp.

23. "Agricultural Education, Review of General Policy," MAF 33/63/TE/19035, PRO; discussions took place between November 1938 and January 1939. The full inquiry was to begin in February 1939.

24. Viscount Astor and B. Seebohm Rowntree, *British Agriculture: The Principles of Future Policy* (Harmondsworth, England: Penguin, 1939), pp. 257–259.

25. "Committee on Higher Agricultural Education in War-time. Appointment. (Loveday Committee)," MAF 33/400, PRO. The charge is V. E. Wilkins to Vice Chancellor [Loveday], 7 March, 1940.

26. "Committee on Higher Agricultural Education in War-time. Proceedings and Report. (Loveday Committee)," MAF 33/401, PRO.

27. "The Luxmoore Report on Agricultural Education—1943," MAF 33/403B, PRO. Quotation of charge in the report, p. i.

28. "Committee on Higher Agricultural Education in War-time. Proceedings and Report. (Loveday Committee)," "Draft Minutes of the fifth meeting held at 23–25, Soho Square, London, W.1. on the 16th Oct., 1941," MAF 33/401, PRO.

29. "The Luxmoore Report on Agricultural Education, 1943," MAF 33/403B, PRO.

30. "HEC 2" in "Joint Advisory Committee on Higher Agricultural Education, Loveday Committee Re-constituted, Papers," MAF 33/407, PRO.

31. R. S. Hudson to Charles Crowthers, 30 May 1944, in "Joint Advisory Committee on Higher Agricultural Education: (Loveday Committee Reconstituted) and Joint Advisory Committee on Agricultural Education: (Junior Loveday Committee), Constitution of Committees," MAF 33/404, PRO.

32. "Joint Advisory Committee on Higher Agricultural Education: (Loveday Committee Reconstituted) and Joint Advisory Committee on Agricultural Education: (Junior Loveday Committee), Constitution of Committees," MAF 33/404, PRO.

33. Agreement within the Ministry of Agriculture on a union member for the committee was difficult. Mr. George Dallas was vetoed by the parliamentary secretary, an MP, so staff came up with the name of Mr. G. Brown, a Labour member of the Hertfordshire War Agricultural Executive Committee and Agricultural Organizer for the Transport and Gen-

eral Workers Union, who was characterized by one trusted informant as " 'very red' but not aggressively unpleasant about it." See C. N. to Secretary, 3 May 1944, in ibid.

34. T. Loveday to Donald Fergusson, 27 March 1944; Donald Fergusson to T. Loveday, 31 March 1944; in ibid.

35. "HEC1," in "Joint Advisory Committee on Higher Agricultural Education, Loveday Committee Re-constituted, Papers," MAF 33/407, PRO.

36. "HEC5," in "Joint Advisory Committee on Higher Agricultural Education, Loveday Committee Re-constituted, Papers," MAF 33/407, PRO. Engledow also produced three other working papers for the committee: "Degree Courses in Agriculture and in Allied Subjects," "Recruitment and Training for Agricultural and Veterinary Appointment in the Public Services and in Industry at Home and in the Colonial Empire," and "A Graduate School in Agriculture." Clearly he was concerned about the role of scientific agriculture as it served to bolster both domestic and imperial production. See "HEC20," "HEC21," and "HEC22," in "Joint Advisory Committee on Higher Agricultural Education Loveday Committee Re-constituted, Papers," MAF 33/407, PRO.

37. "HEC10," in "Joint Advisory Committee on Higher Agricultural Education Loveday Committee Re-constituted, Papers," MAF 33/407, PRO.

38. National Farmers' Union [Publications], *Agriculture and the Nation: Interim Report on Post-War Food Production Policy* (London: National Farmers' Union, 1943), 24 pp.

39. "HEC16" and "HEC15," in "Joint Advisory Committee on Higher Agricultural Education Loveday Committee Re-constituted, Papers," MAF 33/407, PRO.

40. "HEC14," in "Joint Advisory Committee on Higher Agricultural Education Loveday Committee Re-constituted, Papers," MAF 33/407, PRO.

41. "Joint Advisory Committee on Higher Agricultural Education, Loveday Committee Reconstituted, Minutes of Meeting," MAF 33/405, PRO.

42. "Joint Advisory Committee on Higher Agricultural Education (Loveday Committee Re-constituted), Final Papers and Report," MAF 33/408, PRO.

43. Staples "Reconstruction" Digests, Post-war Agriculture, Combined Statement by Principal Bodies, in *Agriculture: A Digest of Fifteen Reports on Post-war Agricultural Policy and Reconstruction* (London: P. S. King and Staples, 1944), p. 1.

44. Staples "Reconstruction" Digests, pp. 13–20, 42–44.

45. House of Commons, Parliamentary Debates, 1944–45 (London: Her Majesty's Stationary Office Hansard, 5th Series), Questions on 14 December 1944, vol. 406, pp. 1348–1349; 6 February 1945, vol. 407, p. 1934; 8 February 1945, vol. 407, p. 2250; 15 February 1945, vol. 408, p. 398; 15 March 1945, vol. 409, p. 388; 19 April 1945, vol. 410, pp. 393–394; 3 May 1945, vol. 410, p. 1591; 17 May 1945, vol. 410, p. 2626; 31 May 1945, vol. 411, p. 359; 5 June 1945, vol. 411, p. 701; and 13 June 1945, vol. 411, p. 1678.

46. Staples "Reconstruction" Digests, pp. 76–83.

47. T. O. Lloyd, *Empire to Welfare State: English History 1906–1985*, 3d ed. (Oxford: Oxford University Press, 1986), pp. 266–269.

48. Tom Williams, MP, *Labour's Way to Use the Land* (London: Methuen, 1935), 120 pp. Clement Attlee served as editor for a series of policy books by Labour, of which Williams's book was one. Christopher Addison wrote the foreword.

49. Biographical details on Williams are from Michael Stenton and Stephen Lees, *Who's Who of British Members of Parliament. Vol. 4, 1945–1979, A Biographical Dictionary of the House of Commons* (Sussex: Harvester Press, 1981), p. 400.

50. Tom Williams, Agricultural Policy, statement by the Rt. Hon. Tom Williams, MP, Minister of Agriculture and Fisheries, House of Commons, Thursday, November 15, 1945, *Agriculture* 52, no. 9 (December 1945): 385–388.

51. Great Britain, Ministry of Agriculture and Fisheries and Ministry of Food, *Post-War Contribution of British Agriculture to the Saving of Foreign Exchange* (London: Her Majesty's Stationary Office, 1947, Cmd. 7022), 9 pp.

52. Great Britain, Ministry of Agriculture and Fisheries, *Agriculture Bill, Explanatory Memorandum* (London: Her Majesty's Stationary Office, December, 1946, Cmd. 6996), 27 pp.; J. Muir Watt, *A Guide to the Agriculture Act, 1947* (London: Sweet and Maxwell, Stevens and Sons, 1947), p. 2.

53. Labour Party, National Executive Committee, *Crisis Quiz* (Sussex: Harvester Press, 1975), microfiche, minutes no. 74, meeting of 24 September, 1947, pp. 16–17. The quote is from the draft of a pamphlet being prepared for promoting the party's views of the foreign exchange crisis of 1947.

54. Astor and Rowntree, *British Agriculture*, p. 254.

55. Agricultural Research Council, *Accounts of the Agricultural Research Council . . . showing the receipts and payments in the year ended 31st March 1942* (London: Her Majesty's Stationary Office, 1942, Cmd. 6434), 5 pp.

56. "Notes for Discussion at the Special Meeting of Council to be held on the 13th December," ARC 7204, 29–11–1943, box 000–8399, 46, Agricultural Research Council Archives, London.

57. Agricultural Research Council, *Post-War Programme*, ARC 8610, 23–5–46, box 8600–8699, 46, Agricultural Research Council Archives, London. A printed, "Confidential" version of this memorandum appeared on 4th June, 1946.

58. Ibid.

59. Ibid. Acreage of field plots available is in G. D. H. Bell, *The Plant Breeding Institute, Cambridge*, The Massey-Ferguson Papers, no. 9/1975 (Massey-Ferguson, 1976), p. 22.

60. John Fryer, The organisation for Agricultural Research, *Agricultural Progress* 22 (1947): 11 pp.; copy in "ARC Constitution Charter," Agricultural Research Council Archives, London.

61. On the interview with the second candidate, a Dr. Philp from South Africa, see Standing Committee on Research Affecting Plants and Soils, "Summary Report of Meeting Held on the 24th of June, 1947," ARC 9394, 14–7–47, Box 9300–9499, 47, Agricultural Research Council Archives. Bell, *The Plant Breeding Institute*, p. 21.

62. G. D. H. Bell, personal communication, 15 June 1987.

63. Bell, *The Plant Breeding Institute*, pp. 23–24; Bell, personal communication, 15 June 1987.

64. Bell, *The Plant Breeding Institute*, pp. 22–23; Bell, personal communication, 15 June 1987.

65. Bell, *The Plant Breeding Institute*, p. 23; Bell, personal communication, 15 June 1987.

66. Bell, *The Plant Breeding Institute*, p. 23; Plant Breeding Institute, *The Plant Breeding Institute, 75 Years, 1912–1987* (Cambridge: Plant Breeding Institute, 1987), pp. 9–10.

67. F. G. H. Lupton, personal communication, 17 October 1986; G. W. Cooke, *Agricultural Research 1931–1981* (London: Agricultural Research Council, 1981), p. 341.

68. Plant Breeding Institute, *The Plant Breeding Institute*, p. 21.

69. Lupton, personal communication, 17 October 1986.

70. John Bingham, personal communication, 4 February 1987.

71. Bell, personal communication, 15 June 1987; John Bingham, personal communication, 4 February 1987.

72. According to many critics of British imperialism, Britain drained resources from India for self-enrichment. Although the "drain theory" has been debated recently, it was generally accepted by nationalists in India and by critics in Britain and America at the outset of World War II. In the midst of the war, a socialist critic in Britain noted that the "drain,"

which was the reason for imperial conquest, had begun to reverse as early as 1931. British efforts to fight Germany resulted in substantial liquidations of British assets overseas, including those in India. As a result, India was for the moment a creditor of Britain. Loss of financial reward, plus pressure from the United States and the Soviet Union against imperialism, doomed both motivation and resources for holding India after the war's end. Britain's economy accordingly had to be restructured. See H. N. Brailsford, *Subject India* (London: Victor Gollancz, 1943), pp. 145–155, for a discussion of the "drain theory" and of the loss of Britain's economic empire. Further discussion of the theory and shifting global balances of power are in Karl de Schweinitz Jr., *The Rise and Fall of British India: Imperialism as Inequality* (London: Methuen, 1983), pp. 197–217; and in Bipan Chandra, *India's Struggle for Independence* (New Delhi: Penguin), pp. 96–101.

73. B. A. Holderness, *British Agriculture since 1945* (Manchester: Manchester University Press, 1985), table 16, p. 174.

Chapter 10

1. T. C. R. White makes the provocative argument that availability of nitrogen is the key chemical limitation in most life forms. In his view, natural selection operates through those individuals who are successful enough in gaining nitrogen to leave offspring. See T. C. R. White, *The Inadequate Environment: Nitrogen and the Abundance of Animals* (Berlin: Springer-Verlag, 1993), pp. 5–14.

2. Janet L. Sprent, *The Ecology of the Nitrogen Cycle* (Cambridge: Cambridge University Press, 1987), pp. 3–28; J. R. Postgate, *The Fixation of Nitrogen* (Cambridge: Cambridge University Press, 1982), pp. 1–4.

3. Sprent, *Ecology of the Nitrogen Cycle*, pp. 3–28; Postgate, *Fixation of Nitrogen*, pp. 9–19.

4. J. R. Postgate, *The Fundamentals of Nitrogen Fixation* (Cambridge: Cambridge University Press, 1982), pp. 163–164.

5. Rowland Edmund Prothero, *English Farming Past and Present* (London: Longmans, Green, 1912), pp. 187–174; Joseph Hutchinson and A. C. Owers, *Change and Innovation in Norfolk Farming: Seventy Years of Experiment and Advice at the Norfolk Agricultural Station* (Morley, Wymondham, Norfolk: Norfolk Agricultural Station, 1980), pp. 37–38; E. John Russell, *A History of Agricultural Science in Great Britain, 1620–1954* (London: George Allen and Unwin, 1966), pp. 44–45, 96.

6. George V. Taylor, Nitrogen Production Facilities in Relation to Present and Future Demand, in *Fertilizer Technology and Resources in the United States*, ed. K. D. Jacob (New York: Academic Press, 1953), pp. 16–25.

7. Ibid., pp. 17–29.

8. Jimmy M. Skaggs, *The Great Guano Rush: Entrepreneurs and American Overseas Expansion* (New York: St. Martin's, 1994), pp. 8–10.

9. E. M. Crowther, Soils and Fertilizers, *Journal of the Royal Agricultural Society of England* 107 (1946): 71–85; see table I, p. 78.

10. In 1938–39, British India produced about 23,000 tons of ammonium sulfate and imported 76,700 tons; see India's Chemicals and Fertilisers: Important Raw Materials Lost to Pakistan, *Chemical Age*, 31 December 1949, pp. 918–919.

11. Taylor, Nitrogen production facilities, pp. 29–31.

12. Ibid., pp. 31–33.

13. Vincent Sauchelli, ed., *Fertilizer Nitrogen: Its Chemistry and Technology* (New York: Rheinhold, 1964), pp. 44–49.

14. Taylor, Nitrogen production facilities, pp. 17, 33–47; percentages are calculated from table 1, p. 17.

15. Sauchelli, *Fertilizer Nitrogen*, table 4.1, p. 55.

16. Bernard Gilland, Cereals, Nitrogen and Population: An Assessment of the Global Trends, *Endeavor* 17, no. 2 (1993): 84–87.

17. Peter M. Vitousek, Beyond Global Warming: Ecology and Global Change, *Ecology* 75, no. 7 (October 1994): 1861–1876.

18. Dana G. Dalrymple, *Development and Spread of High-Yielding Varieties of Wheat and Rice in the Less Developed Nations*, 6th ed. U.S. Department of Agriculture, Foreign Agricultural Economic Report no. 95, September 1978, pp. 10–14.

19. Dalrymple provided a comprehensive review of the origins of the semidwarf genes in a series of reports for the U.S. Department of Agriculture and the U.S. Agency for International Development. These reports are taken as authoritative by plant breeders. See ibid., pp. 10–12, 22–23; and Dana G. Dalrymple, *Development and Spread of Semi-Dwarf Varieties of Wheat and Rice in the United States*, U.S. Department of Agriculture, Agricultural Economic Report no. 455, June 1980, pp. 30–34.

20. Unless otherwise noted, all details about Salmon are taken from his memoirs, *The Odyssey of an Agricultural Scientist, The Memoirs of Samuel C. Salmon*, mimeo, n.d., 116 pp., copy held in library of South Dakota State University. I thank Margaret Rossiter for bringing this autobiography to my attention.

21. H. K. Hayes, E. R. Ausemus, E. C. Stakman, C. H. Bailey, H. K. Wilson, R. H. Bamberg, M. C. Markley, R. F. Crim, and M. N. Levine, *Thatcher Wheat*, University of Minnesota Agricultural Experiment Station, Bulletin 325, January 1936, 39 pp.

22. Salmon, *Odyssey of an Agricultural Scientist*, p. 58.

23. "Norin" is an acronym made from the words designating the Japanese agricultural experiment station network.

24. Salmon, *Odyssey of an Agricultural Scientist*, p. 60.

25. L. P. Reitz and S. C. Salmon, Origin, History, and Use of Norin 10 Wheat, *Crop Science* 8 (1968): 686–689; Dalrymple, *Development and Spread of High-Yielding Varieties of Wheat and Rice in the Less Developed Nations*, pp. 10–23; and Dalrymple, *Development and Spread of Semi-Dwarf Varieties of Wheat and Rice in the United States*, pp. 30–34.

26. Dalrymple, *Development and Spread of Semi-Dwarf Varieties of Wheat and Rice in the United States*, p. 34.

27. Ibid., pp. 34–37.

28. Orville Vogel, personal communication, 12 July 1988; "Orville A. Vogel," Obituary, *Olympian* [Olympia, Wash.], 13 April 1991.

29. Vogel, personal communication, 12 July 1988; Warren E. Kronstad and Norman E. Borlaug, Dedication: Orville A. Vogel, Wheat Breeder, Agronomist, Inventor, in *Plant Breeding Reviews*, vol. 5, ed. Jules Janick (New York: Van Nostrand Reinhold, 1987), p. 1; and "Orville A. Vogel," Obituary.

30. O. A. Vogel, Wheat Development—Getting the Job Done, in Department of Agronomy and Soils and Washington State University Cooperative Extension Service, *O. A. Vogel Honorary Symposium, Wheat in This Changing World* (Pullman: Washington State University, 1973), p. 44; G. D. H. Bell, The History of Wheat Cultivation, in *Wheat Breeding: Its Scientific Basis*, ed. F. G. H. Lupton (London: Chapman and Hall, 1987), p. 32; Mark Alfred Carleton, *The Small Grains* (New York: Macmillan, 1924), pp. 44–45; R. James Cook and Roger J. Veseth, *Wheat Health Management* (St. Paul, Minn.: APS Press, 1991), p. 57.

31. Vogel, Wheat Development, p. 45.

32. Ibid.

33. Burt Bayles to Mr. McCall, 4 August 1935, file: Bayles B.B. 1934–35, General Correspondence (1917–35), Division of Cereal Crops and Diseases, Bureau of Plant Industry, Record Group 54, USNA.

34. O. A. Vogel, personal communication, 12 July 1988.

35. O. A. Vogel, Annual Report for 1937, file: Washington 1937–39, box 141, Annual Research Reports 1930–39, Record Group 54, USNA.

36. Orville Arthur Vogel, Studies of the Relationships of Some Features of Wheat Glumes to Resistance to Shattering and of the Use of Glume Strength as a Tool in Selecting for High Resistance to Shattering (Ph.D. diss., Washington State University, 1939, 68 l., found in library of Washington State University, Pullman, call number 633.11 V862s.

37. O. A. Vogel, A Three-Row Nursery Planter for Space and Drill Planting, *Journal of American Society of Agronomy* 25 (1933): 426–428; O. A. Vogel and Arthur Johnson, A New Type of Nursery Thresher, *Journal of American Society of Agronomy* 26 (1934): 629–630; O. A. Vogel, Wilford Herman, and Loy M. Naffziger, Two Improved Nursery Threshers, *Journal of American Society of Agronomy* 30 (1938): 537–542.

Vogel's colleagues, including Norman E. Borlaug, extolled the importance of Vogel's contributions to the mechanization of plant-breeding operations and noted that these machines enabled substantial increases in the amounts of crosses made each year. See Norman E. Borlaug, International Agriculture's Progress—Everyone's Responsibility, in Washington State University, Department of Agronomy and Soils and Washington State University Cooperative Extension Service, *O. A. Vogel Honorary Symposium, Wheat in This Changing World* (Pullman: Washington State University, 1973), p. 31.

Vogel recounts the distaste he had for the extreme tedium and slowness of handling some of the breeding operations before his inventions made them easier. See Vogel, Wheat Developments, pp. 47–48. Vogel also noted that the reward system within the U.S. Department of Agriculture did not recognize these mechanical contributions because he was not formally assigned to engineering work, a factor that hindered his salary increases before the 1960s (O. A. Vogel, personal communication, 12 July 1988).

38. Vogel, personal communication, 12 July 1988; M. Keith Ellis to Harland E. Priddle, 2 September 1983, letter of nomination and supporting evidence to place Vogel in the Agriculture Hall of Fame, copy supplied by O. A. Vogel.

39. M. Keith Ellis to Harland E. Priddle, 2 September 1983, letter of nomination and supporting evidence to place Vogel in the Agriculture Hall of Fame, copy supplied by O. A. Vogel. At the time, Ellis was the director of the Washington State Department of Agriculture.

40. O. A. Vogel, S. P. Swenson, and C. S. Holton, *Brevor and Elmar—Two New Winter Wheats for Washington*, State College of Washington, Washington Agricultural Experiment Stations, Bulletin no. 525, May 1951, 8 pp.

41. Ibid.

42. Orville A. Vogel, personal communication, 12 July 1988.

43. O. A. Vogel, J. C. Craddock Jr., C. E. Muir, E. H. Everson, and C. R. Rohde, Semidwarf Growth Habit in Winter Wheat Improvement for the Pacific Northwest, *Agronomy Journal* 48 (1956): 76–78. Heights are calculated from table 2, for the years 1953 and 1954.

44. Orville A. Vogel, Annual Report for 1948, file: Washington 44–48, Annual Research Reports 40–49, box 79, Record Group 54, USNA.

45. Nagamitso was a graduate student working with F. C. Elliott. For some reason, Vogel and his colleagues credited Elliott with making the crosses when they published their first paper on the subject. (O. A. Vogel, J. C. Craddock Jr., C. E. Muir, E. H. Everson, and C. R. Rohde, Semidwarf Growth Habit in Winter Wheat Improvement for the Pacific Northwest, *Agronomy Journal* 48 (1956): 76–78. Nagamitso's role was confirmed both in letters from Vogel to Dana Dalrymple and in a personal interview with me. See Dalrymple, *Development and Spread of Semi-Dwarf Varieties of Wheat and Rice in the United States*, p. 35; and Orville A. Vogel, personal communication, 12 July 1988.

46. Vogel et al., Semidwarf growth habit in winter wheat improvement for the Pacific Northwest, 76–78. Yield increases and heights based on data in table 2, p. 77.

47. Ibid., p. 78.

48. Dalrymple, *Development and Spread of Semi-Dwarf Varieties of Wheat and Rice in the United States*, p. 36.

49. O. A. Vogel, R. E. Allan, and C. J. Peterson, Plant and Performance Characteristics of Semidwarf Winter Wheats Producing Most Efficiently in Eastern Washington, *Agronomy Journal* 55 (1963): 397–398. In this paper Vogel and his colleagues noted that under high nitrogen Gaines could yield the farmer as much as a net of $11 per acre more than under conditions of low nitrogen use. Physical yields of Gaines were seventy-five bushels per acre under conditions of high fertility compared with Brevor's fifty-nine bushels per acre.

50. Dalrymple, *Development and Spread of Semi-Dwarf Varieties of Wheat and Rice in the United States*, p. 36–39.

51. Norman E. Borlaug, The Green Revolution, Peace and Humanity, 1970 Nobel Peace Prize ([Mexico]: CIMMYT, [1970]), 31 pp. The first nine pages of this pamphlet contain the remarks of Mrs. Aase Lionaes, president of the Lagting, who introduced Borlaug for his acceptance speech of the prize.

52. Norman E. Borlaug, personal communication, 21 June 1989.

53. Report of the Oficina de Estudios Especiales SAF, 1 February 1943–1 June 1945, Record Group 1.1, series 323, box 6, folder 38, RFA.

54. Ibid.

55. Review of the work of the Office of Special Studies, SAG, for the year 1946–47, Record Group 1.1, series 323, box 6, folder 38, RFA.

56. Report of Mexican Trip with Confidential Supplement Regarding Mexican Agricultural Improvement Project and Personnel, 20 September 1945, by E. C. Stakman, Record Group 1.2, series 323, box 10, folder 60, RFA.

57. Norman E. Borlaug, Oral History Interview, the Rockefeller Foundation Program in Agriculture, June 1967, pp. 147–154, RFA.

58. John H. McKelvey Jr., J. George Harrar, December 2, 1906–April 18, 1982, *Biographical Memoirs of the National Academy of Sciences* 57 (1987): 27–56.

59. Norman E. Borlaug, personal communication, 21 June 1989.

60. Ibid.

61. Norman E. Borlaug, Oral History Interview, June 1967, p. 188.

62. Ibid., pp. 152, 163–164.

63. Norman E. Borlaug, Challenges for Global Food and Fiber Production, *K. Skogs-och Lantbraksakademien Tidskrift*, suppl. 21 (1988): 15–55.

64. Borlaug, Oral History Interview, June 1967, pp. 166–167. Borlaug, personal communication, 21 June 1987.

65. Borlaug, Oral History Interview, June 1967, pp. 166–168, 345; Borlaug, personal communication, 21 June 1989.

66. Borlaug, Oral History Interview, June 1967, pp. 169–170.

67. Report of the Office of Special Studies, S.A.G, 1 September 1949–31 August 1950, Record Group 1.1, series 323, box 6, folder 38, RFA. These varieties included Supremo, Kenya Rojo, Roca mex 481, Yaqui, Roca mex 485, Kentana, Roca mex 483, and Roca mex 484. In a separate report, dated on or after 1 October 1950, Harrar noted that the program had found twelve improved varieties, but he did not mention them by name, nor did he distinguish between selections of new varieties from crosses from the selection of well-adapted varieties from varietal imports. See Report of J. G. Harrar to the Natural Sciences Division and to the Advisory Committee on Agriculture, 30 September 1949–1 October 1950, inclu., Record Group 1.2, series 343, box 10, folder 61, RFA.

68. N. E. Borlaug, J. A. Rupert, B. Ortega C., A. Marino A., and Carlos G. Cavazos, *El trigo como cultivo de verano en los valles altos de Mexico*, Oficina de Estudios Especiales, Secretaría de Agricultura y Ganadería, Mexico, Folleto de Divulgación No. 10, Marzo 1950, 23 pp.

69. Haldore Hanson, Norman E. Borlaug, and R. Glenn Anderson, *Wheat in the Third World* (Boulder, Colo.: Westview Press, 1982), p. 22.

70. Borlaug, Challenges for Global Food and Fiber Production, 15–55.

71. Report of J. G. Harrar to the Natural Sciences Division and to the Advisory Committee on Agriculture, 30 September 1949–1 October 1950.

72. Borlaug, Oral History Interview, June 1967, pp. 185, 194–197; Report of the Office of Special Studies, S.A.G, 1 September 1949–31 August 1950, Record Group 1.1, series 323, box 6, folder 38, RFA.

73. N. E. Borlaug, *Norman E. Borlaug: A Bibliography of Papers and Publications* (Mexico City: CIMMYT, 1988), pp. 21–22, 33–34.

74. Borlaug, Oral History Interview, June 1967, p. 198.

75. Borlaug, Challenges for Global Food and Fiber Production, 15–55.

76. Borlaug, Oral History Interview, June 1967, pp. 199–200; Borlaug, Challenges for Global Food and Fiber Production, 15–55; Borlaug, personal communication, 21 June 1989; Dalrymple, *Development and Spread of High-Yielding Varieties of Wheat and Rice in the Less Developed Nations*, p. 15.

77. Norman E. Borlaug to Glenn Anderson, 31 December 1965, file "IAP Wheat Improvement Admin & Organ, I 1965–66," box 7c 10a, Record Group 6.7, RFA; Borlaug, Challenges for Global Food and Fiber Production, 15–55.

78. Borlaug, Oral History Interview, June 1967, pp. 218–219.

79. Dalrymple, *Development and Spread of High-Yielding Varieties of Wheat and Rice in the Less Developed Nations*, pp. 15–16.

80. Borlaug, Challenges for Global Food and Fiber Production, 15–55.

81. Henry Hopp, Mexico Joins Wheat-Exporting Nations of the World, *Foreign Agriculture*, 8 June 1964, pp. 3–4. Figures are taken from the table on p. 3. Hopp was the agricultural attaché at the embassy.

82. S. Ramanujam, E. A. Siddiq, V. L. Chopra, and S. K. Sinha, eds., *Science and Agriculture: M. S. Swaminathan and the Movement for Self-Reliance* (New Delhi: Indian Society of Genetics and Plant Breeding and twenty other scientific societies, 1980), p. 1.

83. Indian Environmental Society (Desh Bandhu, president), *Economic Ecologists of India* (New Delhi: Indian Environmental Society, [1982]), pp. 2–3.

84. Ramanujam et al., *Science and Agriculture*, pp. 1–2; M. S. Swaminathan, personal communication, 4 July 1987.

85. "Dr. Monkombu Sambasivan Swaminathan, Curriculum Vitae," provided by M. S. Swaminathan, n.d.; M. S. Swaminathan, personal communication, 4 July 1987; Ramanujam et al., *Science and Agriculture*, pp. 3–4.

86. "Dr. Monkombu Sambasivan Swaminathan, Curriculum Vitae," and "Scientific papers published by Dr. M. S. Swaminathan and his research associates," provided by M. S. Swaminathan, n.d.; Swaminathan, Personal communication, 4 July 1987; Ramanujam et al., *Science and Agriculture*, p. 4.

87. "Dr. Monkombu Sambasivan Swaminathan, Curriculum Vitae," and "Scientific papers published by Dr. M.S. Swaminathan and his research associates," provided by M. S. Swaminathan, n.d.; Swaminathan, personal communication, 4 July 1987; Ramanujam et al., *Science and Agriculture*, p. 4.

88. Data in publications are derived from "Scientific papers published by Dr. M. S. Swaminathan and his research associates," provided by M. S. Swaminathan, n.d.; informa-

tion on students who earned their graduate degrees with Swaminathan is in Ramanujam et al., *Science and Agriculture*, pp. 149–154; see also p. 4.

89. Ramanujam et al., *Science and Agriculture*, pp. 131–144.

90. R. K. Agrawal, Development of Improved Varieties, in *Twenty-five Years of Co-ordinated Wheat Research*, ed. J. P. Tandon and A. P. Sethi (New Delhi: Indian Agricultural Research Institute, [1986]), pp. 34–93.

91. M. S. Swaminathan, personal communication, 4 July 1987.

92. M. S. Swaminathan to Dr. Kohli, 18 January 1966, file: "IAP, Wheat Improvement, Admn & Organ, I 1965–55," box 7c10a, Record Group 6.7, RFA. This letter collects a personal statement from Swaminathan about his role in identifying the Norin 10 genes as useful for India and a copy of the formal proposal he prepared to bring Borlaug to India. See also Dalrymple, *Development and Spread of High-Yielding Varieties of Wheat and Rice in the Less Developed Nations*, pp. 15–16.

93. M. S. Swaminathan to Dr. Kohli, 18 January 1966. Borlaug's itinerary is attached to this letter.

94. Norman E. Borlaug, "Wheat production problems in India, Sep 16 1963," Record Group A2 (unprocessed), series 464, box, 961, file: Agriculture 1963, RFA.

95. Ibid.; quote is drawn from pp. 7–8.

96. M. S. Swaminathan, The Impact of Dwarfing Genes on Wheat Production, *Journal of the I.A.R.I. Post-graduate School* 3 (1965): 57–62; Agrawal, Development of Improved Varieties, pp. 34–93. Swaminathan lists the number of additional lines at 613 (table 1, p. 58); Agrawal states that 630 additional lines were sent (p. 38).

97. Figures for yields of improved Indian varieties are taken from information provided by S. P. Kohli of IARI to R. G. Anderson of the Rockefeller Foundation in India. See R. G. Anderson to Dr. Moseman, 30 October 1964, Record Group A2 (unprocessed), series 464, box 961, file: Agriculture 1964, RFA. Kohli provided yield data for eleven different Indian varieties, some of which had been released as early as 1908 by Albert Howard. These eleven varieties were grown in four different locations in three different years: 1961–62 through 1963–64. Yield data on the four Mexican varieties are in Swaminathan, The impact of dwarfing genes on wheat production, 57–62; table 2, p. 59.

Unfortunately, most and perhaps all of the data recorded by Kohli and Swaminathan are not directly comparable because they reported the results of different varieties under perhaps different growing conditions. Three varieties, however, overlapped for the year 1963–64 and thus can be directly compared to the Mexican varieties.

Yields of improved Indian varieties compared with Mexican varieties, in kilograms per hectare, 1961–1964.

Variety	Delhi I	Delhi II	Delhi III	Multiplication	IARI
Sonora 63	3833	3756	4569	4495	—
Sonora 64	4287	—	—	4084	—
Lerma Rojo 64A	3558	—	—	—	—
Mayo 64A	—	—	3650	—	—
N.P. 824	—	—	2800	3737	1714
C. 306	—	2747	—	—	1641
Rs. 31–1	2153	—	—	—	1581

The four Mexican varieties are listed in the top four rows. Delhi I through Multiplication were yield trial results, probably from IARI, reported by M.S. Swaminathan. IARI are from irrigated tests at IARI, reported by S.P. Kohli. Information in the available documents was not sufficient to indicate why the results from Swaminathan and Kohli are different, but presumably they represented different yield trials conducted under different conditions.

98. Figures for yields at the time of independence in 1947 were provided to R. G. Anderson of the Rockefeller Foundation in India by Dr. V. Panse of the Institute of Agricultural Research Statistics. They were based on random sampling of farmers in a number of different states, averaged over five years. Panse's figures indicated little increase in yields per acre from 1947 through 1961. In some areas, such as Maharashtra, with low rainfall and little irrigation, the yields were in the range of 200 to 300 pounds per acre during this period (equivalent to 225 to 337 kilograms per hectare). See R. G. Anderson to Dr. Moseman, 30 October 1964, Record Group A2 (unprocessed), series 464, box 961, file: Agriculture 1964, RFA.

99. M. S. Swaminathan to Dr. Kohli, 18 January 1966, file: "IAP, Wheat Improvement, Admn & Organ, I 1965–66," box 7c10a, Record Group 6.7, RFA. This letter from Swaminathan to Kohli contains an attached letter from Borlaug to Swaminathan, dated 24 June 1964, that reports Borlaug's thoughts after a second visit to India in spring 1964 to see the performance of Mexican wheats planted the previous November. Swaminathan quoted extensively from this letter in his article, The Impact of Dwarfing Genes on Wheat Production, 57–62. See also, N. E. Borlaug, Indian wheat research designed to increase wheat production, 11 April 1964, pp. 3–4, Record Group A2 (unprocessed), series 464, box 961, file: Agriculture 1964, RFA.

100. Borlaug, Indian wheat research designed to increase wheat production, April 11, 1964, p. 6.

101. Ralph W. Cummings to Robert D. Osler, 18 March 1964, Record Group A2 (unprocessed), series 464, box 961, file: Agriculture 1964, RFA.

102. Plans for the Future, A Statement by the Trustees of the Rockefeller Foundation, 20 September 1963, Record Group 3.2, series 900, box 29, folder 158, RFA.

103. Guy B. Baird to Norman E. Borlaug, 29 June 1964, and Guy B. Baird to Norman E. Borlaug, 30 June 1964, both in Record Group A2 (unprocessed), series 464, box 961, file: Agriculture 1964, RFA.

104. Ralph W. Cummings to Norman E. Borlaug, 28 March 1964, Record Group A2 (unprocessed), series 464, box 961, file: Agriculture 1964, RFA.

105. Agrawal, Development of Improved Varieties, p. 38.

106. A brief biographical sketch of Anderson is in Haldore Hanson, Norman E. Borlaug, and R. Glenn Anderson, *Wheat in the Third World* (Boulder, Colo.: Westview Press, 1982), pp. xvi–xviii. Foundation expenditures for the Indian Agricultural Program went from $290,000 in 1963 to $439,350 in 1964. See Rockefeller Foundation, *Annual Report 1964*, p. 101.

107. J. P. Naik to [unspecified form letter], 2 April 1965, and attached minutes and working papers; J. P. Naik to R. W. Cummings, 9 May 1965; Ralph W. Cummings to A. B. Joshi, Kishen Kanungo, 2 July 1965; Ralph W. Cummings to S. Ramanujan, 7 July 1965; all in file: "IAP-Agr Edn, Govt. of India Task Force, 1965–1966," box 2a, Record Group 6.7, RFA. The final report was issued in 1966 with a dissent from Cummings on how funds should be allocated to the agricultural universities.

108. Stanley Wolpert, *A New History of India*, 2d ed. (New York: Oxford University Press, 1982), p. 371.

109. Francine R. Frankel, *India's Political Economy, 1947–1977, The Gradual Revolution* (Princeton, N.J.: Princeton University Press, 1978), pp. 255–256.

110. Union Minister of Food and Agriculture, Speech, to the first meeting of the Committee of National Development Council on Agriculture and Irrigation, 1rst January 1965, in [Directorate of Economics and Statistics], *Agricultural Development, Problems and Perspectives* ([New Delhi: GIPF, 1965]), Appendix 1, pp. 73–83; copy located in National Archives of India, New Delhi.

111. M.S. Swaminathan, personal communication, 4 July 1987. The visits to IARI

probably took place in spring 1965, one year after Shastri and Subramaniam took office.

112. Frankel, *India's Political Economy*, pp. 279, 284.

113. Ibid., pp. 279–280, 284.

114. Wolpert, *New History of India*, pp. 374–375.

115. Glenn Anderson to Norman E. Borlaug, 24 January 1966, file: "IAP, Wheat Improvement, Admn & Organ, I 1965–66," box 7c10a, Record Group 6.7, RFA.

116. Agrawal, Development of Improved Varieties, pp. 38–40; Glen Anderson to Norman E. Borlaug, 25 July 1965, file: "IAP, Wheat Improvement, Admn & Orgsn, I 1965–66," box 7c10a, Record Group 6.7, RFA.

117. B. P. Pal to R. W. Cummings, 2 July 1965, file: "I.A.P., Wheat, Import of Seed from Mexico, 1965–66," box 7c10a, Record Group 6.7, RFA.

118. Cummings to Wellhausen, 30 July 1965 and 5 August 1965, and Ralph W. Cummings to E. J. Wellhausen, 7 August 1965, all in file: "I.A.P., Wheat, Import of Seed from Mexico, 1965–66," box 7c10a, Record Group 6.7, RFA; Norman E. Borlaug to Glen Anderson, 5 August 1965, file: "IAP, Wheat Improvement, Admn & Orgsn, I 1965–66," box 7c10a, Record Group 6.7. RFA.

119. Norman E. Borlaug to R. W. Cummings, 20 August 1965, file: "I.A.P., Wheat, Import of Seed from Mexico, 1965–66," box 7c10a, Record Group 6.7, RFA.

120. R. G. Anderson to Cooperator, 25 September 1965, file: "IAP, Wheat Improvement, Admn & Orgsn, I 1965–66," box 7c10a, Record Group 6.7, RFA.

121. Agrawal, Development of Improved Varieties, p. 40.

122. Borlaug to Anderson, 5 August 1965.

123. Anderson to Cooperator, September 25, 1965.

124. A thorough discussion of the complexities in Indian–United States relations over food and military imports is given by Frankel, *India's Political Economy*, pp. 284–286. See also C. P. Bhambhri, *The Foreign Policy of India* (New Delhi: Sterling, 1987), pp. 36–37; Stanley Wolpert, *Roots of Confrontation in South Asia: Afghanistan, Pakistan, India, and the Superpowers* (New York: Oxford University Press, 1982), pp. 146–148.

125. U.S. Urged to Drop Surplus Exports, *New York Times*, 19 July 1965; Frankel, *India's Political Economy*, p. 286.

126. U.S. Urged to Spur Farm Output to Alleviate World Food Crisis, *New York Times*, 5 December 1965

127. S. M. Sikka to R. W. Cummings, 27 October 1965; Robert D. Osler to RW Cummings, March 1966; I. J. Naidu to Guy B. Baird, 17 April 1966; all in file:" I.A.P., Wheat, Import of Seed from Mexico, 1965–66," box 7c10a, Record Group 6.7, RFA.

128. C. Subramaniam, Message, *Indian Farming* 15, no. 7 (1965): 2.

129. C. Subramaniam, India on the Eve of Breakthrough in Agriculture, *India Farming* 15, no. 8 (November 1965): 7.

130. [Lal Bahadur Shastri], Produce More Food and Preserve our Freedom, Prime Minister Calls the Nation, *Indian Farming* 15, no. 7 (October 1965): 3–4, 52.

131. W. David Hopper to Norman E. Borlaug, 10 February 1966, file: "IAP, Wheat Improvement, Admn & Orgsn, I 1965–66," box 7c10a, Record Group 6.7, RFA.

132. Frankel, *India's Political Economy*, pp. 297–300.

133. Wolpert, *New History of India*, pp. 377–378.

134. Frankel, *India's Political Economy*, p. 292.

135. Norman E. Borlaug to I. J. Naiden [*sic*: Naidu], 26 July 1966, file: "IAP, Wheat Improvement, Admn & Orgsn, I 1965–66," box 7c10a, Record Group 6.7, RFA.

136. N. E. Borlaug to Glenn Anderson, 27 September 1966, file: "IAP, Wheat Improvement, Admn & Orgsn, I 1965–66," box 7c10a, Record Group 6.7, RFA.

137. Norman E. Borlaug to Ralph Cummings, 12 September 1966, file: "IAP, Wheat Improvement, Admn & Orgsn, I 1965–66," box 7c10a, Record Group 6.7, RFA.

138. R. G. Anderson to William I. Myers, 2 March 1966, file: "IAP, Wheat Improvement, Admn & Orgsn, I 1965–66," box 7c10a, Record Group 6.7, Rockefeller Foundation Archives, RFA.

139. R. N. Chopra, *Food Policy in India: A Survey* (New Delhi: Intellectual Publishing House, 1988), pp. 140–141, 162.

140. D. P. Gupta, *Agricultural Developments in Haryana, 1952–53 to 1974–75*, Research Study no. 76/5 (Delhi: University of Delhi, Agricultural Economics Research Centre, 1976), 89 pp. Index numbers for production are from table 1.7, p. 17; fertilizer consumption is from table 2.11, p. 32.

141. Figures on hectares planted in high-yielding varieties of wheat are from Dalrymple, *Development and Spread of High-Yielding Varieties of Wheat and Rice in the Less Developed Nations* table 7, p. 38; figures on food imports are from National Commission on Agriculture, *Report. Part II, Policy and Strategy* (New Delhi: Ministry of Agriculture and Irrigation, 1976), p. 14.

142. B. P. Pal, personal communication, 18 March 1987; Ralph W. Cummings to Robert D. Osler, 6 April 1965, file: "IAP-Admin, Orgn & Policy, 1965–. , V," Box 10c18a, Record Group 6.7, RFA; K. P. A. Menon, *Indian Agriculture: Administrative and Organisational Constraints*, 3d rev. ed. (New Delhi: Sreedeep, 1987), pp. 111–115.

143. "Dr. Monkombu Sambasivan Swaminathan, Curriculum Vitae," copy supplied by Dr. Swaminathan, 1987.

144. John Bingham, personal communication, 4 February 1987.

145. F. G. H. Lupton, History of Wheat Breeding, in *Wheat Breeding: Its Scientific Basis*, ed. F. G. H. Lupton (London: Chapman and Hall, 1987), p. 65.

146. Great Britain, Research Committee on the Increase in Agricultural Production, Rural Reconstruction Association, *Feeding the Fifty Million* (London: Hollis and Carter, 1955), 138 pp. The Rural Reconstruction Association was organized before World War II as a nonpartisan association to restore agriculture to a higher place in British life. Its report argued for the continued importance of balance-of-payments problems and an expanding world population that would soon shrink the wheat surpluses then available. As a result, it concluded that more home production was needed (see pp. 17–18).

147. R. H. Biffen and F. L. Engledow, *Wheat Breeding Investigations at the Plant Breeding Institute, Cambridge* (London: His Majesty's Stationary Office, 1926), 114 pp. This booklet, in retrospect, was remarkably prophetic for the entire breeding program of the high-yielding wheat varieties. Biffen and Engledow outlined the characteristics provided by the wheats with the Norin 10 genes almost perfectly, even though at the time the Norin genes were unknown in the United Kingdom. This pamphlet had a guiding effect on PBI into the 1980s, but it was unfamiliar to American and Indian wheat breeders by that time. The booklet was reviewed in the *Journal of the Agronomy Society of America*, but beyond that notice made no lasting impression in the United States. See JHP, Review of Wheat Breeding Investigations at the Plant Breeding Institute, Cambridge, *Journal of the American Society of Agronomy* 18 (1926): 516–518. JHP was probably John H. Parker of Kansas State Agricultural College, who was the first professor of Paul Mangelsdorf, a member of the Rockefeller Foundation's Survey Commission to Mexico, which recommended establishing the foundation's program in Mexico. Mangelsdorf, however, worked on maize, so it is perhaps not surprising that he seems not to have passed on Biffen and Engledow's work to Borlaug.

148. Britain continued to import wheat as its domestic production climbed, but imports dropped and exports began to grow and eventually surpassed imports. For example, in 1975–77, the United Kingdom imported about 3.8 million metric tons and exported about

0.2 million metric tons. By 1984–86, in contrast, imports were at about 1.5 million metric tons and exports were at 2.7 million metric tons. In 1984, U.K. production of wheat was about 15 million metric tons, approximately 50 to 70 percent more than was consumed domestically. In 1993, U.K. exports of wheat and wheat flour earned £462 million and imports cost £190 million. Thus the United Kingdom earned approximately £272 million by having an exportable surplus of wheat. See United Nations, Food and Agriculture Organization, *Food Balance Sheets, 1975–77 Average* (Rome: Food and Agriculture Organization, 1980), p. 933; United Nations, Food and Agriculture Organization, *Food Balance Sheets* (Rome: Food and Agriculture Organization, 1991), p. 358; United Kingdom, Ministry of Agriculture, Fisheries and Food, *Agricultural Statistics, United Kingdom, 1987* (London: Her Majesty's Stationary Office, 1989), p. 8; United Kingdom, Central Statistical Office, *Business Monitor: Overseas Trade Statistics of the United Kingdom* (London: Her Majesty's Stationary Office, 1993), pp. III 7–8. VI 7–8.

149. F. G. H. Lupton, personal communication, 17 October 1986.

150. Ibid.

151. Lupton's first principal publications concerned disease resistance and whether genes for resistance could be found in species closely related to wheat and transferred into wheat. See G. D. H. Bell and F. G. H. Lupton, Investigations in the Triticine: IV. Disease Reaction of Species of *Triticum* and *Aegilops* and of Amphidiploids between Them, *Journal of Agricultural Science, Cambridge* 46 (1955): 232–246; and F. G. H. Lupton, Resistance Mechanisms of Species of *Triticum* and *Aegilops* and of Amphidiploids between Them to *Erysiphe graminis* D.C., *Transactions of the British Mycological Society* 39 (1956): 51–59. A review of Lupton's work in breeding methodology is in F.G.H.L. and R.N.H.W., Developments in Wheat-Breeding Methods, in *Plant Breeding Institute Cambridge, Annual Report, 1959–60* (Cambridge: Plant Breeding Institute, 1961), pp. 4–18.

152. F. G. H. Lupton and M. A. M. Ali, Studies on Photosynthesis in the Ear of Wheat, *Annals of Applied Biology* 57 (1966): 281–286; and F. G. H. Lupton, Translocation of Photosynthetic Assimilates in Wheat, *Annals of Applied Biology* 57 (1966): 355–364.

153. Lupton, personal communication, 17 October 1986.

154. John Bingham, personal communication, 4 February 1987.

155. G. D. H. Bell, *The Plant Breeding Institute, Cambridge* (Coventry: Massey Fergusson, 1976), pp. 23, 29.

156. Bingham's first published paper was written with Bell on quality of wheat as a trait for plant breeders. See G. D. H. Bell and J. Bingham, Grain Quality—A Genetic and Plant-Breeding character: I. Wheat, *Agricultural Review* 3, no. 6 (1957): 10–22; Bingham, personal communication, 4 February 1987.

157. Keith A. H. Murray, *Agriculture* (London: His Majesty's Stationary Office and Longmans, Green, 1955), pp. 39–40.

158. Viscount Astor and B. Seebohm Rowntree, *British Agriculture: The Principles of Future Policy* (Harmondsworth, England: Penguin, 1939), pp. 79, 108–110.

159. J. K. Bowers and Paul Cheshire, *Agriculture, the Countryside and Land Use: An Economic Critique* (London: Methuen, 1983), pp. 57–61.

160. Ibid., pp. 67–68. Deficiency payments provided the farmer a payment from the government that was based on the difference between a target price set by policy and the average price of wheat sold in the markets.

161. Ibid.

162. B. A. Holderness, *British Agriculture since 1945* (Manchester: Manchester University Press, 1985), p. 49.

163. Ibid., p. 20.

164. Bowers and Cheshire, *Agriculture, the Countryside and Land Use*, pp. 70–71.

165. Bell, *The Plant Breeding Institute*, pp. 26–29.

166. Holderness, *British Agriculture since 1945*, p. 49.

167. Astor and Rowntree, *British Agriculture*, p. 111.

168. Holderness, *British Agriculture since 1945*, p. 49.

169. Bell, *The Plant Breeding Institute*, pp. 26–27.

170. G. D. H. Bell, Crops and Plant Breeding, *Journal of the Royal Agricultural Society of England* 108 (1947): 1–14.

171. F. Earnshaw, A Description of the Recommended Varieties of Wheat and Barley, *Journal of the National Institute of Agricultural Botany* 5, no. 3 (1949): 283–361.

172. E. G. Thompson, Wheat Varieties, *Journal of the National Institute of Agricultural Botany* 6 (1951–53): 289–299. The farmer was a Mr. Cave.

173. Norfolk Agricultural Station, *Thirty-ninth Annual Report 1946–47* (Norwich: Norfolk Agricultural Station, 1948).

174. See the *Annual Report* of the Norfolk Agricultural Station for 1947–48 through 1950–51, all published by the Station, Norwich, in January 1949 through 1952, respectively.

175. J. Bingham and F. G. H. Lupton, Production of New Varieties: An Integrated Research Approach to Plant Breeding, in *Wheat Breeding, Its Scientific Basis*, ed. F. G. H. Lupton, (London: Chapman and Hall, 1987), p. 491 and Table 16.1. Prevalence of Little Joss is from C. C. Brett, The Relative Popularity of Cereal Varieties, *Journal of the National Institute of Agricultural Botany* 6 (1951–53): 55–58.

176. Holderness, *British Agriculture since 1945*, p. 49.

177. Bell, *The Plant Breeding Institute*, p. 34.

178. Plant Breeding Institute, Cambridge, *Annual Report 1959–60* (Cambridge: Plant Breeding Institute, 1961), p. 19; Plant Breeding Institute, *Annual Report 1964–65* (Cambridge: Plant Breeding Institute, 1966), p. 55.

179. Plant Breeding Institute, Cambridge, *Annual Report 1966–67* (Cambridge: Plant Breeding Institute, 1968), pp. 27–28, 31–32, 64.

180. Holderness, *British Agriculture since 1945*, pp. 28–30; Ian R. Bowler, *Agriculture under the Common Agricultural Policy* (Manchester: Manchester University Press, 1985), 255 pp.

181. F. G. H. Lupton, Wheat, *Biologist* 32, no. 2 (1985): 97–105; Plant Breeding Institute, *Annual Report 1972* (Cambridge: Plant Breeding Institute, 1973), p. 67.

182. Bingham, personal communication, 4 February 1987.

183. Bingham and Lupton, Production of New Varieties, pp. 489–492.

184. For example, see Richard Cottrell, *The Sacred Cow: The Folly of Europe's Food Mountains* (London: Grafton, 1987), 192 pp.; and Richard Body, *Farming in the Clouds* (London: Temple Smith, 1984), 161 pp.

Epilogue

1. The stance taken here toward the ability of technology to determine future events is heavily influenced by the historiography of technology that is not strongly deterministic. See, for example, Merrit Roe Smith and Leo Marx, eds., *Does Technology Drive History?* (Cambridge, Mass.: MIT Press, 1994), pp. ix–xiv.

2. Each of these questions has arisen in the numerous lectures and discussions I have engaged in over the past several years. They are in no way exhaustive, but they suggest some of the major themes of interest among scholars of environmental studies.

3. See chapter 1.

4. See chapter 1.

5. World Commission on Environment and Development, *Our Common Future* (New York: Oxford University Press, 1987), pp. 8–9.

6. I am indebted to my colleague Tom Womeldorff of The Evergreen State College, for the clarity of his presentation of these issues.

7. See, for example, James L. Tyson, Falling Wheat Stocks Feed Malthusian Flap, *Christian Science Monitor*, 14 July 1995, pp. 1, 5; and Lester R. Brown, *Who Will Feed China? Wake-Up Call for a Small Planet* (New York: Norton, 1995), 163 pp. For a dissenting view, see, for example, The Food Crisis That Isn't, and the One That Is, *Economist*, 25 November 1995, p. 41.

8. Joel E. Cohen, *How Many People Can the Earth Support?* (New York: Norton, 1995), 532 pp.

Index

Acheson, Dean, 174
Act for International Development (PL 81–535) (United States), 148
Adams Act (United States), 84
Addison, Christopher, 97, 188, 192
Aegilops sitopsis, 26
Aegilops squarrosa, 27, 33
Agrarian Reforms Committee (India), 166
Agricultural Act of 1947 (United Kingdom), 201–204, 247, 250
Agricultural Adjustment Acts (United States), 101
Agricultural College and Research Institute, Lyallpur (British India), 162
Agricultural Improvement Council of England and Wales, 191
Agricultural Prices Commission (India), 184
Agricultural Research Council (United Kingdom), 95, 98, 188, 191, 197, 204, 205, 207, 285n108
agricultural science
  in British India, 168–170
  in India, 78, 79–81, 162–171
  international work in, 142
  in Mexico, 107–108
in United Kingdom, 76–77, 81–83, 188, 192, 200–201, 204–208
in United States, 77–78, 84–85, 89–91
Agricultural Trade Development and Assistance Act of 1954 (P.L. 480) (United States), 156, 175, 180, 183, 255
agricultural universities
  in India, 170, 179–180
  in United Kingdom, 196–199
  in United States, 77–78, 84–85
agriculture
  complexities in, 3
  decline of in the United Kingdom, 77
  expansion of land in, 75
  expansion of production in, 150
  free trade in the United Kingdom, 77
  and the future, 262–264
  and Indian national security, 243–244
  industrialization of, 44, 58–62, 92, 108–111, 114, 171–173, 175–178
  moral purposes of in India, 162–163
  and natural resources, 4
  necessity of, 32
  need for new crop varieties in, 150

agriculture (*continued*)
  organic, 216
  and politics, 59–60
  technology of, 4, 183–186
  and treadmill hypothesis, 13
  use of price controls in India, 164
  use of science to reform, 186
Agriculture Act of 1937 (United Kingdom),
  192
Agriculture (Miscellaneous Provisions) Act
  (United Kingdom), 199
Ali, Imran, 161
All India Coordinated Wheat Improvement
  Project (India), 234, 237
Almazan, Juan Andreu, 105, 111
American International Association for
  Economic and Social Development
  (AIA), 149
American Philosophical Society, 76
American Phytopathological Society, 224
American Society of Agronomy, 141–142
ammonia, 214, 215
ammonium sulfate, 214
Amory, D. Heathcoat, 249
Anderson, Edgar, 57
Anderson, R. Glenn, 238
Argentina, 51, 148
Asquith, Herbert Henry, 81
Atlee, Clement, 128, 131, 201
Australia, 51, 172
Austria-Hungary, 51, 92, 122
Ávila Camacho, Manuel, 105–106, 125

Babcock, Ernest Brown, 66–67
Badische Anilin-und-Soda-Fabrik, 215
Bailey, Liberty Hyde, 41, 56, 71, 75, 84, 156
  career, 52–55
  philosophy of agriculture, 58–62
Baird, Guy, 237
bajra, 158
Baldwin, Stanley, 94, 97, 99
Balfour, Marshall C., 134
Bangladesh, 159
barberry. *See Berberis vulgare*
barley, 20, 24, 33–34, 55, 158, 192, 206, 229,
  249, 250, 251–252, 255
Barnard, Chester I., 134, 135, 137
Barnett, Margaret, 86
Bateson, William, 54, 56, 65, 67, 80, 83, 100
  and Engledow, Frank Leonard, 190
Bayles, Burton B., 222, 230
Bell, G. Douglas H., 251
  career, 206–208
Bengal (British India), 159
Bennett, Henry Garland, 149, 151

Bentley, Amy L., 130
Berberis vulgare, 89–91
bhoodan, 183
Biffen, Rowland Harry, 52, 75, 80, 82, 93,
  246, 248
  and Bell, G. Douglas H., 206
  career, 56–58
  development of wheat varieties, 97
  and Engledow, Frank Leonard, 190, 191
  and Pal, B. P., 178
  retirement of, 188, 206
Bihar, 174
Bingham, John, and Plant Breeding Institute,
  208, 246, 249, 253
biometrics, 55–56, 66
black rust. *See Puccinia graminis tritici*
Blyn, George, 158
Board of Agriculture (and Fisheries) (United
  Kingdom), 77, 86, 94
Board for the Encouragement of Agriculture
  and Internal Improvement (United
  Kingdom), 77
Bolley, Henry L., 89
Borgstrom, Georg, 6
Borlaug, Margaret Gibson, 224
Borlaug, Norman E., 246
  career of, 223–232
  and Hayes, Herbert, 224
  joins Mexican Agricultural Program,
  107
  photo of, 178, 223
  and population issues, 137
  publications of, 230
  and resignation from Mexican Agricultural
  Program, 228
  and Stakman, Elvin, 224
  and Swaminathan, M. S., 235
  and travel in India, 235–237
  and Vogel, Orville, 222
  and wheat breeding in Mexican
  Agricultural Program, 226
Bosch, Carl, 215
Boveri, Theodor, 69
Bradfield, Richard, 107, 142
Brannan, Charles F., 156
Britain. *See* United Kingdom
British East India Company, 76, 163
British West Indies, 190
Broekema, 41
Brooks, F. T., 208
Bryan, William Jennings, 84
Bulgaria, 48
bunt, 221
Bureau of Plant Industry. *See* U.S.
  Department of Agriculture

Burma, 159
Buxton, Noel, 97

calcium cyanamide, 215
Calles, Plutarco Elías, 110, 228
Calles, Rodolfo Elías, 110, 228
Cambridge University, 56, 58, 82, 93
    and devolution of Plant Breeding Institute,
        205
    and Engledow, Frank Leonard, 188–190,
        196
    and Pal, B. P., 178
    School of Agriculture, 86
Campoy, Aureliano, 228
Canada, 48, 51, 91, 127, 172, 215, 229, 238
capital, 12, 163, 171
capitalism, 39, 58, 61, 81, 111, 119, 181,
    241
Capron, Horace, 217
Cárdenas, Lázaro, 104–105, 108, 110
Carleton, Mark Alfred, 64, 71, 89
carrying capacity. See ecosystems
Cattle Plague Department (United
    Kingdom), 77
Cavendish, Richard, 95
Central Landowners' Association (United
    Kingdom), 198, 199
Central Rice Research Institute (India), 234
Centro Internacional de Mejoramiento de
    Maíz y Trigo. See CIMMYT
Cercosporella herpotrichoides, 248, 252, 253
cereal crops, changing yields of, 3
Ceylon. See Sri Lanka
Chartered Surveyors' Institution (United
    Kingdom), 199
chemurgy, 156
Chiang Kai-shek, 173
Chile, 214, 254
China, 103, 125, 134, 135, 154, 173, 174,
    180, 182, 242
Chopra, R. N., 158
Chorleywood Baking Process, 253–254
chromosome theory of inheritance, 69–70
Churchill, Winston, 127, 250
CIMMYT (Centro Internacional de
    Mejoramiento de Maíz y Trigo), 108,
    115, 231
Clark, Colin, 6
Clausen, Roy Elwood, 66–67
Clifford, Clark, 149
co-dependency of wheat and people, 33–35,
    213, 257
co-evolution of wheat and people, 30–35
Cohen, Joel, 265
Coimbatore Agricultural College, 233

coke, 214
cold war, 145
Columbia University, 69, 123, 136
Commission on Country Life (United States),
    59–60, 84
Committee on Agricultural Research
    Organisation (United Kingdom), 95
Committee of the Privy Council for
    Agricultural Research (United
    Kingdom), 95
Common Agricultural Program, 254, 255
Commoner, Barry, 5
communism
    containment of by United States, 144–
        145
    and India, 147, 154, 172, 181–182
    and origins of Point Four Program, 149
    and theories of overpopulation, 138, 260
    and theories of U.S. national security, 119,
        172, 174–175
Communist Party (United Kingdom), 192,
    200
Communist Party of India, 181, 184
Community Projects Administration (India),
    176–177
Compton, Karl T., 138
Connecticut Agricultural Experiment
    Station, 123, 137
Conservation Foundation, 136
Conservative Party (United Kingdom), 94,
    97, 192, 200, 208, 250
Coordinator of Inter-American Affairs
    (United States), 148
corn. See maize; see also wheat
Cornell University, 52–53, 61, 90, 106, 107,
    124, 142, 146
Corn Laws (United Kingdom), 44–45, 58,
    75, 77, 85, 191, 208
Corn Production Act (United Kingdom), 86
Correns, Karl, 35, 55
Cotter, Joseph, 108
Cotton Supply Association of Manchester,
    78
Council of Agriculture for England (United
    Kingdom), 199
Council of Agriculture for Wales (United
    Kingdom), 199
Creighton, Harriet, 69
Creole Petroleum, 148
Crosby, Alfred W., 46
Crosby, F. M., 91
Cummings, Ralph, 235, 238
currency devaluation
    in India, 164, 165, 172, 185, 244
    in United Kingdom, 165

Curzon, George Nathaniel, 79–81, 99
Czechoslovak Academy of Sciences, 234

Daniels, Josephus, 105, 106, 110
Darwin, Charles, 43, 63, 66
Defense of the Realm Consolidation Act
    (United Kingdom), 86
Democratic Party (United States), 84
Denmark, 43
Department of Agriculture (British India),
    168
Department of Agriculture for Punjab
    (British India), 170
Department of Agriculture of Scotland
    (United Kingdom), 95
Department of Food (British India), 160,
    164, 172
Department of Revenue, Agriculture, and
    Commerce (British India), 78
Department of Science and Education
    (United Kingdom), 209
development, 15
Development Commission (United
    Kingdom), 82–83, 85, 94, 95, 96
Development and Road Improvement
    Funds Act (United Kingdom), 82,
    286n113
Dey, S. K., 176
Díaz, Porfirio, 104
disease resistance, 210
Donaldson, Peter J., 133
Drosophila melanogaster, 68, 69
Dyer, R. E. H., 99

East, Edward Murray, 64, 66, 68, 137
    career, 123
ecology
    and carrying capacity, 8, 264–265
    and photosynthesis, 8, 257
    and political ecology, 9, 256–267
    and nitrogen cycles, 212–213, 215
    and north-south balances, 266–267
    and rural-urban balances, 265–266
Economic Cooperation Administration
    (United States), 150
ecosystems
    and carrying capacity, 8, 32, 136
    and food webs, 8
Ehrenfeld, David, 6
Ehrlich, Anne, 5
Ehrlich, Paul, 5
einkorn, 27, 34
Eisenhower, Dwight D., 156
El Bajío (Mexico), 109, 226–228
    map of, 227

electricity, 214–215
Elliot, Walter, 98
emmer, 27, 33–34
enclosure, 38, 214
England, 27–30, 37–39, 43–44, 57, 120–121,
    178. See also United Kingdom
Engledow, Frank Leonard, 246, 248
    and Bateson, William, 190
    and Bell, G. Douglas H., 206
    career, 188–192
    and Loveday Committee Reconstituted,
    196
    and Pal, B. P., 178
    photo of, 189
    and postwar expansion of agricultural
    science, 204
Ensminger, Douglas, 151, 179, 181
environment
    and agriculture, 39–40, 61
    and capitalism, 39–40
    and science, 39–40
    and technology, 4
Ernle, Lord. See Prothero, Rowland
Erysiphe graminis tritici, 248
Etawah project (India), 153, 176–178
European Community, 209, 254
eyespot. See Cercosporella herpotrichoides

famine, 124–125, 150, 164, 184, 201,
    240
    Great Bengal Famine, 159–160
Famine Emergency Committee (United
    States), 130, 146
Famine Inquiry Commission (British India),
    159
Farrer, William, 41
Ferrell, John A., 106
fertilizers, 86, 101, 155, 179, 181, 210,
    242
    nitrogen, 211–216, 235, 241, 251
First World War, 85–92, 142, 190
Fitzgerald, Deborah, 113
Food Control Act (United States), 88
Food Corporation of India (India), 158, 184,
    238
Foodgrains Policy Committee (India), 165
Food Grains Policy Committee (British
    India), 164
Foodgrains Prices Committee (India),
    184
Food for Peace (United States), 156
food self-sufficiency
    India, 158, 171
    United Kingdom, 96–97, 188, 191, 204,
    243, 246, 248, 255

Ford Foundation, 114, 231
  and assistance to India, 151–152, 154,
    176–180, 181, 185
  and *India's Food Crisis and Steps to Meet
    It*, 181–182
  and Intensive Agricultural Districts
    Programme, 185, 235, 240
  and Point Four Program, 145
foreign assistance, 130
  requests for, from India, 146–147, 242
foreign exchange
  and India, 165, 171, 185, 242, 244
  and Mexico, 112
  and United Kingdom, 202–204, 246,
    250
Formosa, 134, 216
Fosdick, Raymond B., 106, 133
France, 48, 51, 217, 252
Frankel, Francine, 180, 184, 185
Freeman, E. M., 91
Fryer, John, 83

Gaines, Edward Franklin, 220
Galton, Francis, 56
Gamble, William K., 114
Gandhi, Indira, 183, 186, 244
Gandhi, Mohandas K., 98, 166, 170, 232
General Mills, 91
genes
  origin of word, 69
  semi-dwarfing in wheat, 211–255
Germany, 43, 48, 51, 69, 85, 87, 92, 97, 98,
    122, 127–128, 141, 195, 215
Ghadar movement, 87
Gilbert, Arthur W., 66
Gilbert, Joseph Henry, 77
Gomez, Marte, 107, 125
Gold Coast, 190
Government of India (British India), and
    Great Bengal Famine, 159–160
Government of India Act (British India), 99,
    170
gram, 158
Gray, Asa, 52
Great Bengal Famine, 159–160, 162, 185
Great War. *See* First World War
green revolution, 92, 101, 163, 182, 183–186,
    187, 209, 211, 223, 232, 238–246, 256–
    258
Griswold, A. Whitney, 83
Group of Peers (United Kingdom), 199
Grow More Food (India), 153, 160, 165,
    167–168
guano, 214
Gunn, Selskar M., 103

gunpowder, 213
Gupta, R. L., 172–173

Haber, Fritz, 215
Haber-Bosch process, 215
Hall, Alfred Daniel, 83, 93–94, 95, 96, 100,
    188, 191, 192
Hallett, F., 40
Hardy, Ben, 149
Harrar, J. George, 107, 152, 191, 224
  and India, 237, 243
  photo of, 233
  and plant breeding in Mexican
    Agricultural Program, 225
  as president of Rockefeller Foundation, 225
Harvard University, 52, 106, 123, 220
Haryana, 245
Hatch Act (United States), 77
Hayes, Herbert Kendall, 71, 137, 217, 224,
    226
Hays, W. M., 41
high yielding varieties, 16, 92, 181, 185, 210
Hinduism, 162
Hoffman, Paul G., 151, 177, 179
Holmes, Horace C., 151, 176
Hoover, Herbert, 88, 91, 130, 146
Hope, Victor Alexander John, 100
Hopkins, Cyril G., 123
Houston, David, 89
Howard, Albert, 80–81, 82, 191, 216, 281n27
Howard, Gabrielle L. C., 80–81, 191
Howard, H. W., 233
Hudson, R. S., 195
Hull, Cordell, 105, 126
Hume, Allan Octavian, 78
Hungary, 48
Hunter, Howard, 206
Hussey, Obed, 50
Huxley, Thomas H., 56

Illinois Agricultural Experiment Station, 123
Immer, Forrest Rhinehart, 71
Imperial Agricultural Research Council
    (British India), 100, 168. *See also* Indian
    Council for Agricultural Research
    (India)
Imperial Agricultural Research Institute
    (British India), 80, 87, 100, 170. *See
    also* Indian Agricultural Research
    Institute (India)
India, 18, 30, 51, 138, 190, 209
  affects of British rule, 158–159
  agrarianism in, 170–171
  agricultural education in, 238
  agricultural price reform, 184, 238–239

India (*continued*)
  agricultural production in, 164, 180–181, 185, 245
  agricultural science in, 78, 79–81, 98–101, 162–171, 211
  agricultural yields in, 236, 318n97
  assistance from Rockefeller Foundation, 180
  assistance from United States, 154, 241
  castes in, 162–163
  community development program in, 176–178, 179, 185
  currency devaluation in, 164, 165, 172, 185, 244
  drought in, 174, 180, 185, 245
  expansion of agricultural land in, 174
  famine in, 78, 125
  and first five year plan, 167, 171, 175–177
  and First World War, 87
  food production in, 158–159, 170–171, 172–174
  and food shortage after Second World War, 128
  and import of Mexican wheats, 236–238, 241–244, 245
  and import of wheat from the United States, 238
  independence of, 161–162
  industrialization of, 163
  irrigation in, 80, 241
  map of wheat-growing areas, 29
  moral purposes of agriculture in, 162–163
  and movement to intensify agricultural production, 181–183, 245
  national security planning in, 232, 243–244, 261–262
  partition of British India, 161–162
  and politics of the green revolution, 238–246
  rebellion of 1857, 78
  reconstruction of agriculture, 100, 170
  relations with Pakistan, 242–243
  relations with United States, 146–147, 171–175, 185, 242, 244
  and second five-year plan, 171, 175–177, 180–181
  and third five-year plan, 241
  use of nitrogen fertilizer in, 214, 245
  use of price controls, 164
  use of science to reform agriculture, 239
  use of social reform in agriculture, 178–179
  war with China, 182
  war with Pakistan, 164–165, 241–243

Indian Agricultural Research Institute (India), 162, 169
  appointment of B. P. Pal to direct, 178–179
  assisted by Rockefeller Foundation, 153, 180
  and import of Mexican wheats, 244
  and Shastri, Lal Bahadur, 240
  and Subramaniam, C., 239
  and Swaminathan, M. S., 233–234
Indian Civil Service, 163
Indian Council for Agricultural Research (India), 101, 163
  directed by B. P. Pal, 179, 245
  and import of Mexican wheats, 237
Indian National Congress Party, 87, 98, 163–164, 166, 171, 175, 181, 182, 183–184, 244
Indian National Science Academy, 234
Indonesia, 135
industrial revolution, 44–45, 58, 75
  in India, 155, 163, 171–173, 175–178, 180
  in Mexico, 108–111
Intensive Agricultural District Programme (IADP) (India)
  adoption of, 185–186
  conflicts of, with Gandhianism, 182
  conflicts of, with socialism, 182
  operation of, 240–241
  origins of, 182
International Center for the Improvement of Maize and Wheat. *See* CIMMYT
International Development Advisory Board (United States), 148
International Education Board. *See* Rockefeller Foundation
International Institute of Agriculture, 143
International Rice Research Institute, 246
International Union for the Conservation of Nature and Natural Resources, 246
Iran, 151, 154
irrigation, 101, 210
  in India, 160, 161–162, 177, 179, 181, 235, 241
  in Mexico, 109–111, 226
Irwin, Lord. *See* Wood, Edward
Islam, 162
Italy, 51, 143, 195, 217

Jain, H. K., 232
Jallianwala Bagh, 98–99
Japan, 70, 127, 134, 135, 141, 159, 195, 201, 211, 216–218
Jardine, W. M., 142
Java, 134

Johannsen, Wilhelm, 67, 69, 71
John Innes Horticultural Institute (United
    Kingdom), 82, 206
Johnson, Sherman, 181–182
Joint Advisory Committee on Higher
    Agricultural Education (United
    Kingdom), 191
Jones, Donald F., 123
jowar, 158
Junior Loveday Committee, 196

Kahn, Abdul Rahman, 80
Kansas State College of Agriculture, 217
Kennedy, John F., 133
Kerala, 181, 184–185, 232
Kevles, Daniel, 123
Keynes, John Maynard, 121–122
Kirk, Dudley, 132
Kloppenburg, Jack, 16
Knight, Henry, 159
Knight, Thomas Andrew, 40
Kohli, S. P., 235, 237, 238, 244
Korea, 135, 216
Korean War, 149
Kosygin, Aleksei, 244
Krishnamachari, V. T., 177
Kumarappa, J. C., 166

Labour Party (United Kingdom), 94, 97, 127,
    192, 200, 204, 208, 246, 250
Ladejinsky, Wolf I., 147
Lakshmi, Vijaya, 174
Lamarck, Jean-Baptiste de Monet de, 53, 73
Land Agents' Society (United Kingdom), 199
land grant university system (United States),
    77
land reform
    in India, 164, 166–167, 175, 181, 183
    in United Kingdom, 93, 96, 200
Land Settlement Association (United
    Kingdom), 199
land tenure
    in British India, 166–167
    in United Kingdom, 192
Land Union (United Kingdom), 199
Law, Bonar, 94
Laws, John Bennet, 77
League of Nations, 124, 143
LeCouteur, John, 40, 65
Leiss, William, 6
Leonard, Warren H., 217
Liberal Party (United Kingdom), 81–83, 94,
    97, 192, 200
Linnaeus (Carl von Linné), 63
Lloyd George, David, 81–83, 86, 92

Long Ashton Research Station (United
    Kingdom), 82
Loveday Committee (United Kingdom),
    194–195
Loveday Committee Reconstituted (United
    Kingdom), 196, 197–198. *See also*
    Junior Loveday Committee
Loveday, Thomas, 194
Ludmerer, Kenneth, 123
Lupton, Francis G. H.,
    and Plant Breeding Institute, 208, 246
Luxmoore Committee (United Kingdom),
    194–195, 197

MacArthur, Douglas, 217
MacDonald, Ramsey, 94–95
Mahalanobis, P. C., 180
maize, 18, 19, 36, 70, 112, 123, 125, 153,
    158, 170, 185, 224, 226, 229, 237
Malaya, 190
Malenbaum, Wilfred, 47–48
Malthus, Thomas R., 16, 121
Malthusian models, 6
Malthusianism, 120, 123
Mangelsdorf, Paul C., 106, 152
Mann, Albert R., 106
Mao Tse-tung, 173
Marshall, George, 130
Marshall Plan, 144
Massachusetts Institute of Technology,
    138
Matthaei, Gabrielle L. C. *See* Howard,
    Gabrielle
Mayer, Albert, 176
Mayr, Ernst, 63
McClintock, Barbara, 69
McCloy, John J., 182
McCormick, Cyrus, 50
McKinley, William, 84
mechanical reaper, 49–51
mechanization, 49–51, 75, 86, 92, 155, 181
    of wheat breeding, 221
Medieval period, 37
Mehta, N. C., 163
Mendel, Gregor, 23, 35, 43, 54–55, 80
Merchant, Carolyn, 6
Messersmith, George, 125
Mexico, 18, 30, 91, 92, 209
    agricultural production in, 229
    agricultural science in, 107–108, 211
    food shortage in, 125–126
    land reform in, 104–106
    national security planning in, 262–262
    revolution in, 104
    Spanish conquest of, 104

Mexico (*continued*)
  wheat exports from, 231
  wheat imports to, 231
  wheat in diet of, 112–113
  wheat yields in, 229, 231
Miami University, 123
Michigan State Agricultural College, 52
Middleton, Thomas H., 86, 95, 100, 191
Milbank Memorial Fund, 123
Miller, Henry M., 107
millet, 158
Ministry of Agriculture (India), 146, 172,
  174, 178, 182, 239
Ministry of Agriculture (Japan), 217
Ministry of Agriculture (Mexico), 229
Ministry of Agriculture (and Fisheries)
  (United Kingdom), 95, 96, 97, 188, 193,
  197, 207, 253
Ministry of Agriculture and Irrigation (India),
  246
Ministry of Community Development and
  Cooperation (India), 182, 239
Ministry of Education (United Kingdom),
  193
Ministry of Food (India), 172–173, 174
Ministry of Food (United Kingdom), 127
Ministry of Food and Agriculture (India),
  174, 184
Ministry of Irrigation (India), 239
Moe, Henry A., 134
Montgomery, Bernard, 128
Morgan, Thomas Hunt, 68, 69
Morris, Joe Alex, 149
Mugica, Francisco, 105, 111
Mukherjee, J. N., 178
Munshi, Kanaiyalal M., 174, 178
Museum of Natural History, 136
Myanmar. *See* Burma

Nagamitsu, Dick, 222
National Academy of Sciences (United
  States), 234
National Agricultural Advisory Service
  (United Kingdom), 199, 201, 203, 204
National Association of Audubon Societies,
  136
National Development Council (India),
  239
National Extension Service (India), 179
National Farmers' Union (United Kingdom),
  196, 197, 199
National Federation of Women's Institutes
  (United Kingdom), 196
National Institute of Agricultural Botany
  (United Kingdom), 252, 253

national security planning
  and agricultural production, 156
  in British India, 79–81, 87
  in United Kingdom, 85–87, 209
  in United States, 87–89, 92, 119–120,
    149–150
National Seed Distribution Organization
  (United Kingdom), 209
National Union of Agricultural Workers
  (United Kingdom), 197, 199
Nehru, Jawaharlal, 151, 153
  and community development program,
    176, 240
  death of, 183, 238
  and Ford Foundation, 179–183
  and Great Bengal Famine, 160
  at Indian independence, 163, 166
  and industrialization of India, 171
  and land reform, 167
  and modern Indian state, 155
Neo-Europes, 46, 62, 220
Neolithic period, 27, 32, 36–37, 213
neo-Malthusianism, 121–122, 137
  ecological, 123
  political, 123
Netherlands Agricultural University,
  233
New York Zoological Society, 136
Nigeria, 190
Nilokheri project (India), 176–178
Nilsson, Hjalmer, 41, 71
Nilsson-Ehle, Herman, 41, 67–68
nitrogen, 211–216, 226, 229, 235
nitrogen fixation, 212–215
Nobel Prize
  and Borlaug, Norman, 223
  and Bosch, Carl, 215
  and Haber, Fritz, 215
  and Morgan, Thomas Hunt, 69
Norfolk Agricultural Station (United
  Kingdom), 252
Norfolk Four Course System, 253
Norin 10, 218, 222, 224, 232, 235, 254
North Atlantic Treaty Organization,
  144
North Dakota Agricultural College, 89
Notestein, Frank, 124, 134
nuclear weapons, 173
Nuffield Foundation, 191

oats, 19, 192, 249, 251
Office of the Coordinator of Inter-American
  Affairs, 136
Office of Population Research. *See* Princeton
  University

Office of Special Studies. *See* Rockefeller Foundation, Mexican Agricultural Program
Oficina de Estudios Especiales. *See* Rockefeller Foundation, Mexican Agricultural Program
Oklahoma State College of Agricultural and Mechanical Arts, 149
organic agriculture, 216
Osborn, Fairfield, 135–136
Osborn, Henry Fairfield, 136
Osoyo, Roberto, 113
Ottoman Empire, 99
overpopulation theories, 123, 213
Oxford University, 199

Pakistan, 161, 185, 241
Pal, Benjamin Peary, 153
    and agricultural education, 238
    as director of Indian Agricultural Research Institute, 178–179, 245
    and Engledow, Frank Leonard, 190
    photo of, 178
    and Swaminathan, M. S., 233, 235
pangenesis, 43
Parker, Marion, 245
Patel, Vallabhbhai, 163
Pearson, Karl, 56, 65, 67
Peru, 214
Phaseolus vulgaris, 67
Philippines, 134, 135, 185, 246
Phipps, Henry, 80
photosynthesis, 8, 32, 212, 257, 267
P.L. 480. *See* Agricultural Trade Development and Assistance Act of 1954
Planned Parenthood Federation of America, 136
Planning Commission (India), 176, 179–181, 184, 185, 240, 244, 246
plant breeding
    as affected by Darwin-Mendel, 62–63
    assumptions of, 223
    in British India, 170
    and capital, 14
    and classification of organisms, 63–64
    and development, 14–15
    embraced in India, 186
    and evolution, 63–64
    and industrialization, 44
    and inexpensive fertilizer, 155
    and international vision, 141–142
    mechanization of, 221
    and Mendelian genetics, 35
    and nation states, 12–14
    and national security planning, 120, 138

and nitrogen fertilizer, 216
origins of, 35–41, 52, 62–74
as shuttle breeding, 226
and statistical methods, 65, 73
systematic programs for, 70–74
and technology, 14
and war, 76
and yields, 10–13, 35–40
Plant Breeding Institute (PBI) (United Kingdom), 24, 58, 93
    and Biffen's retirement, 188
    and Engledow, Frank Leonard, 190
    and independence from Cambridge University, 205
    origins of, 82
    and postwar expansion of, 205–208
    and postwar research program, 246–255
    and Swaminathan, M. S., 233–234
Plant Varieties and Seeds Act (United Kingdom), 208–209
PNST. *See* population–national security theory
Point Four Program (United States), 117
    and community development program in India, 176
    development of program, 148–150
    origins of, 144–145
    and Rockefeller, Nelson A., 148–150
Poland, 98
political ecology
    and agriculture, 7
    analytical framework of, 7–10
    central mission of, 10
    definition of, 5
    and Neolithic period, 32–33
    and plant breeding, 14–15
    and yields, 17–18, 32–33
population, 16, 32, 37–38, 40, 75, 258–262
    expansion of European, 46
    in India, 181–182
    issues of, 122–123
    and Malthus, 121
    and nitrogen fixation, 213
    in United Kingdom, 187
    and U. S. national security planning, 119–120, 154, 186
Population Association, 123
population–national security theory, 119–120, 138, 154, 181, 258–262
Population Reference Bureau, 123
Populist Party (United States), 84
Populist revolt (United States), 59
Portugal, 51
postwar food shortage, 127–131, 172, 201
powdery mildew. *See Erysiphe graminis tritici*

Princeton University, Office of Population
    Research, 123, 124, 132
Privy Council (United Kingdom), 95
Progressive Era, 84, 91
Prothero, Rowland, 86, 94
Provine, William, 55
Puccinia glumarum, 248
Puccinia graminis tritici, 88–91, 217, 225,
    226, 229
Puccinia striiformis, 82, 237
Punjab, 98–99, 160, 170, 237
pulse, 158

ragi, 158
Raja, S. T., 147
Reading University, 249
Research and Marketing Act of 1946 (United
    States), 156
revenue extraction (British India), 161
rice, 18, 19, 36, 152, 158, 185, 229, 234, 237,
    245
Richey, Frederick D., 141–142
Rimpau, Wilhelm, 41
Rockefeller Foundation, 92, 103–104, 143
    and assistance to India, 146, 152–154, 180,
        185, 237
    criticisms of agricultural programs, 135
    and efforts to establish study of human
        ecology, 134–135
    and Engledow, Frank Leonard, 191
    Indian Agricultural Program, 117, 139
    International Education Board, 103
    International Health Division, 134
    Mexican Agricultural Program, 106–108,
        115–117, 125, 138, 142, 191,
        222–232
    and neo-Malthusianism, 137
    Office of Special Studies, 108, 115
    and overpopulation theories, 133–135, 138
    and Point Four Program, 145
    and population studies, 124
    and postwar population survey, 134
    program in China, 103, 142
    and Vogel, Orville, 222
    and "World Food Problem", 138
Rockefeller, John D., III, 117, 133, 134, 237
Rockefeller, Nelson A., 106, 136, 148–150
Romania, 48
Roosevelt, Franklin D., 101, 106, 125, 126,
    143
Roosevelt, Theodore, 59, 84
Roper, Mildred Emmeline, 190
Rothamsted Station (United Kingdom), 77,
    82
Rowlatt Acts (British India), 98

Royal Agricultural Society of England
    (RASE) (United Kingdom), 76–77, 199–
    200, 208
Royal Agri-Horticultural Society (British
    India), 78
Royal Botanic Gardens (Kew Gardens)
    (United Kingdom), 76
Royal Commission on Agriculture (British
    India), 99
Royal Society (United Kingdom), 58, 76,
    191, 208, 234
Rudolph, Lloyd, 167
Rudolph, Susanne, 167
Rupert, Joe, 230, 254
Rural Electrification Administration (United
    States), 101
Russel, E. John, 83
Russia, 48, 122
rye, 19–20, 24, 55

Sahay, Vishnu, 172
Sakamura, Tetsu, 70
Salmon, Samuel Cecil, 217–218, 222, 224
Saunders, William, 41
Schumpeter, Joseph, 14
science, 39
Scottish Society for Research in Plant
    Breeding, 206
Scripps, E. W., 123
Scripps Foundation for Research in
    Population Problems, 123
Second World War, 98
    and India, 146, 159
    and Japanese agriculture, 217
    postwar food shortage, 127–131
    and United Kingdom, 192–195, 200, 249–
        250
    U.S.-Mexico agricultural trade, 126
Sen, Amartya, 159
Shastri, Lal Bahadur, 184–186, 238, 243
    photo of, 240
Shirreff, Patrick, 40–41, 65
Shull, George H., 123
Singh, Bahadur Datar, 163
Singh, D. P., 176
Sinoloa, 114, 226–228
Smith, Adam, 121
Smith, Ben, 129
smut, 220
socialism, 163
sodium nitrate, 214
Solbrig, Dorothy, 6
Solbrig, Otto, 6
Solvay Process Company, 214
Sonora, 110, 114, 226–228

sorghum, 158
South Dakota State College of Agriculture, 217
Southern Rhodesia. *See* Zimbabwe
Spain, 51
Spillman, William Jasper, 220
Sri Lanka, 134, 190
Stakman, Elvin Charles, 137, 217
 and Borlaug, Norman, 224
 career, 89–91
 and Mexican Agricultural Program, 107
 photo of, 90
Standard Oil, 105, 148
stem rust. *See Puccinia graminis tritici*
Stopford, John S. B., 195
stripe rust, 241
Strode, George K., 134
Studebaker Corporation, 151
submarine warfare, 86, 129
Subramaniam, C., 184–186, 238, 243
 photo of, 239
sustainability, 263–264
Sutton, Walter, 69
Svalof plant breeding station (Sweden), 67–68
Swaminathan, Monkombu Sambasivan
 career of, 232–238, 245–246
 as Director General of Indian Council of Agricultural Research, 245–246
 and Pal, B. P., 233
 photo of, 178, 233
 and Subramaniam, C., 240
 and Vogel, Orville, 234
 and work on potato, 234
Sweden, 51, 67–68, 71
Swedish Seed Association, 234
Switzerland, 51

Taiwan, 135, 185
Tamil Nadu, 232
Taylor, Clifford C., 147, 172–173
Technical Cooperation Administration (United States), 147, 149, 150, 153
technology, 4, 14
Tennessee Valley Authority (United States), 101
Thompson, Warren S., 123–124, 132
Topley, W. W. G., 205
Transport and General Workers' Union (United Kingdom), 196, 197, 199
Travancore University, 233
treadmill hypothesis. *See* agriculture
Triticum aestivum, 23, 27, 33
Triticum dicoccoides, 26
Triticum dicoccum, 26–27
Triticum durum, 23, 34

Triticum monococcum, 27
Triticum urartu, 26
Truman, Harry S., 117, 128, 130, 131, 144–145
Tschermak, Erich von, 35, 55

Union of South Africa, 190
Union of Soviet Socialist Republics, 173, 242, 244
United Kingdom, 12, 18
 and agrarianism, 83
 agricultural education in, 193–194
 agricultural imports to, 249
 agricultural prices policy, 203
 agricultural production in, 86, 97, 249–250
 agricultural reconstruction, 86, 93–98, 195–205
 agricultural science in, 76–77, 81–83, 93–98, 188, 200–201, 204–208, 211, 286n113
 barley production in, 251–252
 and Chorleywood Baking Process, 253–254
 in contrast to India, Mexico, and United States, 247
 control of empire by, 161, 163
 and Corn Laws, 44–46
 currency devaluation in, 165
 and enclosure, 38, 214
 environmental criticisms of agriculture in, 255
 and First World War, 85–87
 food shortage in, 127–131
 and free trade in wheat, 51
 land nationalization, 93, 96, 192, 200
 map of wheat-growing areas, 28
 national security planning in, 261–262
 Norfolk four-course rotation, 213
 postwar Labour government, 201–205
 relations with United States, 201
 use of nitrogen fertilizer in, 214
 use of price policy in, 250–251, 253
 use of wheat in, 248
 wheat exports, 255
 wheat imports, 48
 wheat prices in, 188
 wheat production in, 209, 255
 wheat yields in, 255
United Nations
 Division of Population, 124
 Food and Agriculture Organization (FAO), 143, 168, 182, 191
United States, 12, 18, 30, 43, 69
 and agrarianism, 83
 agricultural depression, 59, 101
 agricultural production in, 88–90, 156

United States (*continued*)
  agricultural science in, 77–78, 84–85, 89–
    91, 211
  agricultural trade with United Kingdom,
    127
  changes in agricultural technology, 155–156
  and First World War, 87–90
  food supplies in wartime, 126
  map of wheat-growing areas, 31
  national security planning in, 149–150,
    242, 261–262
  production of nitrogen fertilizer in, 215
  promotion of modern agriculture in, 101
  promotion of wheat exports, 51
  and *Puccinia graminis tritici* in, 229
  relations with India, 146–147, 171–175,
    185, 242, 244
  relations with Mexico, 105–106
  relations with Pakistan, 242
  relations with United Kingdom, 201
  wheat exports to Europe, 48
United States Agency for International
  Development, 117
United States Department of Agriculture
  (USDA), 54, 77, 146
  Bureau of Plant Industry, 84, 89, 91, 126,
    141, 219
  Cooperative Extension Service, 77
United States Department of State, 146, 149,
  150, 172
United States Food Administration, 88
United States Overseas Economic
  Administration, 150
University of Aberystwyth, 206
University of Bristol, 194
University of California, 66
University Education Commission (India),
  170
University of Manchester, 195
University of Minnesota, 71, 89, 90, 91, 107,
  137, 217, 224
University of Nanking, 103, 146
University of Nebraska, 219
University of North Wales, 206
University of Wisconsin, 234
Uppal, Badri Nath, 153
Uttar Pradesh, 78, 153, 176

variation, 62–69
Vavilov, N. I., 57
Venezian, Eduardo L., 114
V. I. Lenin All-Union Academy of
  Agricultural Sciences, 234
Vilmorin, Henrie de, 40–41

Vogel, Orville Arthur, 246
  career of, 218–223
  photo of, 219
  and Swaminathan, M. S., 234
Vogt, William, 135–137
Vries, Hugo de, 35, 54–55, 65, 67, 69, 80

Wallace, Henry A., 106, 111, 125, 148
War Agricultural Executive Committees
  (United Kingdom), 86, 97, 127, 128,
  129, 192–193, 196
War Emergency Board of Plant Pathology
  (United States), 89
Washington state, 6, 218–223
Washington State College of Agriculture,
  218, 220
Watson, J. A. Scott, 199
Waynick, Capus M., 147
Weaver, Warren, 106, 116, 134, 138, 152,
  153
Webber, Herbert J., 54
Weldon, W. F. R., 56, 65, 67
Wellhausen, Edwin J., 107, 226
Welsh Plant Breeding Station (United
  Kingdom), 206
West Bengal, 159
Western Regional Cooperative Wheat
  Improvement Program (United States),
  220
wheat, 18, 19–35, 55, 152, 158, 169, 172,
  185, 192, 193, 224, 250
  anatomy of, 20–22
  barter for manganese from India, 174
  breeding of in Washington state, 220
  classification of, 23–25
  color of, 25
  and day-length sensitivity, 229
  decline of production of in England,
    45
  diagrams of, 21–22
  evolution of, 26–27
  excess land given to, 47–48, 220
  expansion of area in, 46
  and free-threshing trait, 27
  global production of, 25
  hard, 24
  high yielding varieties of, 92, 190, 240,
    318n97
  and laborsaving machinery, 49–51
  life-cycle of, 20–22
  lodging under high fertilizer use, 155,
    216, 220–221, 241, 251
  and mythology, 26
  and nitrogen fertilizer, 216

and plant diseases, 220, 221, 225
and ploidy, 70
policies for in United Kingdom, 204
prices of in India, 184
prices of in United Kingdom, 188
production during First World War,
   88
research in Mexico, 107–108
research in United Kingdom, 248
semi-dwarfing genes in, 211–218, 230
semi-dwarfing varieties of, 217, 254
shattering in, 220
and shuttle breeding, 226
soft, 24
spread of, 27–30
spring, 24, 48
straw as a crop of, 216
supplies in First World War, 85–86
traditional varieties of, 216
uses of, 19–20, 24, 211
winter, 24, 220
varieties of, 82, 220
yields of, 20–23, 175, 236
Wheat Act of 1932 (United Kingdom),
   188
wheat trade, 45, 47–48
   in British India, 161–162
   within India, 166
   and Mexico, 112–113
   from Pakistan to India, 165
   promotion of export trade, 51
   tariffs on, 51
   from United States to India, 146, 165, 168,
      172–175, 242
wheat varieties
   Brevor, 221, 222
   Capelle Desprez, 252, 253, 254
   CIANO 67, 231
   club, 220
   Elgin, 221
   Elmar, 221, 222
   Gaines, 218, 222
   Hobbit, 255
   Holdfast, 246, 253
   Hybrid 46, 246
   INIA 66, 231
   Jaral 66, 231
   Kentana, 229
   King Red Chaff White, 40
   Lerma Rojo, 229
   Lerma Rojo 64, 231, 241, 244

Lerma Rojo 64A, 236
Little Joss, 58, 97, 252
Marfed, 221
Maris Ranger, 254
Maris Widgeon, 246, 253, 254
Mayo 64, 236
Nord Desprez, 252
Noreste 66, 231
Norin 10, 218, 220–223, 230
Norteno 67, 231
Orfed, 221
Penjamo 62, 230–231
Pitic 62, 230–231
Pringle, 40
Redman, 246
Ridit, 220
semi-dwarf, 211–255
Shirreff's Bearded Red, 40
Shireff's Bearded White, 40
Shirreff's Squarehead, 40
Sonora 63, 236, 239
Sonora 64, 231, 236, 241
Talevara, 40
Thatcher, 217
Tobari 66, 231
Yeoman, 97
Whelpton, Pascal K., 124
White, Richard, 6
Whitehouse, R. N. H., 249
Williams, Thomas, 201–203
   photo of, 202
Wilson, Edmund B., 69
Wilson, James, 84
Women's Land Army (United Kingdom),
   196
Wood, Edward, 99
World Bank, 185, 240
World War I. *See* First World War
World War II. *See* Second World War
Wye College of Agriculture, 80

Yankton College, 219
Yaqui River Valley, 110, 228
yellow rust. *See Puccinia striiformis* or
   Puccinia glumarum
yields, 11–12, 15–17, 35–40, 45–46, 190,
   210–211, 221, 231, 234
Yugoslavia, 48

zamindar abolition, 166–167
Zimbabwe, 190